Robert Hogg, Henry Graves Bull, Edith G. Bull, Alice B. Ellis

The Herefordshire Pomona

Containing Coloured Figures and Descriptions of the most Esteemed Kinds of

Apples and Pears

Robert Hogg, Henry Graves Bull, Edith G. Bull, Alice B. Ellis

The Herefordshire Pomona
Containing Coloured Figures and Descriptions of the most Esteemed Kinds of Apples and Pears

ISBN/EAN: 9783337523800

Printed in Europe, USA, Canada, Australia, Japan

Cover: Foto ©berggeist007 / pixelio.de

More available books at **www.hansebooks.com**

THE

HEREFORDSHIRE POMONA,

CONTAINING

COLOURED FIGURES AND DESCRIPTIONS OF THE MOST ESTEEMED KINDS OF

APPLES AND PEARS.

WITH ILLUSTRATIONS DRAWN AND COLOURED FROM NATURE BY MISS ELLIS AND MISS BULL.

TECHNICAL EDITOR: ROBERT HOGG, LL.D., F.L.S.,

Honorary Member of the Woolhope Naturalists' Field Club; Vice-President of the Royal Horticultural Society; Author of 'The Fruit Manual'; 'British Pomology'; 'The Vegetable Kingdom and its Products', &c., &c.

" *Hope on. Hope ever.*"

" Ζεφυρίη πνείουσι τὰ μὲν φύει ἄλλα δέ πέσσει
ὄγχνη ἐπ' ὄγχνῃ γηράσκει, μῆλον δ' ἐπὶ μήλῳ,
αὐτὰρ ἐπὶ σταφυλῇ σταφυλή, σύκον δ' ἐπὶ σύκῳ."

Homer Odyssey vii. 119-22.

"THE BALMY SPIRIT OF THE WESTERN GALE,
ETERNAL BREATHES ON FRUITS UNTAUGHT TO FAIL;
EACH DROPPING PEAR, A FOLLOWING PEAR SUPPLIES,
ON APPLES APPLES, FIGS ON FIGS ARISE."

Pope.

GENERAL EDITOR: HENRY GRAVES BULL, M.D., &c., J.P. for the City and County of Hereford.

VOLUME I.

HEREFORD : JAKEMAN AND CARVER, HIGH TOWN.

LONDON : JOURNAL OF HORTICULTURE OFFICE, 171, FLEET STREET, E.C.

1876—1885.

DEDICATED

BY PERMISSION TO

THE RIGHT HONOURABLE LORD BATEMAN,

OF SHOBDON COURT,

LORD LIEUTENANT AND CUSTOS ROTULORUM

OF THE COUNTY OF HEREFORD,

BY THE MEMBERS OF

THE WOOLHOPE NATURALISTS' FIELD CLUB,

AS A TRIBUTE TO THE CORDIAL INTEREST HE HAS SHOWN

IN THIS EFFORT TO IMPROVE THE PRODUCTIONS OF

HEREFORDSHIRE.

. . . . Post hanc Malusque Pyrusque,
Illustres veneræ domus, numeroque potentes;
Non illas aspernatur vel Citria regum
Mensa superborum, non aspernantur et illæ
Vel tripedes inopum mensas at terrea vasa.
Subsidium vitæ, luxusque paratus egenæ
Et non ullius desertrix copia mensis.
Hinc sibi vina parant gentes quas igne remotus,
Languidiore fovet non vitibus æquus Apollo.
Principis illa quidem munusque vicemque liquoris
Non indigna ferunt; sic curas illa metusque
Pauperiemque domant, sic lætitiisque jocisque
Speque nova et liquido perfundunt lumine venas,
Et pulchrum accendunt Veneris Martisque furorem
Hoc contenta mero lætum Normannia degit
Vicinasque tuens Gallorum haud invidet uvas.

> COWLEY. Plant: v., 606-621.
> (c. 1665.)

"The tribe of Pears and Apples next succeed,
Of noble families and numerous breed;
No monarch's table e'er despises them,
Nor they the poor man's board on earthen dish contemn.
Supports of life as well as luxury,
Nor like their rivals a few months supply;
But see themselves succeeded e'er they die. .
Where Phœbus shines too faint to raise the vine,
They serve for Grapes, and make the Northern wine;
Their liquor for th' effects deserves that name,
Love, valour, wit, and mirth, it can inflame;
Care it can drown, lost health, lost wealth restore,
And Bacchus' potent juice can do no more.
With Cyder stor'd, the Norman province sees
Without regret the neighbouring vintages.

> COWLEY's "Six Books of Plants," 1689.
> Book V.

TABLE OF CONTENTS.

VOLUME I.

GENERAL INTRODUCTION.

"SUNT NOBIS MITIA POMA
CASTANEÆ MOLLES, ET PRESSI COPIA LACTIS."
Virgil. *Ecl.* I., 81-2.

"Now cheese and chestnuts are our country fare,
With mellow apples for your welcome cheer."
Dryden.

"AND SUCH THE GROVES
WHICH BLITHE POMONA REARS ON VAGA'S BANKS."
Akenside.

"WHAT SOIL THE APPLE LOVES; WHAT CARE IS DUE
TO ORCHARDS; TIMELIEST WHEN TO PRESS THE FRUITS,
THY GIFT, POMONA."
Philips. *Cider.*

The Fungus Forays, commenced by the Woolhope Naturalists' Field Club in the year 1867, and carried on with so much success every Autumn since that period, have given considerable renown to the Club, and have attracted annually many scientific men, British and foreign, to its meetings. These Forays could not fail to impress upon the members the sad state of neglect into which the orchards of Herefordshire had been allowed to fall, since their decaying trees formed a rich field for many interesting varieties of the Fungus tribe. On one of these Fungus Forays (in the year 1872), when the conversation had turned upon the apples cultivated here, the observation was made, that "celebrated as Herefordshire is for its orchards, it was very remarkable that so few of the best varieties of apples should appear in the markets, or in the fruit shops, of Hereford." This remark, though not followed up, or commented on at the time, was not lost. The celebrated mycologist, the Rev. MILES J. BERKELEY, M.A., F.R.S., &c., was present at the Foray. At that time he was the scientific authority of the Royal Horticultural Society of London, and in the following March (1873), with thoughtful kindness, he induced the Council of the Society to send to the

Woolhope Club, grafts of all the most esteemed varieties of apples grown in their garden at Chiswick.   Large bundles of excellent grafts of ninety-two different varieties came down, and were distributed with much care amongst the members.   This generous gift was highly appreciated, and it encouraged very greatly the study of Pomology in the county.   In the first instance it tended to improve very much the private gardens, by creating the desire to grow good fruit, and to learn the right names of those that were growing there; next it gave rise to the formation of a Special Committee in the Club; to the establishment of Apple and Pear Shows under its influence; and led eventually to the publication of the present work.

The study of Pomology is rather beyond the domain of a strictly Scientific Society; but the members of the Woolhope Club had become strongly impressed with the necessity of some great effort to restore Herefordshire to its true fruit-growing supremacy; to call the attention of the growers to the best varieties of fruit for the table and the press; to improve the methods followed in the manufacture of Cider and Perry, and the quality of these products; and thus to increase in every way the marketable value of its orchard products.   The Herefordshire Agricultural Society was unable to take up the subject, and no other Society, or persons, could be found to undertake the great labour and expense required to carry it to a practical issue.   For these reasons, the members unanimously consented that the organisation of the Woolhope Club should be rendered available for the purpose, impelled, not only by the excellent beginning just mentioned, but also by the desire to render some useful return to the landed proprietors of the county, for the very kind way in which their gardens, parks, woods, and fields, have ever been thrown open to the researches of the Club.

In the production of this important work, the Club has been especially fortunate in obtaining the gratuitous assistance of the learned pomologist, Dr. HOGG—a man of European reputation—who, in accepting the position of Technical Editor, gave a guarantee to the scientific correctness of the Pomona.   Dr. HOGG met the wishes of the Committee in a cordial and generous spirit, and placed the stores of his knowledge—the fruits of life-long study—entirely at its disposal.   To him, therefore, the Woolhope Club has the pleasure of offering its best and most grateful thanks.

The Club has also been greatly favoured by the gratuitous services of two ladies who undertook the preparation of the illustrations for the Pomona, thus removing a difficulty, which, under less fortunate circumstances, might almost have proved insurmountable.   Miss ELLIS had just come to reside in Hereford when the work was under consideration, and on learning the object which the Club had in view, and the disinterested spirit in which it was undertaken, she most kindly and most generously offered her services.   Miss ELLIS is a Queen's gold medalist of the Bloomsbury School of Art, and possesses talents of no mean order. She was joined by Miss BULL, and together, these two ladies have worked most perseveringly for the benefit of the Club.   It is not too much to say, that in this their labour of love, they have spent all the sunshiny hours of eight autumnal seasons in succession.   How admirably they have succeeded, the Plates of the Pomona demonstrate.   The correct representations of the fruit, so essential in a work of this nature, under their skilful hands, have become pictures; and the great beauty of these illustrations, which are superior to anything hitherto produced, is the charming result of their labours.

It should also be mentioned that one of the original drawings made for this work, viz.—that of the EGGLETON STYRE on Plate xxix.—was made by Mrs. STACKHOUSE ACTON, of Acton Scott, Salop. This lady was the eldest daughter of Mr. THOS. ANDREW KNIGHT, and his associate in many of the experiments carried out at Elton, and Downton Castle, at the commencement of the present century. These experiments, especially such as related to the growth and hybridization of fruits, may be said to have created a fresh era in Scientific Horticulture, and thus to have given rise to the great increase in the variety of apples and pears which we now enjoy. Mrs. STACKHOUSE ACTON took a very lively interest in the present work ; she supplied much of the material from which the life of her father was written ; and having made drawings for his " Pomona Herefordensis " in the year 1808, it was a source of great pleasure to her after an interval of seventy years, to be able in 1878 to make a drawing for " The Herefordshire Pomona." Mrs. STACKHOUSE ACTON died at Acton Scott, on January 24, 1882, in her eighty-seventh year.

The letter-press of the work throughout has also been gratuitously afforded. A large amount of original information is distributed throughout its pages from beginning to end, for which the Woolhope Club is deeply indebted to the various writers. Dr. BULL, the general Editor, gives the introductory essays on " The Early History of the Apple and Pear " ; " Modern Apple Lore " ; " Thomas Andrew Knight and his Work in the Orchard "; and " The Life of Lord Viscount Scudamore "; all of which possess the highest interest to the county of Hereford. Due recognition must also be made of the excellent practical monograph on " The Cordon System of Growing Pears," contributed by the Right Hon. the EARL OF CHESTERFIELD (then SIR HENRY H. S. SCUDAMORE STANHOPE, Bart.), giving the results of his own experience at Holme Lacy. Mr. EDWIN LEES, F.L.S., F.G.S., &c., of Worcester, has also contributed a learned and highly interesting paper on the " Crab and its Associations." There is also the able and exhaustive paper on " The Orchard and its Products, Cider and Perry," by the Rev. C. H. BULMER and the Pomona Committee. The labour bestowed on this treatise, embracing so extensive a subject, was necessarily very great. Mr. J. GRIFFITH MORRIS worked at it cordially, and to him is chiefly due the lucid description of M. PASTEUR's views of Fermentation, and much also of the statistical work it contains.

The examination of the juice of the several Vintage Fruits has been made with much care by Mr. G. H. WITH, F.R.A.S., F.C.S., of Trinity College, Dublin, and the accuracy of the results may be relied on. It will be observed that the density, or amount of sugar contained in the juice, has a much lower range than that of the Norman fruits now introduced ; and much lower also than that taken from Herefordshire apples by Mr. THOS. ANDREW KNIGHT at the beginning of the present century. This fact is due, beyond doubt, to the succession of ungenial summer seasons experienced during the last seven years. The Club is also indebted to Mr. WITH for the valuable recipe for Orchard Manure, which will be found on page 160.

The Chromo-lithographs for the work have been executed by Monsieur G. SEVEREYNS, of Brussels. The Plates show their own excellence, and it is satisfactory to know that the artistic skill he has displayed in reproducing the beautiful original drawings so faithfully, together with the testimonials sent by the Club, have aided materially in obtaining for him the high honour of being made a " Chevalier de l'Ordre de Léopold," a distinction which has just been awarded to him by the King of the Belgians

The researches made into Orchard Literature, in addition to their own personal investigations, soon made the Committee aware of the fact, that many of the most valuable fruits of the last two centuries, which are mentioned by Dr. BEALE, 1657 ; WORLIDGE, 1675 ; EVELYN, 1706 ; PHILIPS, 1706 ; HUGH STAFFORD, of Pynes, 1753 ; MARSHALL, 1796 ; KNIGHT, 1808, and other writers, were either altogether lost, or had almost disappeared from the orchards. The following varieties, which formerly were very highly esteemed, may be mentioned as examples, viz : *Woodcock ; Friar ; Pawson ; Oaken Pin ; Arier ; Olive ; Coleing ; Whitesour ; Blackamore ; Mydiate ; Dufflin ; Meriot Ysnot ; Lings ; Dean's Apple ; Peleasantine ; Heming ; Westbury Crab ; Bromsberrow Crab ; The Stocking Apple ; Underleaf ; Best Bache ; Great White Crab ; The Redstreaks, Summer, Winter, Yellow, Moregreen,* and *Red ; White, Red,* and *Green Must ; Summer* and *Winter Fillets ; Elliott ; Devonshire Royal Wilding ; Gennet Moyle ; Hagloe Crab ; Forest Styre ; Skyrme's Kernel ;* and *Foxwhelp.* The neglect to cultivate these valuable varieties is, doubtless, very much to be attributed to the prevailing belief, that "Sorts die out of necessity," or, as Mr. THOMAS ANDREW KNIGHT expressed it, "There was no renewal of vitality by the process of grafting, but that the scion carried with it the debility of the tree from which it was taken." (See further on pp. 31-3, where the subject is discussed and this opinion shown to be opposed to the laws of Vegetable Physiology.)

It was resolved, to put the question once again to the test of practical experience. Mr. RICHARD SMITH CARINGTON, of St. John's Nursery, Worcester, at the request of the Pomona Committee, kindly undertook to conduct the experiment with three good old varieties of fruit which were almost gone. The result is shown by the following "Special Report" to the members of the Club, which was issued in June, 1883 :—

"The Pomona Committee have the great satisfaction to inform the Members, that the experiments they have caused to be carried on during the last four years, for the restoration of those valuable orchard fruits, the FOXWHELP and SKYRME'S KERNEL Apples, and the TAYNTON SQUASH Pear, have completely succeeded. They have now upwards of 800 young trees in vigorous health, viz :—

	Foxwhelp.	Skyrme's Kernel.	Taynton Squash.
One year maidens, about 3 ft. high	500	100	30
Two year's old trees, 4 to 5 ft. high ...	80	30	18
Standard FOXWHELP trees, 5 to 6 ft. high	100		

These young trees have been distributed through the county, and so far as can be judged at present are doing well."

The difficulty of procuring true grafts of the old noted varieties is often very great ; for example, it was not until 1883 that the Committee were able to obtain grafts of those valuable fruits, FOREST STYRE and HAGLOE CRAB. They were obtained at last through the kindness of that excellent practical pomologist, Mr. WILLIAM VINER ELLIS, of Minsterworth, near Gloucester, who sent both fruit and grafts, and these excellent varieties are now being propagated by Messrs. CRANSTON & Co., of King's Acre Nursery, Hereford.

The new varieties of Cider Apples which have been introduced into our orchards during the present century are very numerous. Several of the most valuable bear the name " Norman " and have been represented on Plate xi. of the work, published in 1878. The doubt is there expressed, whether these so called " Norman " apples were really apples from Normandy ; and every effort has since been made to ascertain their history and origin.

Marshall in his book on " Rural Economy " (1796), in the chapter on " Herefordshire Orchards," first notices the fact, of the name " Norman " having been given to a *Wilding* growing in a hedgerow near Ledbury. He very properly points out the error ; but from that time, notwithstanding, the custom seems to have prevailed, more and more, until of late years, all seedlings, or other unknown fruits, especially if they are " Bittersweets," have had the name " Norman " attached to them. The absurdity is very glaring, when the varieties are named after Englishmen, as *Barnett's Norman, Hawkins Norman, Phillips Norman;* or from English villages, as *Cummy Norman, Didley Norman, Marden Norman, &c.;* and equally self evident is the anomaly when they bear such names as " *American Norman,*" *Duke of Normans, Pride of Normans, &c.*

There are, in our orchards, nearly twenty of these so called Norman apples, and several of these varieties have become well-known throughout the county, and are highly esteemed. It was resolved to compare them with the real apples of Normandy.

Last year (1883), through the great kindness of Monsieur FÉLIX DENNIS, a merchant at Hâvre, a very fine collection of cider apples was obtained direct from Normandy and sent to Hereford. Eighty-five of the best Norman varieties were exhibited in the Woolhope Room, at the Free Library ; but not a single one of them was similar to any of the Herefordshire fruits. In order to complete the experiment, it was necessary to take the first opportunity of placing the " Norman " apples of this county upon the tables in Normandy.

This year (1884), a grand Congress of the Pomological Societies of France was announced to be held at Rouen. THE SOCIÉTÉ POMOLOGIQUE DE FRANCE, in conjunction with the ASSOCIATION POMOLOGIQUE DE L'OUEST decided to hold its Session at Rouen from October 2nd to the 12th ; with the co-operation of the SOCIÉTÉ CENTRALE D'HORTICULTURE DE LA SEINE-INFÉRIEURE, and other kindred Societies from the Departments of LA MANCHE, ILLE ET VILAINE, &c. Exhibitions of Table Fruits and Vintage Fruits were also held, including Cider, and all other Orchard products and Orchard Machinery. An invitation was sent to the Woolhope Club to attend the Congress, and a Committee consisting of Dr. HOGG, of London ; Mr. GEO. H. PIPER, of Ledbury ; and Dr. BULL of Hereford, was appointed to represent the Club at Rouen and to compare the Fruits of Herefordshire with those of Normandy. The ability and energy with which these gentlemen carried out their duties, may be said to have added an international feature to the Congress. The Report of the Committee is attached to this paper. Its success was most gratifying, and, as will be seen, a Gold Medal was awarded to the Herefordshire Table Fruit; a Bronze Medal to the Orchard Fruit ; a Silver Gilt Medal to the Cider from mixed fruit, and a Silver Medal to Cider made from a single variety of Apples, and a large Silver Medal was also given to a bunch of Black Alicante Grapes from Eastnor Castle. To the Parts already published of the present work, the high reward of a " Diplôme d'Honneur " was given from each of the Societies under whose auspices the Exhibition for Table and Orchard fruits were held. The very high personal compliment of a Gold Medal, was also given to Dr. Hogg for the great services he has rendered to Pomology.

The comparison of the Orchard Fruits of the two countries more immediately concerns THE HEREFORDSHIRE POMONA. The labours of the Committee here, have also been very effective and practical. They have proved as far as possible, that the so called Norman apples of Herefordshire are not really Norman fruits ; and it may be added, that the result of a long series of enquiries renders it almost certain, that they are merely local seedlings. The conclusion therefore is, that wherever the name " Norman " has hitherto been attached to a descriptive prefix, it should at once be changed into " Hereford," and where it is attached to the name of an English person or an English place, it should be changed to " Kernel" or "Seedling." The following varieties, which were exhibited at Rouen will therefore lose their Norman appellation, and assume the following names :—

BLACK HEREFORD.	RED HEREFORD.
BROADLEAVED HEREFORD.	SPREADING HEREFORD.
BROWN HEREFORD.	SHORTJOINTED HEREFORD.
CHERRY HEREFORD.	SQUARE HEREFORD.
GREEN HEREFORD.	STRAWBERRY HEREFORD.
HANDSOME HEREFORD.	SWEET HEREFORD.
HEREFORD BITTERSWEET.	UPRIGHT HEREFORD.
HEREFORD REDSTREAK.	YELLOW HEREFORD.

The right name of the apple hitherto called *White Norman*, is WHITE BACH, which it must retain ; *Phillips' Norman* should be PHILLIPS' KERNEL ; *Marden Norman*, MARDEN SEEDLING, and so on for all varieties bearing the names of English persons, or English places.

The great care with which the Committee carried out their next very important duty, that of selecting some of the best Norman apples to introduce into Herefordshire, is shewn by the Report. The apples they have selected are ROUGE BRUYÈRE, BRAMTOT, MÉDAILLE D'OR, BÉDAN-DES PARTS, MICHELIN, ARGILE GRISE, DE BOUTTEVILLE, and FRÉQUIN AUDIÈVRE. Sections of them have been taken, and in the order they are named here, they are placed on a plate from the drawings Miss BULL has made of them. Their descriptions, analyses, and characters are also fully given.

The Orchards throughout Normandy—as observed in long journeys in several directions— are well cared for, and the great extent of young orchards planted was very remarkable. It is said that the supply of Cider Fruits is still much below the demand for them, owing in great measure to the extent of the phylloxera disease amongst the vines; whether from this cause, or from changes in Agricultural commerce, the increase of apple trees appears to have become a necessity on the estates of landed proprietors of the West of France, and more particularly of Brittany.

The following official report shows that the increase is not so rapid in England. The AGRICULTURAL RETURNS published by Parliament show that the amount of orcharding in England, that is, "The acreage of arable, or grass land, but used for fruit trees of any kind," was

In 1877	...	...	159,095 acres
„ 1880	..	..	175,200 „
„ 1883	...	...	185,782 „

The following counties stand highest in the list :—

	1877	1880	1883
Herefordshire ...	24,885	26,683	27,081
Devon ...	24,776	25,758	26,348
Somerset ...	20,921	22,993	23,407
Kent	13,097	14,685	17,417
Worcester	14,621	15,854	16,804
Gloucester	11,965	14,178	14,926

Then with a wide difference—

	1877	1880	1883
Cornwall	4,497	4,678	4,869
Dorset	3,814	3,716	4,073
Monmouth	2,932	3,618	3,919
Salop ...	2,944	3,248	3,718
Middlesex ...	3,051 ...	3,249	3,467

The remainder is divided between 39 other counties.

In Herefordshire, and chiefly also in Devonshire and Somersetshire, the hardy fruits grown are almost confined to apples and pears. The increase in the fruit tree acreage for these counties is steady though not great ; but there is ample room to improve the orchards that already exist by supplying the place of the worthless varieties of fruit with those of value.

It has been one great object of this work, to encourage as far as possible the increased growth of the best varieties of Table Fruit. The fact that a Gold Medal was this year awarded to Herefordshire Apples in the centre of Normandy bears out this recommendation, and proves most satisfactorily that both the climate and soil here are well adapted to their cultivation. The demand for an increased supply of Table Fruit must always exist so long as foreign apples are imported so largely as is shown by this Return :—

"The quantity of apples imported into this country from various countries in 1882, amounted to 2,386,805 bushels, valued at £783,906. In 1883 the quantity was 2,251,925 bushels, the value being stated in the Agricultural Returns as £553,488."

American apples are brought in considerable quantities into our own markets, and the careful manner in which they have been sorted out and packed, to make the sample equally good throughout, affords an example that Herefordshire fruit growers would do well to follow, in sending their fruit for sale. Much more might be said on this subject, but it is scarcely necessary, for the practical conclusions must be self-evident to everyone.

VIII. GENERAL INTRODUCTION.

It only remains to give, on behalf of the Woolhope Club, a brief but cordial expression of their obligation to those ladies and gentlemen, who, in various ways, have rendered assistance in the preparation of the Pomona; who have supported the Apple and Pear Shows; and who have supplied fruit of the first quality for illustration. Some names must be given.

The Right Hon. LORD BATEMAN, the Lord Lieutenant of the County, from the first mention of the Pomona, did all he could to encourage the work, and had indeed at that time more faith in the ability of the Club to carry it out, than the Club had in itself. The Right Hon. the EARL of CHESTERFIELD, besides the valuable monograph on "The Cordon System of Growing Pears" already mentioned, has supplied many of the specimens of fruit drawn and coloured on the plates of the Pomona. The LADY EMILY FOLEY, from the commencement, has most kindly placed at the disposal of the Committee any fruit that could be supplied from the extensive collection at Stoke Edith. Mr. HENRY HIGGINS has supplied many excellent varieties from his fine collection of young trees at Thing-hill. The Right Hon. the EARL of DUCIE was good enough to direct Mr. SHINGLES, the gardener, to supply any fruit required from Tortworth. The Committee of the Royal Horticultural Society of London kindly authorised Mr. A. F. BARRON, the Superintendent, to send typical specimens of the varieties cultivated at Chiswick.

The Rev. Prebendary PHILLOTT, and the Rev. THOS. WOODHOUSE, have at all times placed their rich stores of learning at the service of the Committee, and have done much for the Pomona without their work being apparent.

The following gentlemen have also aided the work in various ways : The Rev. SIR GEO. H. CORNEWALL, Bart.; Messrs. J. H. ARKWRIGHT; H. C. BEDDOE; THOS. BLASHILL, of London; THOS. CAM; the late Rev. JAMES DAVIES, of Moor Court; JAMES DAVIES, of Hereford; Dr. CHAPMAN; Rev. Canon DOLMAN; Rev. W. D. V. DUNCOMBE; WM. VINER ELLIS, of Minsterworth, Gloucester; J. T. OWEN FOWLER; Rev. F. T. HAVERGAL; F. H. HERBERT; ARTHUR HUTCHINSON; HUGH JENNER; the late THOS. MASON, of Wellington; Dr. MASTERS, F.R.S., &c., of the "Gardeners' Chronicle"; the late Dr. McCULLOUGH, of Abergavenny; J. GRIFFITH MORRIS; GEO. H. PIPER, of Ledbury; JAMES RANKIN, M.P.; the Very Rev. PRIOR RAYNAL, St. Michael's, Belmont; JAMES RENNY, The Almners, Surrey; and WORTHINGTON G. SMITH, F.L.S., of London.

The following fruit growers in the county have also rendered good service in supplying information, and fruit from their orchards on many occasions : Messrs. W. H. APPERLEY, Withington; JOHN BOSLEY, Lyde; HENRY DAVIES, Venns Green, Marden; CHARLES JONES, Bell Orchards, Ledbury; J. H. HALL, Garford; WM. HILL, Lower Eggleton; the late WM. JAY, Lyde; WM. SIVELL LANE, Bosbury Farm; JOHN TAYLOR OCKEY, Leddon Court, Bishop's Frome; J. H. SMITH, Hall House, Ledbury; the late WILLIAM WARD, Stoke Edith Gardens; CHARLES WATKINS, Wilcroft; JOHN WATKINS, Pomona Farm, Withington; and HERBERT YEOMANS, of Canon Pyon.

The Club has also been indebted for fruit on several occasions to the following nurserymen : Messrs. GEO. BUNYARD, Maidstone; CHEAL and SONS, Crawley, Sussex; CRANSTON and Co., King's Acre; RIVERS and SON, Sawbridgeworth; RICHARD SMITH and Co., Worcester; and WHEELER and SON, Gloucester.

In the first instance a General Committee was formed, to superintend the publication of the work, with the president of the Club, Mr. J. GRIFFITH MORRIS, as Chairman ; and a Sub-committee, with the Rev. C. H. BULMER, as Chairman, both of which included many of the names given above ; but it was soon found that the delay caused by this arrangement was too inconvenient, and that a small Central Committee, residing in Hereford, was necessary. This Committee consisted of DR. BULL, MR. THOMAS CAM, and MR. J. GRIFFITH MORRIS.

The management of the Apple and Pear Shows, with all the necessary trouble entailed, was kindly undertaken and carried out by Mr. H. C. MOORE, with the assistance of Mr. D. R. CHAPMAN, of the Free Library, and very well the work was done.

Lastly, throughout the work of the Pomona and Fruit Shows from beginning to the end, Mr. J. REGINALD SYMONDS has been at the helm as Honorary Secretary and Treasurer.

To one and all of these gentlemen, the thanks of the Woolhope Club are due, and are hereby gratefully tendered for the assistance they have so freely given, in bringing this work to a satisfactory conclusion.

HENRY G. BULL.

January, 1885.

"The Members of the Woolhope Club desire to offer their special and grateful thanks to DR. BULL for the ability and perseverance with which he has carried out the publication of 'The Herefordshire Pomona.' In addition to his own valuable papers, he has edited and superintended the Work from beginning to end. He originated it, and without the untiring diligence and zeal he has displayed for so many years, it could not have been so successfully completed."

J. Griffith Morris 1877
A. W. Whifott 1878
Arthur Armitage 1879
J. H. Knight 1880
Augustin Ley 1881

Thomas Blashill 1882
George H. Piper 1883
Charles Burrough 1884
C. G. Martin 1885

Theophilus Lane Secretary.

March, 1885.

(APPENDIX TO THE INTRODUCTION.)

REPORT OF THE COMMITTEE APPOINTED TO ATTEND THE CONGRESS OF THE POMOLOGICAL
SOCIETIES OF FRANCE, ON BEHALF OF THE WOOLHOPE CLUB, HELD AT
ROUEN, from OCTOBER 2nd TO THE 12th, 1884.

Your Committee, having obtained the Schedules of the Exhibition to be held at Rouen, thought it best to compete in the classes open to strangers. A collection of Table Fruit was therefore obtained from the gardens of Stoke Edith, Holme Lacy, Thing-hill, and other places. It consisted of fifty-seven varieties of Dessert Apples, fifty-seven varieties of Culinary Apples, and thirty-six varieties of Pears. This collection was very fine. It formed the leading attraction at the Exhibition in the Hall of the Hôtel de Sociétés Savantes, and a Gold Medal was awarded to it by the Société Centrale d'Horticulture de la Seine Inférieure.

A fine bunch of Black Alicante Grapes from the garden at Eastnor Castle was also taken, and received from the same Society a large Silver Medal.

The collection of Orchard Fruits exhibited by the Woolhope Club consisted of fifty-six varieties of Cider Apples, and forty-two varieties of Perry Pears. To this collection the Association Pomologique de l'Ouest awarded a Bronze Medal.

Two varieties of Cider made from mixed fruits, and four varieties of Cider made from a single variety of fruit; with two varieties of Perry, were also exhibited. To the Cider from mixed fruits a Silver Gilt Medal was given, and to that from a single fruit, a Silver Medal. Prizes were not offered for Perry, of which very little is made in Normandy.

The first six Parts of the present Work, THE HEREFORDSHIRE POMONA, were also exhibited, and a " Diplôme d'Honneur" was awarded to the Woolhope Club, from the Société Centrale d'Horticulture de la Seine Inférieure, for the Table Fruits represented in the Work; and a second was also given by the Association Pomologique de l'Ouest for the Vintage Fruits.

A Gold Medal was also specially awarded to Dr. Hogg, for his life-long work in Pomology.

The receipt of these high honours did not cause your Committee to forget that the chief objects of their visit to Rouen were, first, to ascertain whether the Apples called " Norman " in Herefordshire were really Norman varieties; and secondly, if they were not so, to select a few of the most valuable varieties from the Norman orchards to introduce into Herefordshire.

Eighteen of the best so-called Norman Apples of Herefordshire were placed together on the exhibition tables at Rouen. Your Committee carefully compared them with the three thousand plates of Vintage Fruits present: the attention also of the leading exhibitors from Normandy and Brittany was specially called to them; but, with one exception, they were quite different to all others there, and were unknown to the Norman nurserymen and growers. The exception was the "*Foley Norman*" which local tradition states to have been introduced into Herefordshire by Mr. Edward Thomas Foley of Stoke Edith (c. 1810-20). This Apple was the same as the *Blanc Doux* of the Rouen Catalogue, but it is one that has not borne well the modern test of exact analysis, and it has therefore lost much of its repute in the Norman Orchards.

Your Committee next proceeded to select a few of the best real Norman varieties to be introduced into Herefordshire. They decided that the following characteristics were essentially necessary: 1. The fruit must possess the very best quality of juice; 2. The trees must be hardy, vigorous and fertile; 3. They must blossom at varying intervals; 4. The fruit must attain maturity in late autumn, or winter; and 5. They must have obtained the highest repute in the Norman orchards.

With the kind assistance of Monsieur A. Hauchecorne (one of the distinguished authors of the great French work "Le Cidre"); M. Michelin, of Paris (one of the original promoters of the Congress appointed by the French Government for the study of Cider Fruits); Monsieur Héron (President of the SOCIÉTÉ CENTRALE D'HORTICULTURE DE LA SEINE INFÉRIEURE); M. Legrand, Nurseryman at Yvetot; M. Lesueur, of Rouen; and other Norman growers of Cider Fruits; your Committee have selected eight varieties, which meet all the requirements laid down for their guidance.

Specimens of their several fruits have been procured from Normandy and they have been drawn and coloured by Miss Bull on the accompanying plate to illustrate their detailed description.

HISTORY AND DESCRIPTION OF THE NORMANDY CIDER APPLES INTRODUCED BY THE WOOLHOPE CLUB INTO HEREFORDSHIRE, 1884.

[N.B.—The numbers refer to the fruits as represented on the Plate; and a Table at the end gives the summary of their habits, analysis, and character, as published in the Rouen Catalogue.]

1.—ROUGE BRUYÈRE.

An old variety, whose history is not known. Its name is often given to other Apples (varieties of *Argile* or *Fréquin*, &c.), a fact which proves the general esteem in which it has long been held.

2. Bramley

1 Rouge Bruyère

3 Médaille d'or

4 Bédan-des-parts

5 Micheim

Angloe grise

7 de Boutteville

6 Fréquin Audièvre

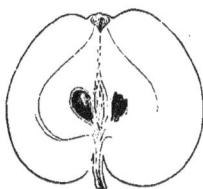

1.—ROUGE BRUYÈRE.

Fruit: small and symmetrical, broad at the base, becoming slightly angular at the upper third. Skin : almost entirely carmine, deeper on the sunny side, and having small grey spots scattered over the surface. Eye : small and closed, set in a very shallow depression. Stalk : short and woody, inserted in a narrow cavity, lined with russet, which also spreads over the base of the apple. Flesh : whitish yellow, firm, with a sweet juice, a bitter, pleasant taste and an excellent aroma.

This is a very favourite apple throughout the orchards of Normandy. "It is superior," says M. Hauchecorne, "to all others bearing its name, and makes excellent cider without mixture with other fruits." Its esteem is only equalled by the *Argile grise.* The abundance of tannin in the juice, renders it very valuable to give good keeping qualities to the cider from mixed fruits. The density of the juice is 1.075 to 1,080. In 1,000 parts it contains of alcoholisable sugar 175; tannin 7; mucilage 8; acidity (as compared with monhydrous sulphuric acid) 1 ; salts, &c. 9; and water 800.

2.—BRAMTOT.

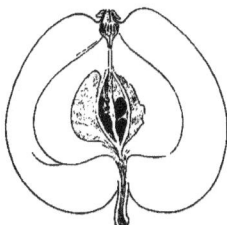

2.—BRAMTOT.

A seedling grown by M. Legrand, of Yvetot, Seine Inférieure. It first fruited in 1856, and was named after M. Bramtot, a manufacturer of Yvetot. It is thought to be a seedling from the old variety, *Martin Fessard.*

Fruit: of middle size, symmetrical, but sometimes with unequal sides, wide and flattened at the base, but contracted towards the eye. Skin: clear yellow, with a touch of carmine towards the sun, its surface being scattered over with numerous grey spots. Eye: small and closed, with long reflected sepals, and placed in a very narrow cavity with grooved sides. Stalk: short, thin, and woody, set in a narrow, deep cavity. Flesh : whitish yellow, and tender, with an abundant juice of a sweet and pleasant though slightly bitter flavour.

"This excellent variety," says M. Hauchecorne, "both in tree and fruit, possesses virtues as an apple for the press, which are rarely united in so high a degree." The juice is of good colour, and has a pleasant aroma. Its density is so high as 1,092, and in good seasons it reaches 1,105. A kilogramm contains 226 grammes of sugar, which gives an alcoholic strength from 13 to 14 per cent. There are also 6 grammes of tannin and 1,070 of acidity in each kilogramm of juice.

3.—MÉDAILLE D'OR.

A seedling raised by M. Goddard, of Boisguillaume, Rouen. A Gold Medal was awarded to its fruit in 1873 for its superior properties by the SOCIÉTÉ CENTRAL D'HORTICULTURE DU DÉPARTEMENT DE LA SEINE INFÉRIEURE.

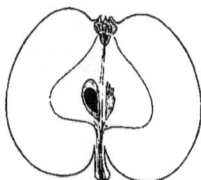

Fruit: small, oblate, broad at the base, often irregularly spheroidal. Skin : golden yellow, almost completely covered with a marble work of thin brown russet, which often concentrates in patches, and becomes continuous round the eyes ; there is often a slight touch of rose colour on the side next the sun.

3.—MÉDAILLE D'OR.

Eye : large and closed, sunk in a deep cavity, with slightly grooved borders. Stalk : thin and woody, about half an inch long, and inserted in a deep depression. Flesh : yellowish and tender. Juice : very sweet, with a strong, rough, astringent flavour, and not unpleasant.

The tree is very fertile and bears its fruit in clusters. In general appearance, and lightness of structure, this fruit resembles the old English variety *Forest Styre*. As a vintage fruit it takes the very highest rank. The juice attains the very high density of 1,102 ; and each kilogramm contains 238 grammes of sugar, giving 14 to 15 per cent. of alcohol ; 5.509 of Tannin ; and 1,428 of acid, as compared with monhydrous sulphuric acid.

4.—BÉDAN-DES-PARTS.

The *Bédan, Bédengne, Bec d'Âne*, with other varieties in name has held a high repute in the Norman Orchards from time immemorial. This particular variety is superior to all the *Bédans* in the richness and colour of its juice. It is a seedling grown by M. Legrand, of Yvetot, which first bore fruit in 1874.

4.—BÉDAN-DES-PARTS.

Fruit: small, broad at the base, often larger on one side. Skin: pale yellowish green, with a clear red cheek on the side next the sun; small grey spots are scattered over the surface, and sometimes brown patches. Eye; small and closed, set in a shallow, irregular cavity, with grooves and small tubercles between them. Stalk: strong, half an inch long, inserted in a shallow, narrow cavity, which is lined with russet, which russet extends, more or less, over the base of the apple. Flesh: yellowish, tender, and juicy, slightly bitter in taste, but with good flavour. Juice: highly coloured.

"This new variety," says M. Hauchecorne, "takes a high place among fruits of the first quality, from the fertility of the tree, the high colour of its juice, and its richness in sugar, tannin, and aroma." The density of the juice is 1,084. One thousand parts contain of alcoholisable sugar 195; tannin 5; mucilage 10; acidity 1,070; salts, &c. 1,030; and water 776.

5.—MICHELIN.

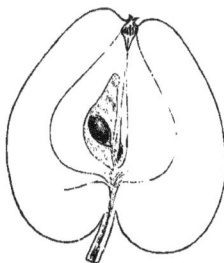

5.—MICHELIN.

A seedling raised by M. Legrand, Yvetot. It first bore fruit in 1872, and was dedicated by him to M. Michelin, of Paris, Chevalier de la Légion d'Honneur, Member of the Société Centrale d'Horticulture de France, et de la Seine Inférieure, and one of the original promoters of the Congress appointed by the French Government for the study of Cider Fruits, and who attended all its meetings.

Fruit: of middle size, conical, with obtuse angles, becoming more marked as the fruit becomes more narrow towards the eye. Skin: green throughout, becoming yellowish green as it ripens; it presents a slight blush of red on the sunny side, and numerous small specks over the surface, with here and there a streak of russet. Eye: small and closed, almost level with the surface, and surrounded by a patch of light grey russet. Stalk: half an inch long, and inserted in a shallow cavity lined with russet, which spreads in streaks over the base of the apple. Flesh : white, tender, sweet, and rich.

"This is an apple of the highest merit," says M. Hauchecorne, "and is well worthy of extensive cultivation." The juice has a high colour and a density of 1,083. In 1,000 parts there are of alcoholisable sugar 194 ; tannin 5,509 ; mucilage 11 ; acidity 1,071 ; salts, &c. 8,420 ; and water 780.

6.—ARGILE GRISE.

The *Argile* is one of the oldest varieties in the Norman orchards. Its origin is unknown, but it has been long highly esteemed in all the Departments of the North-west of France in which Cider is produced. Its name is so popular that it has been given to many varieties, and often to those of inferior value. The *Argile Grise* is the best of all the varieties.

Fruit : rather below the middle size, ovoid, with obtuse angles as it narrows towards the eye ; often fuller on one side than the other. Skin : greenish yellow, more or less covered with a thin grey russet ; it sometimes takes a pale tinge of red colour on the sunny side. Eye : small and closed, with short, broken sepals, seated in a shallow cavity, with folded margins and small tubercles between the folds. Stalk : small and short, frequently connected with the fruit by a fleshy prominence on one side. Flesh : yellow and tender. Juice : plentiful, slightly bitter but still sweet and pleasant.

6.—ARGILE GRISE.

"The *Argile Grise* belongs to the *Fréquin* group of Cider Fruits," says M. Hauchecorne ; "it is equally valued in the orchard with *Rouge Bruyère*," and is believed to make cider of the best quality. The juice has a good colour and a density of 1,075, and sometimes more. One thousand parts contain of alcoholisable sugar 194 ; tannin 5,509 ; mucilage 15 ; acidity 0.920 ; salts, &c. 3,571 ; and water 781.

7.—DE BOUTTEVILLE.

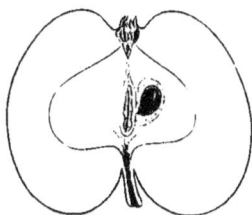

7.—DE BOUTTEVILLE.

A seedling raised at Yvetot, by Monsieur Legrand. It first fruited in 1873, and was dedicated to Monsieur L. De Boutteville, Honorary President of the Société Centrale d'Horticulture de la Seine Inférieure, and Author, with Monsieur A. Hauchecorne, of the celebrated work "Le Cidre," published at Rouen, in 1875.

Fruit : of middle size, oblate, smooth and round, without angles. Skin : pale yellow, with an orange blush on the sunny side, more or less spotted over the surface, and the spots often become dark and tinged with red under the sun's influence. Eye : closed, seated in a narrow, deep cavity, with folded margins. Stalk : short, placed in a broad and deep cavity, lined with a thin russet that radiates over the base of the apple. Flesh : yellowish, with a sweet and pleasant flavour, free from bitterness. Juice : of a high colour, sweet, and pleasant.

" This apple," says M. Hauchecorne, " is one of the best varieties for making a good cider that will keep well. The apple is firm in flesh and travels well. Its juice is well coloured with excellent perfume and taste." It has a density of 1,083. One thousand parts contain of alcoholisable sugar 193 ; tannin 6 ; mucilage 11 ; acidity 2.14 ; salts, &c. 7.86 ; and water 780.

8.—FRÉQUIN AUDIÈVRE.

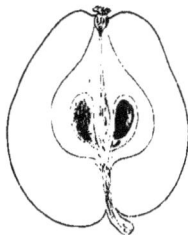

8.—Fréquin Audièvre.

A seedling raised by Monsieur Audièvre, treasurer of the Société d'Horticulture d'Yvetot, in 1868. It is thought to have been a seedling from *Petit-Fréquin,* or *Fréquin Rouge,* with greatly improved qualities to either of these varieties.

Fruit : very small, flattened at the base, but contracting rapidly towards the eye. Skin : with a pale yellow ground, almost entirely covered with red carmine, and frequently with many fine white spots on the surface. Eye : small and closed, set in a narrow cavity with sulcated borders. Stalk : variable, generally very short, and set obliquely in a small and shallow cavity. Flesh: yellowish white, and firm. Juice: sweet, slightly bitter, but with good perfume and flavour.

"This valuable variety," says M. Hauchecorne, "possesses the highest merit of the *Fréquin* tribe. It contains all the elements for making a strong, pleasant, and healthy Cider." The juice has a very high colour, and a density of 1,079. One thousand parts contain of alcoholisable sugar 180; tannin 5,509; mucilage 12; acidity 1,320; salts, &c. 11,171; and water 790.

The table on the next page presents, in a condensed form, the valuable qualities of all these Norman apples. It is taken from the "Catalogue" of the Société Centrale d'Horticulture de la Seine Inférieure.

Trees of all these valuable varieties of true Norman apples have been sent to Messrs. Cranston and Co., King's Acre, Hereford, who will propogate carefully from them. It is believed that they will prove very valuable in the orchards of Herefordshire.

Your Representatives, in conclusion, desire to express their great sense of the kindness and courtesy shown to them during their visit to Rouen.

ROBERT HOGG.
GEO. H. PIPER.
HENRY G. BULL.

October, 1884.

NORMAN CIDER APPLES, INTRODUCED BY THE WOOLHOPE CLUB, 1884.

HART.—CHARACTER—ANALYSIS OF JUICE, &c., FROM THE CATALOGUE OF THE SOCIÉTÉ CENTRALE D' HORTICULTURE DE LA SEINE-INFÉRIEURE.

No. on Plate.	Name.	Time of Blossoming.	Ripe.	Flavour of Fruit.	Quality of the Juice.	Density of Juice (Sp. Gr.)	Useful Properties contained in 1 Kilogram (35 oz., 100 gr.) of Juice.				Character of the Tree.
							Sugar Grammes (15½ gr.)	Alcohol per 100.	Tannin Grammes (15½ gr.)	Acidity Grammes (15½ gr.)	
1	Rouge Bruyère	Beginning of May.	November	Slightly bitter, excellent.	High in colour, very good flavour.	1,075 to 1,080	175	10 to 11	7,000	0,001	Healthy and fertile, round headed growth.
2	Braintot	Beginning of May.	November	Slightly bitter sweet, excellent.	Good in colour, perfume, and taste.	1,092 to 1,105	226	13 to 14	6,000	1,070	Healthy, vigorous and fertile, of handsome growth.
3	Médaille d'Or	Beginning of June.	November	Bitter, excellent.	Good in colour, and perfume.	1,102	238	14 to 15	5,599	1,428	Very vigorous and fertile, with upright growth.
4	Bédan-des-Parts	End of April.	December	Bitter sweet, excellent.	Very good in colour, taste, and perfume.	1,084	197	12	5,000	1,070	Horizontal growth, healthy, vigorous, and fertile.
5	Michelin	End of May.	December	Sweet and good.	Good in colour, flavour, and perfume.	1,080 to 1,083	190	11 to 12	5,599	1,071	A round headed tree, healthy, vigorous, and fertile.
6	Argile Grise	Beginning of May.	November and December	Slightly bitter, very good.	Good in colour, perfume, and taste.	1,075 to 1,080	194	10 to 11	5,599	1,071	Upright in growth, vigorous, and fertile.
7	De Boutteville	Beginning of May.	December	Sweet and good.	High in colour, perfume, and flavour.	1,083	183	11 to 12	6,000	2,142	Upright in growth, healthy, and very fertile.
8	Fréquin Audièvre	End of May.	December	Bitter sweet, excellent.	High in colour, good taste, and perfume.	1,079	180	10.5	5,599	1,320	Horizontal growth, vigorous, and fertile.

PART I.   PRICE 16s.

THE

HEREFORDSHIRE POMONA,

CONTAINING

COLOURED FIGURES AND DESCRIPTIONS OF THE MOST ESTEEMED KINDS OF

APPLES AND PEARS.

EDITED BY

ROBERT HOGG, LL.D., F.L.S.,

Honorary Member of the Woolhope Naturalist's Field Club; Secretary of the Royal Horticultural Society;
Author of 'The Fruit Manual'; 'British Pomology'; 'The Vegetable Kingdom and its Products', &c., &c.

" Hope on. Hope ever."

" Ζεφυρίη πνείουσα τὰ μὲν φύει ἄλλα δὲ πέσσει,
ὄγχνη ἐπ' ὄγχνη γηράσκει, μῆλον δ' ἐπὶ μήλῳ,
αὐτὰρ ἐπὶ σταφυλῇ σταφυλή, σῦκον δ' ἐπὶ σύκῳ."
Homer Odyssey vii. 119-22.

LONDON: HARDWICKE AND BOGUE, 192, PICCADILLY, W.
HEREFORD: JAKEMAN AND CARVER, HIGH TOWN.

1878.

THE HEREFORDSHIRE POMONA.

IMPORTANT CORRECTIONS.

PLATE IV.—Erase "Monarch," and write, 1. "An unnamed Seedling Pear of Mr. T. A. Knight's." See PLATE LI. (51) for the true KNIGHT'S MONARCH.

[The figure given here represents the variety which Mr. Knight's Gardener distributed in error.]

PLATE X —Erase from the *letterpress* No. 4, the synonym "New Hawthornden."

PLATE XI.—In the names to figures 2, 3, 4, 5, 6, and 7, and also in the letterpress, erase the word "Norman," and write "HEREFORD."

PLATE XII.—For "Cok's Pomona, read COX'S POMONA.

PLATE LIII.—For Bergamotte "Dietrich," write BERGAMOTTE HERTRICH

" Ζεφυρίη πνείουσι τὰ μὲν φύει ἄλλα δέ πέσσει.
ὄγχνη ἐπ᾽ ὄγχνη γηράσκει, μῆλον δ᾽ ἐπὶ μήλῳ,
αὐτὰρ ἐπὶ σταφυλῇ σταφυλή, σῦκον δ᾽ ἐπὶ σύκῳ."
Homer Odyssey vii. 119-22.

" THE BALMY SPIRIT OF THE WESTERN GALE,
ETERNAL BREATHES ON FRUITS UNTAUGHT TO FAIL ;
EACH DROPPING PEAR, A FOLLOWING PEAR SUPPLIES
ON APPLES APPLES, FIGS ON FIGS ARISE."

Pope.

THE EARLY HISTORY OF THE
APPLE AND PEAR.

Ὄγχιας μοι δῶκας τριςκαίδεκα καὶ δέκα μηλέας
συκέας τεσσαράκοντα."   *Homer. Odyssey XXIV. 339.*
              (c. 700 B.C.)

" Twelve pear trees bowing with their pendant load
And ten, that red with blushing apples glowed,
Full fifty purple figs."   *Pope.*

" Pomaque et Alcinoi Silvæ : nec surculus idem
Crustumiis Syriisque piris, gravibusque volemis."
                        Virgil. Geo. II. 87–9.
                          (c. 40 B.C.)

" So apples and Phæacian orchards gleam
With divers hues : and pears diversely team
Crustumian, Syrian, and the big Voleme."
                         Blackmore.

" Pomis proprietas pyrisque vini."
           Pliny. Nat. Hist. LXV. c. 15.
                  (c. A.D. 50.)

" Wine may be made from apples and from pears."

" Kent doth abound with apples of most sortes, but I have scene in the pastures
and hedgerowes about the grounds of a worshippfull Gentleman dwelling two miles from
Hereford, called *M. Roger Bodnome*, so many trees of all sortes that the seruants drinke
for the most part no other drinke but that which is made of Apples. The quantitie is
such, that by the report of the Gentleman himselfe, the Parson hath for tithe many
hogsheads of Syder."
           Gerarde. Herball. (1597.)

THE fruits that have ever been most widely cultivated are the Apple and the Pear. The trees
grow freely and without trouble in all temperate climates. The beauty and fragrance of the
blossom give a charm to Spring, and in Autumn the fruit is varied, delicious, and abundant. It is

excellent when fresh, keeps well, and may be cooked in a variety of ways. The Apple and the Pear are the most wholesome and the most generally useful of all fruits.

From the earliest records the Apple has ever been held in the highest esteem, and its merits are so great, that we cannot wonder that it should be surrounded with many poetical and super-stitious fancies. Imagination has indeed been busy with it. In every age and in almost every country some poetical legend or some mystical allusion concerning the apple is to be found. It was an apple that Paris awarded to Aphroditê as the prize of beauty. It was the golden fruit of apples that the dragon watched in the garden of the Hesperides. This is the healing fruit of the Arabian tales. In Greece its name, strangely enough was the same as that of sheep, and it thus became the symbol of all manner of wealth. Ulysses enjoys it in the garden of Alcinous. Tantalus gasps for it vainly in Hades. The golden apple is the golden ball which the Frog-prince brings up from the water: the golden egg which the red hen lays in Teutonic story: the gleaming sun which is born of the morning. Golden apples are frequently used in Grecian mythology. They do not always mean golden fruit of any kind, but sometimes denote the golden tinted cloud-flocks, or herds of Helios. The fact that the same word "μῆλα" means apples, as well as sheep, accounts for the transformation of numerous myths of cloud-flocks, into stories of boughs of golden apples—notably in the case of the golden apples which Gaia gave to Hêrê when she became the bride of Zeus, these apples being the golden herds of Helios, the apples of the Hesperides guarded by the fierce dragon Ladon, which never slept. *Cox. "Aryan Mythology"*.

It was one of the labours of Hercules to procure this golden fruit, or these fine herds; but at the Heracleia, or festival held in honor of Hercules, by the Thisbians and Thebans in Bœotia, by another play upon the word, the sheep again became apples, which were offered at his altars. The custom arose in this way. It was always usual to offer sheep to Hercules, but the overflowing of the river Asopus, on one occasion, having rendered it impassable, the sheep could not be con-veyed across the stream, and some youths remembering the ambiguity of the Greek word, offered apples instead, with much sport and festivity. To represent the sheep they raised an apple upon four sticks as the legs, and two more were placed on the top to represent the horns of the victim. Hercules was delighted at the ingenuity of the youths, and the festivals were ever after continued with the offering of apples.

The Scholiast on Hesiod (Theog. 215,) says "The golden apples are the stars. The even-ing hours are said to tend them, because it is at that time that we see the stars. But Hercules is the sun; and when the sun rises the stars disappear."

The golden apple itself is also used for the new born sun, as in the story of the marriage of Peleus with the sea-sprung Thetis, when Eris the goddess of discord, the only one of the gods not invited to the wedding feast, cast the golden apple on the banquet table, inscribed "a gift for the fairest." It was forthwith claimed of right by the three queens of heaven, as goddesses of the dawn, Hêrê, Athênê, and Aphroditê—Hêrê as queen of the blue sky, Athênê as cold queen of the morning light, and Aphroditê as queen of the warm young rays. Paris bestowed the apple on Aphroditê, and henceforth the wrath of Hêrê and Athênê dwelt always on the city of Ilion.

It was by the aid of the three golden apples from the garden of the Hesperides, or as some versions of the story run, from an orchard in Cyprus, that Hippomenes won the race with the fleet Atalanta. He artfully threw them down separately at some distance from each other, and the fair nymph, charmed at the sight, stopped to pick up the apples: Hippomenes hastened on the course, arrived first at the goal, and obtained Atalanta in marriage.

As the Grecian, so also do the myths of the Scandinavian and the Sclavonic nations abound in references to the apple, often very curious, and always interesting. The poetical fancies repeat themselves again and again in varied terms, and the mythical apples are usually golden. Where however a touch of reality appears, the apple may be the citron, the orange, the quince, the peach, the apricot, or the pomegranate; or it may even be the plum, the almond, or the nut.

In the prose Edda we are told that the goddess Iduna had the care of the apples which had the power of conferring immortality. They were, consequently, reserved for the gods who ate of them when they began to feel themselves growing old. It is in this manner that they will be kept in renovated youth until Ragnarok, the general destruction. The evil spirit Loke, it is said, took away Iduna and her apple tree, and hid them in the forest where they could not be found by the gods. In consequence of this malicious theft everything went wrong in the whole world. The gods became old and infirm; enfeebled both in mind and body. They no longer paid the same attention to the affairs of the earth; and men having no one to look after them, fell into evil courses and became the prey of mischief and distress. At length the gods finding matters getting worse and worse every day, roused their last remains of vigour, and, combining together, forced Loke to restore the tree. In the prose Edda also, Skirnir offers eleven golden apples and the ring Draupnir—from which on every ninth night eight equally heavy rings drop—to Gerda, if she will return Freyr's love.

A Polish legend, given by Mannhart, says, There is a glass mountain, on the top of which stands a golden castle, and before it is a tree of golden apples. In the castle lives the enchanted daughter of a prince. Many vainly try to get on the mountain; but at last the youth who has fastened the claws of a lynx to his hands and feet is successful. With the golden apple, he calms the dragon he finds at the entrance; and finally having broken the spell that bound the princess, he must remain with her and not return to the lower earth.

In the curious Polish myth, "Madey," a Bishop, passing through a dense forest, smells the sweet odour of apples, and his attendants discover a tree full of fruit, which they are unable to gather, and under the tree an old gray-haired man is kneeling. It is the desperate robber Madey, full of repentance and sorrow for the past. He entreats the Bishop to hear his confession and give him absolution. His request is granted, and during the confession, the apples on the tree, one after another, are changed into snow white doves, fly up, and disappear. Soon there is only one apple on the tree. It is the soul of Madey's father, whom he had murdered, but could not bring himself to confess the deed. At length he does so, and the last apple, as a beautiful white dove, flies away to heaven. The Bishop in the end grants him absolution, and the body of Madey crumbles into dust. "Sclavonic Fairy Tales." Naake.

A Hanoverian legend says that a girl was asked by the dwarfs to be godmother to one of their children. On the appointed day she was led down a beautiful staircase which was under an apple tree in the court of a superb garden, whose trees were laden with fruits. She was repaid for coming with an apron full of apples, which when she returned to the earth's surface were found— like the fruit of Aladdin—to be of solid gold.

The golden apples are often met with in Northern Mythology. The Golden Bird seeks the golden apples in the king's garden in many a Norse legend; and when the tree bears no more, the good mother goddess " Frau Bertha " reveals to her favourite, that it is because a mouse gnaws at the tree's root. These golden apples, it is related in some legends, may be taken from a tree growing over a fountain of holy water with a rejuvenating power; myths traceable to the tree and fountain of Urd, one of the Naunir.

In a pretty Sclavonian myth,—the Russian Cinderella,—a small apple, rolled round a small silver plate, shews everything its fair owner wishes to see. Her sisters choose fine dresses and are evil-minded; but she is all goodness, and becomes of course the Czarina.

The Prince Ahmed, in the Arabian Nights' Entertainment, discovers the wondrous apple which has the power of restoring all sick persons who smell it, to immediate health, and he thus cures the princess Nouronnihar. The Bohemian version of the same legend makes the one apple into three, which must be eaten, and thus the king, the queen, and the princess Libena are all snatched from the grave.

In the Danish story of " Svend's Exploits," his father gives Svend three magic apples grown from a pip brought from a dragon's island. Two of the apples were large and red, but the third was small, shrivelled and green. " Don't eat the apples yourself," said his father, " and take especial care of the least, for though it looks the worst it is the best of them all, and can cure any injury caused by the others." A certain knight, Peter, afterwards robbed Svend of all his magic treasures, among them the two fine looking apples; and being through gross duplicity, on the eve of espousing a fair princess whose hand had been pledged to Svend, he gave one apiece to the princess and the king, her father. The king and princess had no sooner eaten them, than their royal noses began to grow to such a frightful length that nobody in court could refrain from laughing. No one but Svend could restore those royal features to their normal propriety. He denounced knight Peter as an impostor, fetched the shrivelled, green apple, and dividing it, gave half to the father and half to the daughter, whose noses immediately resumed their natural proportions, and Svend was rewarded with the hand of the princess and the half of her father's kingdom. *Thorpe.* "*Yuletide Stories.*"

The apple is again said sometimes to make the nose grow so large that the sacred pear can alone restore it to moderate size. In the goddess Holla's garden the favourite fruit trees are the apple and the pear—the latter of which fruits retained its sanctity in France long after the introduction there of Christianity.

The apple however is not always beneficent. Azrael, the angel of death, accomplished his mission by holding an apple to the nostril: and in the Northern folklore, Snowwhite is tempted to her death by an apple, half of which a crone has poisoned, but she recovers life when the apple falls from her lips.

A present of apples is ofttimes the symbol of a matrimonial proposal.

" Malo me Galatea petit." (*Virgil. Ec.*)

In the Servian legend of " The Three Brothers," whose only property was a pear tree, from which they shared the fruit, each of them in turn gives a pear to a poor beggar. The good angel on whom they had thus bestowed charity unconsciously, grants to each of them his wish. One asks for rivers of wine; the second for countless flocks of sheep; and about these no difficulty is made.

But the third asks for a pious wife; which it seems was a hard thing to find in those days. He was directed however to propose for a certain king's daughter. Two princes were before him, and had put their apples on the table, he puts his apples by the side of the others, and aided again by the good angel, he obtains the princess, but is sent with her to live in poverty in the midst of a forest. The charity of the elder brothers is quickly destroyed by their worldly wealth, and they are reduced to their pear tree again; whilst that of the younger one, and his good wife is increased, and meets with the customary reward of fairy wealth and happiness. *"Sclavonic Tales."* *Naake.*

The Volsunga Saga tells us how King Rerir (son of Sigi) and his wife having grown old and childless, cried without ceasing to the gods to give them a child. And Freyja of the golden hair, taking pity on them, called Ljod, her handmaiden, and gave her an apple to drop in the queen's lap. The queen ate the apple, and in due season bare a son, the famous Volsunga.

The apple tree has been supposed to be of Eastern origin, in consequence, probably, of the frequent occurrence of the word "apple" in our translation of the sacred writings. It has been said to be the tree of knowledge from which Eve ate the fruit in Paradise; it is mentioned by the prophet Joel; and it is thought to be the tree alluded to by Solomon.

"As the apple tree among the trees of the wood so is my beloved among the sons."—*Cant. ii 3.*

"Comfort me with apples."—*Cant. ii. 5.*

"A word fitly spoken, is like apples of gold in pictures of silver."—*Proverbs xxv.*

The best authorities in the Hebrew language believe that the citron is the fruit spoken of here, and not the apple. It is right, however, to say, that Dr. Thompson, the Eastern Missionary, in his well known work *"The Land and the Book,"* states that when at Askelon in June he saw "quite a caravan" start for Jerusalem, loaded with apples that would not have disgraced an American orchard. He thinks that the translation which represents the Hebrew word "taffûah" by *citron* is not correct. "The Arabic word," he says, "is almost the same as the Hebrew, and it is as definite, to say the least, as our English word." It means apple. Citrons are large, weighing several pounds each, hard and indigestible. The tree must be propped up, or the fruit will bend it to the ground. It is scarcely a tree at all, and too small and straggling to give any shade, and no one would dream of sitting under it. And he concludes, "as to smell and colour, all the demands of the Biblical allusions are fully met by these apples of Askelon; and no doubt, in ancient times and in royal gardens, their cultivation was far superior to what it is now, and the fruit larger and more fragrant. Let "taffûah," therefore, stand for "apple," as our noble translation has it."—p. 546.

Learned authorities have discussed, with great minuteness, the question whether the forbidden fruit was really an apple, but the word "apple," in its early use, comprehended so many different sorts of fruit, that it is not possible to determine accurately the precise nature of

"that alluring fruit . . . . . .
. . . that crude apple that diverted Eve."

It is however curious that the protuberance in front of the throat should be called *"Pomum Adami,"* or Adam's apple, from a whimsical supposition that the forbidden apple had not been swallowed, but had stuck there.

The old ecclesiastical writers abound in fancies, similies, and typical allusions to the apple-tree and its fruit. It was believed by some writers to be the tree of life, that it miraculously survived the deluge, and that it was this identical tree which was used to make the cross on which our Saviour was crucified. It is said the apple tree takes the form of the cross in growth, its trunk, the stem; and its transverse boughs, the limbs of the cross. By others, that Christ fastened to the cross gives the odour of our redemption. Again, it is our Saviour himself who is compared to the apple tree; as this tree excels all others in the beauty and freshness of its blossom, in the sweetness of its odour, and in its abundant fruitfulness, so does Christ draw all to himself: Then again, Christ in the Sacrament is like the fruit of the apple tree, both food and drink: the colour of the juice of the apple may be white and red, typifying the water and the blood. Again, as there are different species of apples, all good in their sorts, so Christ has many graces, &c., &c. Apple trees were often planted in the monastery gardens, in the form of a cross, and a certain number of the trees were put under the charge of each monk. There are also abundant allusions through all ancient writers, sacred and profane, as to the medicinal virtues of the apple, which it is unnecessary specially to notice.

The Druids are said to have paid particular reverence to the apple tree. It was the next most sacred tree to the oak, and orchards of apple trees are said to have been planted in the vicinity of the sacred groves. It is easy to believe that this may have been the case as well from the great usefulness of the fruit, as from the fact of the mistletoe growing so frequently upon the apple tree. The apple tree was cultivated in Britain from the earliest ages of which we have any record, meagre and unsatisfactory as such records may be, especially with reference to apple trees. At the time of the invasion of the Romans, Glastonbury is said to have borne the name of "Avallonia," or the "apple island," from the quantity of apples grown there, though modern writers think there is no proof whatever that Glastonbury even existed at that time. If the Druids had apples, as doubtless they had, they grew them for some centuries after the Roman invasion in the more secluded districts of the country, where they still practised their rites amongst the people.

There is a poem by Merddin, said to have been a bard of the sixth century, entitled the "*Avallenau*," or the "Apple Trees." Merddin's existence, and the age in which he lived, is called in question by matter of fact people, but the poem "*Avallenau*" remains. The Rev. Edward Davies, in his work on "*The Mythology and Rites of the British Druids,*" (1809) says that "Mr. Sharon Turner has proved this poem to be the genuine production of Merddin."—(p. 480.) The purport of the poem Mr. Davies states to be "a tribute of gratitude for the sight of an orchard containing one hundred and forty-seven delicious apple trees, which had been privately exhibited to the bard by his lord Gwenddleu. This number being the square of seven multiplied by the mystical three, carries with it an allegorical meaning, and affords one proof amongst many, that the rites of Druidism were practised in Britain as late as the close of the sixth century." Mr. Davies gives the following translations from the poem "*Avallenau.*"

"To no one has been exhibited at one hour of dawn what was shewn to Merddin before he became aged, namely, seven score and seven delicious apple trees, of equal age, length, and size, which sprang from the bosom of Mercy. One bending veil covers them over. They are

guarded by one maid with crisped locks.   Her name is Olwedd with the luminous teeth."
—*Stanza* I.

   " The delicious apple tree, with blossoms of pure white and widespreading branches, produces sweet apples for those who can digest them, and they have always grown in the wood that grows apart," &c.—*Stanza* II.

   " The sweet apple tree has pure white sprigs, which grow as a portion for food," &c.—*Stanza* IV.

   "'The fair apple tree grows on the border of the vale, its yellow apples and its leaves are desirable objects," &c.—*Stanza* V.

   "Thou sweet and beneficent tree ! not scanty is the fruit with which thou art loaded," &c.—*Stanza* VI.

   " The proper place of this delicate tree is within a shelter of great renown, highly beneficent and beautiful ; but princes devise false pretences with lying, gluttonous, and vicious monks, and pert youngsters, rash in their designs," &c.—*Stanza* VII.

   " The tree with delicate blossoms which grows in concealment amongst the forests," &c.—*Stanza* XIV.

   " The sweet apple tree with delicate blossoms grows upon the sod amongst the trees," &c.—*Stanza* XVI.

   This is the Merddin called the Caledonian, to distinguish him from the better known Merddin of South Wales. He is styled supreme Judge of the North. " The *white blossoms* seem to imply the robe of the Druid ; the *spreading branches*, his extensive authority ; the *fruit*, his doctrine and hopes ; and the sequestered wood which had always produced this fruit, his sacred grove." (p. 484.)

   " The use of *Tallies* or *Sprigs* cut from a fruit-bearing tree, which Tacitus ascribes to the Germans, was probably common to them and the Druids. They were the omen sticks or points of sprigs so often mentioned by the Bards." *Davies*, p. 43.

   The ancient Welsh Bards were rewarded for excelling in song by " the token of the apple spray :" thus Gualchmai, the golden tongued, the son of Meilyr, in a poem called " Gorhofedd," or " the Boast," (end of 12th century) speaking of the mystical sprigs, says :

> " Gorwyn blaen avall, bloden vagwy,
> Balch caen coed—
> Bryd pawb parth yd garwy."

   " The point of the apple tree supporting blossoms, proud covering of the wood, declares, every one's desire tends to the place of his affections." *Davies*.

   And again in the song by Hywell, the son of Owen, he says :

> " My shield remains white upon my shoulder ; the wished for achievement I have not obtained, though great was my desire ; another has worn the token of the apple spray."
>   *Davies*, p. 284.

   The apple still holds its place in the customs of many nations—always the emblem of fertility, and usually the symbol of happiness. It is oracular in love matters—an omen of love—a love charm, or a token of affection. Auguries are to be drawn from the cutting of an apple in half—the number of its pips—throwing the peel over the shoulder—sleeping with an apple, or the half of one under the pillow, &c., &c. If the tree blossoms out of season, it betokens some joy, or some unlooked for sorrow, as the wit of the observer is enabled to solve the enigma. In some countries, it may also be added, it has been the custom to place an apple in the hand of a child when buried, that it may have it to play with in Paradise.

It is time however to leave the myths of apple lore, and see what a somewhat more trustworthy history has to say about the Apple and Pear.

The oldest systematic writer on the subject in Latin is the Elder Cato, (B.C. 234-149) who in his treatise on Husbandry gives lists of fruits, and directions about grafting, planting, propagating and storing them. He is often quoted by Pliny. Other writers may be mentioned, as Varro (B.C. 116-27) ; Columella (A.D. c. 42) who treats the subject carefully and at some length ; the Greek physician and botanist Diascorides of Anazarbus in Cilicia, (A.D. 64) and others of later date : but the one who enters most fully into the subject is the Elder Pliny, (A.D. 23-79) who in his wonderful Encyclopœdia of knowledge, the Natural History, has given a descriptive list of the different sorts of Apples and Pears, taken in great measure from Cato and other writers. This account is so often referred to, is so full of quaint humour, and is so thoroughly practical, that the three chapters on the Quince, the Apple and the Pear will be here translated in full.

PLINY, BOOK XV, CHAPTER 11.

"ON THE QUINCE APPLE.

"Next to these (fir-cones) in size are Quince Apples, which we call *Cotonea*, and the Greeks *Cydonia:* they were imported from the island of Crete. The branches which they draw from the stem are bent inwards, and hinder its free growth. There are many sorts of them, *chrysomela* (golden apples) strongly marked with grooves, of a colour inclined to golden. Those which are lighter in colour are called by names belonging to our own country, and are possessed of a most exquisite odour. The Neapolitan sort also are not undeserving of praise. A smaller sort of the same kind called *Struthea* (sparrow-apples) emit a still stronger fragrance as they quiver on the stalk ; they are late in appearing, while those which are called *Mustea* (new-wine apples) are very early. The *Cotonea* grafted on the *Struthea* produce a special sort called *Mulvian*, the only sort out of those already named which are eaten raw. At the present time all of them are commonly placed in men's reception rooms, upon effigies of deities placed there as guardians of the night season. There are besides woodland apples of small size, which are the most highly scented of all next to the *Struthea;* they grow on hedges. The name *mala* (apples), though the fruits belong to a different race, we give to peaches and pomegranates, of which nine sorts are named, growing on trees which came originally from Carthage. The distinguishing feature of the latter is seed beneath the rind, that of the former a solid stone in the body of the fruit. There are moreover some of the pear tribe called *Libralia* (pound-pears) from their great weight."

CHAPTER 14.

"OF THE VARIOUS SORTS OF APPLES, 29 SORTS.

"There are many sorts of apples. Of citrons and their parent tree we have already spoken, but to another sort of fruit the Greeks give the name Median, from the country of its origin. Equally foreign are the *Zizypha* (jujubes) and *Tubercs*, both of which have only lately been introduced into Italy, the latter from Africa, the former from Syria. It was Sextus Papinius, whom I remember as Consul, who introduced both of them in the latter days of Augustus of divine memory. They were planted in fortified places, and are more like berries than apples, but are particularly ornamental to raised mounds, seeing that it is the custom now even for the roofs of our houses to be overgrown with wood. There are two kinds of *tuberes*, a white one, and one called from its colour *Syriace.* There is a sort which may almost be called foreign, for it grows in one place alone in all Italy, namely in the territory of Verona; it is called *lanata* (woolly) for a downy wool covers the fruit, and is most abundant in the *struthea* (sparrow-apples) and the *persica* (peaches). It gives them their peculiar name, and they are not worthy of any other praise. There is no reason why we should not point out the other sorts by name, seeing that they have conferred an immortality on those who first produced them, as if on account of some illustrious achievement. Unless I am deceived, the skill shewn in grafting, will be sure to make itself manifest, and also the fact that nothing is so trifling in itself as to be incapable of conferring credit on some one: and thus there are some apples which derive their origin from Matius, and Gestius, and Manlius, and also from Claudius. Those on which the *cotoneum* (quince) was grafted by Appius, of the Claudian family, are called *Appiana* from him: they have the quince odour, the size of the Claudian apple, and a ruddy colour. And lest any one should think they came into use from the influence of an illustrious family, there are also some apples remarkable for their roundness, called *Sceptiana* from their originator, who was only a freed-man. Cato adds the *Quiriana* and the *Scantiana*, which he says may be stored in jars *(Cato R. R. 7)*. But the most recently introduced are the small apples of most exquisite flavour and scent, which are called *Petisia*. The *Ameria* apples, and the *Græcula* (small Greek apples), have reflected credit on their native soils (*i. e.* Ameria and Etruria, and probably the Italian district called Magna Græcia). The rest derive their name from physical causes, *c. g.* from *fraternity*, such as cohere and are found in pairs, never single in their growth: those called *Syria* from their colour (deep red): the *Melapia* (apple-pear), from their affinity to the pear: *Mustea* (new-wine apples), from the rapidity with which they ripen: *Melimela* (honey-apples), from their honey flavour: *Orbiculata* (globe-apples), from their round shape, like that of a revolving globe; the Greek name for these, *Epirotica*, proves that they made their first appearance in Epirus: *Orthomastica* (mammary-apples), from the resemblance in their form to the female breast: *Spadonia* (barren apples), so called by the Belgians because their seed produces no plants: those called *Melofolia* (leaf-apples), have one leaf, but sometimes a double one, growing on their side: the *Pannucea* (ragged-apples) shrivel very quickly into wrinkles: the *Pulmonea* (lung-apples) bulge out to an absurd degree: and there are some of the colour of blood, deriving this peculiarity from the mulberry being grafted on their stock, but it is common to all apples to have the side towards the sun of a red colour. There are some woodland apples with little that is pleasant in the way of taste, and even less so in point of smell, which is highly pungent. They have a wicked peculiarity of bitterness, and enough acidity to turn the edge of a sword. *Farina* (corn-flower) apples are the worst of all, though they are the first to appear, and are early fit for gathering."

CHAPTER 15.

"ON PEARS, AND THE VARIETY OF GRAFTED SORTS."

For the same reason in the case of Pears, those which ripen early are taxed with the name of pride (proud pears): they are small but very early. All agree in liking the Crustumian Pears very much. Next to them come the Falernian Pears, so called from the abundance of their juice which is called milk, and which people drink; and among them others possess the dark colour of Syria. Of the rest, the names vary in various places. Some by their names at once betray and ennoble the names of their originators and the place to which they belong, as the *Decimiana* pears, and a sort derived from them called *Pseudo-decimiana*; the *Dolabelliana* pears with a very long stalk; the *Pompeiana* which are called *Mammosa* (breast-like); the *Liceriana*; the *Severiana* and their immediate offspring the *Turraniana*, differing from them in the length of their stalk; the *Favoniana* which are red, and a little larger than the early proud pears; the *Lateriana*; and next the *Auiciana*, autumnal pears with a pleasant acidity of flavour. The pears which were special favorites with the Emperor Tiberius are called *Tiberiana*; except that they are more deeply coloured and attain a greater size under the influence of the sun, these would be the same as the *Liceriana* pears. The following pears take the name of their native localities: *Ameriua*, the latest of all pears; *Picentina*; the *Numantina*; the *Alexandrina*; the *Numidiana*; the *Graeca*, and among these the *Tarentina* pears; the *Signina* which from their colour some people call *testacea* (brick-dust), just as others are called onyx-coloured, and others purple. Some get their names from their smell, as the *myrapia* (musk pears); *laurea* (laurel pears); *nardina* (nard pears); some from the season at which they ripen, as the *hordearia* (barley pears); some from having a neck like a jar, are called *ampullacea*, and also *Coriolana*; some are called *cucurbitina* (gourd pears) with a rough skin and a slightly acid juice. The origin of the names of those called *Barbaric* and *Venereau* is uncertain, the latter are also called *coloured pears*; so also is it the case with the *Regal* pears, which are sessile having a very small stalk; with the *Patrician* and *Voconian* pears which are green and oblong in shape. Besides these there are the *Volema* pears mentioned by Virgil, who borrowed the name from Cato *(Cato R. R. 7)*, who also mentions the *Sementina* (late pears gathered in sowing time), and the *mustea* (new-wine pears). This branch of natural economy has indeed already reached its climax, for there is no experiment which men have not tried, seeing that Virgil tells us of the arbutus grafted with the nut, the plane tree with the apple, and the elm tree with the cherry. Nor can any further discovery be made. It is certainly a long time since any new fruit has been discovered; nor indeed is it right that all grafts should be made without distinction, as for example that thorn-trees should be used as stocks, seeing that lightning strokes are not easily expiated, and it is laid down by the priests, that whenever a tree has been struck, as many lightning strokes must be accounted for, as there are kinds of grafts made. The form of pears is more turbinate than apples. Among them the late pears hang on the tree even until winter, and ripen with the cold, as do the Greek, the jar-shaped, and the laurel pears, and this is the case also with the *Amerina* and *Scantiana* apples. Pears are stored in as many ways as grapes. Except plums, it is only pears which are kept in jars. Wine may be made both from apples and pears, and doctors take advantage of this property in their treatment of the sick. They are prepared with wine and water, and serve the purpose of a cooling drink, which no other fruit but *cotonea* (quinces) and *Struthea* (sparrow apples) will give."

There can be little doubt that the Romans on their arrival in Britain found apples growing there; for the crab is indigenous to the soil, and the apple is but a cultivated crab. Cæsar makes no allusion to fruit. He describes the Britons as more a pastoral than an agricultural people. But some parts of the island, he says, were already fruitful in corn, especially, as it appears, in the districts inhabited by the Belgæ; who had recently crossed over from the Continent *(De Bello Gallico,* v. 14). Tacitus, who in his Life of Agricola, gives the most interesting and trustworthy account of Ancient Britain, which has come down to us, does not mention either apples or pears, but expressly says: "The soil is adapted for produce of all kinds, except the Olive and the Vine, and other things (fruit trees) accustomed to grow in warmer countries, (and is) fruitful: they are quick in coming, slow in ripening: both effects arising from the same cause, the excessive dampness of the soil and climate." *(Vita Agric.* 12). There can be still less doubt that the Romans as they settled down in Britain, brought with them the varieties of fruit they had been accustomed to use in Italy, though there is no distinct record of their having done so. They ever loved to surround themselves with the plants of their own country, and it is to them we owe the introduction of the elm, the box, the walnut, the cherry, and the pear. The coarse pot-herb Alexanders, *(Smyrnium olusatrum)* is generally found in the neighbourhood of Roman earthworks, and unwittingly they brought the Roman nettle, *(Urtica pilulifera)* which still haunts some of the ruined Roman stations in England. From the country in which the Romans settled, the fruit there, would gradually spread through the country. In the third Century the Romans obtained permission, it is said, of the Emperor Probus to introduce the Vine into Britain, and soon made wine from the fruit.

It is natural to suppose, also, that as the native inhabitants receded before the invaders, they too would carry with them their own varieties of apples into the most remote districts of the country. The Druidical legends, for such evidence as they may afford, support this idea. At a later period, during the fifth and sixth Centuries, there is some indirect evidence to show that this was the case. The native Britons sought refuge from the Saxons amongst the mountains of Wales, and many of them fled from thence to the North-Western coast of France, called Armorica, which in consequence of this emigration, received the name of Brittany, which it has since retained. From Wales they carried with them their apple trees, and one remarkable instance of their having done so is recorded in the *Liber Landavensis*. It is thus mentioned by Montalembert in his "*Les Moines d'Occident*" which, translated says : " When St. Brieuc and his eighty monks from Great Britain landed in Armorica (Brittany), and marked the site on which the town which bears his name was afterwards built : they acted just as the soldiers of Cæsar did in the forests sacred to the Druids. "They first surveyed the ancient woods with curiosity," says the chronicle : "they hunted every-where through them, and finding a branching valley with pleasant shade and a stream of clear water running through it : they all set to work   .   .   .   . frequently replacing the trees of the forest by fruit trees ; like the British monk Teilo who planted with his own hands, aided by St. Samson, an immense orchard, a true forest of fruit trees, three miles in extent in the neighbourhood of Dôl." (*Book* vi. p. 394). Teilo was the son of Tegwedd, who was also the mother of Bishop Afan of Bualle (Builth). She is said to have suffered martyrdom at Llandegveth, near Caerleon. Teilo received his religious education at this college of Iltutus, or Illtyd, situated at the village now known as Llantwit Major, in Glamorganshire. Teilo succeeded Dubricius in the see of Llandaff; and on the death of St. David he was appointed to the see of Menevia (St. David's). He placed his nephew Ishmael there as suffragan, and continued himself at Llandaff, with the title of Archbishop. " He is known as one of the three blessed visitors of the Isle of Britain, Dewi (St. David's) and Padarn, being the other two. They were so called from the zeal with which they preached the faith in Christ, to rich and poor alike, without fee or reward, and from their deeds of charity." (*Williams' Cymry* p. 133.)

Samson was Bishop of Dôl, but it seems there were two Bishops of Dôl of the same name, and both were educated at the college at Iltutus. In early life Teilo passed over to Armorica and spent some years with his old fellow-student, Bishop Samson of Dôl. The *Liber Landavensis* says : " St. Teilo also left there another testimony of his patronage, for he and the aforesaid Samson planted a great grove of fruit-bearing trees to the extent of three miles, that is from Dôl as far as Cai, and these woods are honoured with their names until the present day, for they are called the "Groves of Teilo and Samson." (*Liber Landavensis. Llfr. Teilo. Welsh M.S. Soc. Chap. III. p.* 346). This orchard still existed in the twelfth Century, under the name of " Arboretum Teliavi et Samsonis." (*Vie St. Brieuc, by the Canon of La Devison,* 1627 ; *cited by La Bordérie*). Tradition states that the planting of this orchard first led to the manufacture of cider in Normandy, and certainly no notice of it is to be found until some centuries afterwards, when the cider of Normandy began to attain the celebrity it afterwards gained. Teilo died A.D. 540, and was succeeded at Llandaff by his nephew Oudoceus, also a person of eminent sanctity. Samson died A.D. 599. There are twelve churches in the Diocese of St. David's, founded by St. Teilo, or dedicated to him, of which Llandeilo Fdwr in Carmarthenshire is the principal. In the diocese of Llandaff the Cathedral is dedicated to St. Teilo and St. Peter, and there are five other churches, including Llanarth, Llandeilo Cressency and Llandeilo Pertholly in Monmouthshire.

In the Sarum Missal there is a special blessing for apples, which is appointed to be used on St. James' Day, July 25th, but this form does not appear either in the Missal, or Breviaries of the Hereford use.

It must also be mentioned, that it is a common belief in the Midland Counties, that apples are not fit to be cooked until they have been christened by the showers of St. Swithin on July 15th.

A shrivelled apple is amongst the remains of the Lake cities of Switzerland.

> " Time rolls on his ceaseless course. The race of yore,
> *   *   *   *   *   *   *
> *   *   *   *   *   *   *
> How are they blotted from the things that be."
>
> *Scott—Lady of the Lake*, iii. 1.

From the time of the conversion of the Anglo-Saxons to Christianity, and for many succeeding centuries, even as late as the fourteenth century, the cultivation of fruit was chiefly carried on by the Ecclesiastics. The monks were men of peace and study, and living in retired spots, depended upon their gardens for much of their food. Through ages of war and bloodshed they pursued their peaceful avocations and cultivated the soil with sedulous industry. Many a monk like Scott's Abbot Bonniface of Kennequhair, has found great pleasure in the pears and apples he had grafted with his own hands. The " Abbey Garden" is always observed to occupy the best and most sheltered situation that could be found, and by their foreign connections the monks were enabled to obtain, from more favourable climates, not only better kinds of vegetables and more choice fruits for their own delectation; but also valuable medicinal herbs for the treatment of the sick poor in their neighbourhood. The ruins of most of the old Abbeys afford, to this day, proofs of the care bestowed by their former inhabitants in introducing foreign plants. From the gardens attached to these Institutions, they have often been found by botanists to wander into the neighbouring fields and woods; Asarabacca (*Asarum Europæum*) recently found by the Woolhope Club in the Forest of Deerfold, is one of these medicinal plants: Thorn apple (*Datura Stramonium*); Belladonna (*Atropa Belladonna*) ; stinking Groundsell (*Senecio squalidus*) ; the plant always grown in nunnery gardens, (*Aristolochia clematitis*) are other examples ; and more might be mentioned. As early as 674, there is a record that Brithnot the first Abbot of Ely, laid out extensive gardens and orchards, which " he planted with a great variety of herbs, shrubs, and fruit trees. In a few years, the trees which he planted and engrafted appeared at a distance like a wood, loaded with the most excellent fruits, in great abundance, and added much to the commodiousness and beauty of the place." *Hist. Eliens. apud. Gale, L.* ii. c. 2.

" *The ancient laws and Institutes of Wales,*" published by the Commissioners of Public Records in 1841, which comprise the laws supposed to have been enacted by Howel Dda, about the early part of the tenth Century, modified by subsequent regulations under the Princes of Wales, previously to the subjugation of Wales by Edward I. (1283), give several references to the great value of apple trees.

In the Dull Gwynedd, the Venedotian, or North Wales Code, Book III., Chapter 20, is entitled, as rendered in the English translation, "On the worth of Trees this treats :"

Section 8.—" Every tree that shall bear fruit is of the same worth as the entire hazel grove, excepting the oak and the apple tree." (*Mem.* A hazel grove was valued at twenty four pence).

Section 9.—" A graft four pence without augmentation until the calends of winter after it is grafted."

Section 10.—" And thenceforward an increase of two pence is added every season until it shall bear fruit, and then it is three score pence in value, and so it graduates in value as a cow's calf."

Section 11.—A sour crab tree is four pence in value until it bear fruit."

Section 12.—" And after it bears fruit it is thirty pence in value."

In the Dull Dyfed, the Dimetian, or West Wales Code, Book II., Chapter 35, entitled "of furniture."

Section 81.—"An apple tree is three score pence in value."
Section 81.—"A crab tree is thirty pence in value."

In the Dull Gwent, or Gwentian Code, Book II., Chapter 18, entitled "The worth of petty utensils."

Section 61.—"An apple tree three score pence." "A crab tree thirty pence."

In the Cyforithian Amrywial, or Anomalous Laws, Book XIV., Chapter 31, "of land."

Section 9.—"All lands are to be shared but these, a bog, oak wood, and a quarry. And these erections are to be in common among Brothers, an orchard, a mill, and a wear."

In the Leges Walliæ, Liber II., Cap. 28, "De Arboribus."

Section X.—"Pomus dulcis IX. a denarii."
Section XI.—" "Pomus amara XXX. a denarii."

CAP. XXXVII.

"De preciis domorum et aliarum rerum."

Section XCVIII.—"Pomus dulcis IX. denarios valet."
Section XCIX.—"Ramus ejus XXX. denarii."
Section C.—"Pomus amara XXX. denarii."

Leges Howeli Boni.
"ri dulcis pomi + LX.
"ri peris pomi XXX.

There is no reference to the apple tree in "*The Ancient Institutes of England.*"

In an alleged account of the "Antient Saxon rite of coronation as recorded in the time of Edgar" (959-975), the following passage is given as forming part of the blessing pronounced by the Archbishop or Bishop at a Saxon coronation :—

"May the Almighty Lord give thee, O King, from the dew of heaven and the fatness of the earth, abundance of "corn, and wine, and oil. Be thou the Lord of thy brothers and let the sons of thy Mother bow down before thee; let the "people serve thee, and the tribes adore thee. May the Almighty bless thee with the blessings of heaven above, and the "mountains and the valleys; with the blessings of the deep below; *with the blessings of grapes and apples.*"

Wild apple trees were not uncommon in this reign. William of Malmesbury, says, (lib. ii. cap. 8), that King Edgar in 973 while hunting in a wood and separated from his followers was overcome by an irresistible desire to sleep, and alighting from his horse, he lay down under the shade of a wild apple tree.

Henry of Huntingdon, in describing a quarrel that arose at the Court of Edward the Confessor, between two of the sons of Earl Godwin, represents one of them as departing in a rage to Hereford, where his brother had ordered a royal banquet to be prepared. "There he seized his brother's attendants and cutting off their heads and limbs, he placed them in the vessels of wine, mead, ale, pigment, morat, and cyder." (*Henry of Hunt.* Vol. vi. p. 367). It must, however, be admitted that although this history was written in 1154, it was not published until 1576, by Saville, when it is possible that the last word "Cyder," by that time the common drink of Herefordshire, may have been added.

It is difficult to find any record of the different sorts of apples grown in England, at an early period, though it is well known that there were many varieties both cultivated and wild. Malmesbury in the early part of the twelfth Century, speaking of Gloucestershire, says, " *Cernas tramites publicos vestitos pomiferis arboribus, non insitiva manûs industriâ, sed ipsius solius humi naturâ.*"

Different circumstances, however, have preserved the names of two apples—the Pearmain, and the Costard apple. The Pearmain appears in a legal deed. So early as the 6th of King John (1205), Robert de Evermue was found to hold his Lordship of Redham and Stokesley, in Norfolk, by petty serjeantry, the paying of 200 pearmains, and four hogsheads (*modios*) of wine made of pearmains into the exchequer, on the Feast of St. Michael yearly." (*Bloomfield's "Norfolk," 4th Edit.* xi. p. 242 ; *quoted by Cullum, "Hist. and Antiq. of Hawsted,"* p. 117 *note*).

The Pearmain is therefore a very old apple in England, though its name, and Drayton, pronounce it not English :—

> " The Pearmain which to France long ere to us was known ;
> Which careful fruiterers now have denizen'd our own."
> *Polyolbion,* s. 18.

The other well known apple of the thirteenth Century, was the Costard apple. It is mentioned under the name of " Poma costard" in the fruiterers bills of Edward I. (1292), at which time it was sold for one shilling the hundred. William Lawson who speaks of it in 1597, says quaintly, " Of your apple trees you shall finde difference in growth. A good Pipping will grow large and a Costard tree." It must have been extensively grown and appreciated at one time for it has given the name of " costard-monger," or as it has now become "coster-monger" to the retail sellers of fruit and vegetables ; this apple, in the earliest history of the trade, being probably the only thing they sold. The beneficent qualities of the fruit however, seem ever to have been in marked contrast with the roughness of the sellers.

> " He'll rail like a rude coster-monger."—(*Beaumont & Fletcher's " Scornful Lady.*")
> " Yonder are two apple women scolding and just ready to uncoif one another."—(*Arbuthnot & Pope.*)

The history of the apple during the middle ages, is chiefly to be gleaned from the incidental notices with reference to cider, which have come down to us. From these scant notices it would appear that the manufacture of cider was not confined to certain districts as it now is, since but little was known of the influence of the soil, or its quality, in those days. Where apples grew, and drink was scarce, cider was made. The first distinct notice of it as being made in England was in Norfolk ; the next we have is in Yorkshire. " In 1282 the bailiff of Cowick, near Richmond, stated in his account, that he had made sixty gallons of cider from three quarters and a half of apples. (*Hudson Trowers' Archæological Journal,* vol. v.) In these days no one would think of making it so far north. In Scotland it seems never to have been made, or used, to any extent. In the annals of the Lord High Treasurer of Scotland, edited by Thomas Dickson (1877), cider is only once mentioned, and then as being obtained for Peter Warbeck's English Followers in 1497." (*Compotum Thesaurum regum Scotorum,* vol. i., 1473-98).

In a tract on Husbandry, written early in the XIV. Century, it is stated under the rubric in old Norman French, "*Coment hom doit mettre le issue de sun estor a ferme,*" that ten quarters of apples or pears ought to yield a ton (tonel) of cyder as rent." (*Add. M.S.* 6159, p. 220).

Most of the early notices of cider refer chiefly to its mode of manufacture, but that it was made very widely, there can be no doubt.

The mention of cider occurs in the "*Roll of the Manners and household expenses of the Countess of Leicester*" (thirteenth Century)—but in the "*Roll of the household expenses of Bishop Swinfield,*" so carefully edited by the Rev. John Webb, there is no mention whatever of cider. The apple itself is only once named, and then as being purchased with a lemon. In this minute detail of the expenses of the Bishop in his progress through the Diocese of Hereford from village to village, commencing at Michaelmas, 1289, and continued until nearly the same time in 1290, if apples had been abundant, if orchards had existed—and certainly if cider had been made at that time there must have been some mention of it. The Bishop visited, again and again, the districts of the county now most celebrated for their orchards, but the Roll is silent on the subject. On the "endorsement" of the Roll, the proceedings of the visit of the steward Kemseye to an estate, held by the Bishop in his native county of Kent, at Womenswould, near Wingham, are recorded. Here "other particulars" says the Editor, "proclaim the forming of a homestead : a virgate of land was was bought, and some apple trees were planted—*plantis pomiferis,* grafted stocks ; and it may seem strange that while throughout the Roll no mention is made of orchards in Herefordshire one should be forming in Kent." (*Editor's note,* p. 121.)

The "Roll" mentions white wine from Ledbury several times, but this was a grape wine, and not the Perry it is so celebrated for now. "*Vineâ de Ledebur.*" This vintage had yielded during the preceding Autumn (1288) seven pipes (*dolia*) of white wine and nearly one of verjuice. It was valued at eight pounds the pipe, or about half the price of the foreign wine got from Bristol, and brought up the Severn to Hawe (p. 59). This vineyard was planted, or renewed by the preceding Bishop Cantilupe. A farm in the parish of Ledbury, on the Gloucester-road, still bears the name, and in after times the descendants of Bishop Skipp had a vineyard on their estate of Upper Hall, in the parish of Ledbury. Towards the end of the seventeenth Century, George Skipp, Esq., made both white and red wine from his plantation. He died in 1690. The Editor has often seen the site on which the vines grew." (*Roll of Bishop Swinfield,* vol. ii., *note by Editor,* p. cxxvii.)

There is also a "Vineyard" estate on the banks of the river Wye, one mile east of Hereford, which might well have existed before this time. The property was left to the Trinity Hospital Charity, in the city of Hereford, in 1607, by Mr. John Kerry. The vines here grew on terraces, supported by stone walls, built at considerable expense, and one or two very aged vinetree stocks exist there at this time.

The Roll of Bishop Swinfield is an authentic and trustworthy document, and since both beer and wine are very frequently mentioned in it, it affords the strongest negative testimony against the existence of orchards and the making of either cider or Perry in Herefordshire at that time.

Another Century had scarcely passed, however, before cider must have become a well known drink of the people, since it is a curious fact that in the Hereford Wycliffe M.S. Bible, now under a glass case in the Cathedral, the passage in the 15th verse of the 1st Chapter of St. Luke's Gospel, where the angel is speaking to Zacharias, respecting the birth of his son John the Baptist, rendered thus in the authorised version :—

"For he shall be great in the sight of the Lord and shall drink neither wine nor strong drink."

Is thus given :—

"For he schal be gret bifore the Lord, and he schal not drinke wyn ne sider."

"The date when this Wycliffe Bible was written is believed to be about 1420. (Havergal —*Fasti Herefordiensis*), and thus it may very possibly have been written by Wycliffe's companions and followers who are known to have lived for several years in seclusion in the wilds of Deerfold Forest, North Herefordshire, to escape the persecution which set in against them on the death of John of Gaunt in 1399. It must, however, be admitted that the word "sider" may be a translation of "σικερα," the equivalent for "strong drink," though not adopted in any of the later translations.

The apple plays an important part in the story of William Tell, the renowned champion of Swiss liberty. The story goes that (in 1307) when Tell refused to uncover his head before the hat of the Austrian governor Gessler, set up on a pole for the purpose, he was condemned, in derision of his reputed skill as an archer, to shoot an apple from the head of his own son. Tell does so successfully but openly states, that a second arrow which he had concealed on his person, was destined for the heart of the tyrant if he had injured his son. Eventually, the story says, Tell does shoot Gessler and thus begins the insurrection, which at length succeeds in emancipating his country. What matters it, that this same apple story had been told before, or that this particular version is a myth. It is related by Saxo Grammaticus, the Danish historian as occurring in 950, to a certain Palnatoki, a celebrated archer in the bodyguard of the ruthless King Harold Bluetooth. In Norway, it is told of Pansa, the splayfooted, and Hemingr, the Norse archer, a vassal of Harold Hardrada, who invaded England in 1066. In Iceland, there is a kindred legend of Egil, brother of Wayland Smith, the Norse vulcan. In England, there is the ballad of William of Cloudeslee, which supplied Sir Walter Scott with many details of the archery scene in Ivanhoe.

> "I have a sonne seven years old ;
>   Hee is to me full deere
> I will tye him to a stake—
>   All shall see him that be here.
>
> "And lay an apple on his head
>   And goe six score paces him froe,
> And I myself with a broad arrowe
>   Shall cleave the apple in towe."
>
> *(Bell's Early English Ballads).*

It is told of Puncher, a famous magician on the Upper Rhine. It is common to the Turks and Mongolians. It is a legend of the wild Samoyeds : and finally it is told in the Persian poem of Farrid-Uttin Attar, born in 1119. The facts are the same in all the stories. It is always an unerring archer, who at the capricious command of a tyrant, shoots from the head of some one dear to him an apple, a nut, or a coin ; and he always has a second arrow, and when questioned as to the use for this, the invariable reply is, "To kill thee tyrant, had I slain my son."

The early English ballad, " The Jew's Daughter," which is supposed to refer to the murder of a child at Lincoln in 1256, for which some Jews were tried and executed, and their goods confiscated, says :—

> " She pulled an apple red and white
> To entice the young thing in ;
> She pulled an apple white and red
> And that the sweet bairn did win."
>
> *(Bell's Early English Ballads).*

This ballad has, however, also been supposed to refer to an Italian or German legend.

Chaucer mentions cider (1360).

> " This Sampson never Sider drank ne wine."
>   " *The Monke's Tale*," 14061.

Shewing that cider was a well known drink in his day.

Good Master William Langland, of Cleobury Mortimer, in 1362, evidently knew what a good apple was, for in that curious, quaint poem of " *Piers' Ploughman*," he says :—

> " I preide Piers tho to pulle a down
> An appul and he wolde,
> And suffre me to assaien
> What savour it hadde."
>
> *(Vision of Piers the Ploughman).*

The inference is clear, that bad apples must have been common, and that he did not mean to eat this one, if he didn't like it.

Fuller states that Pippins were first introduced into England in the 16th year of Henry VIII. (1525) by Leonard Mascal, who brought them " from over the sea" and planted them at Plumstead, in Sussex, a small village on the north side of the South Downs, near the Devil's Dyke." This statement is quoted by most fruitgrowing authorities, and it is extremely probable that Mascal did introduce a lot of Pippins ; but as all plants from pips are called " Pippins"—it is quite possible that these " Pippins" were mere stocks for grafting. A seedling whose fruit proves good enough to grow often retains its name of "Pippin" and Mascal's Pippins may therefore have been approved kinds of apples—but if so, many varieties existed in England long before his time. Mascal is also said " to have brought the first carp to England, and thus to have furnished at one time our orchards and our ponds with the rarest variety of each kind."

Noakes in his " Monastery and Cathedral of Worcester," (1866) referring to a date *circa* 1533, says, " There is no mention of cider, or home-made wine, or vineyards at Worcester Monastery, though we know that the vine was much cultivated in this county and neighbourhood," p. 180. And again, cider is not alluded to in the Worcester Rolls till a comparatively late period, though we know the beverage was so called in Chaucer's Monks' tale, middle of fourteenth Century. It probably existed from a very early period *under some other name or as wine*," p. 300. And again, *circa* 1662, a hogshead of cider then cost £1 14s 0d., while ditto of strong beer was but £1 4s. 0d. Bottled cider was sold at 6d. a bottle in 1720, equal to 3s. of present money," p. 305.

The simple fall of an apple has done good service in the cause of science :—

> "Lo ! sweetened with the Summer light
> The full juiced apple, waxing over mellow,
> Drops in the silent Autumn night."
>
> *Tennyson.*

It was the falling of an apple that attracted Sir Isaac Newton's attention, and led that profound Mathematician to the discovery of the law of Gravitation, which forms the very foundation of the Newtonian philosophy.

———————————

The early history of the Pear offers much less of interest than that of the apple, It is one of the trees which Homer describes as forming the gardens of Alcinous and of Laertes, the father of Ulysses (Odyssey vii. 119, and xxiv. 337). It is named by several of the old writers ; who mention the great age to which the tree lives, and Theophrastus especially notes the fact of the fruit from old trees being of better quality than from young ones. We have seen that Pliny mentions numerous varieties and the great esteem in which they were held. There is no mention of Pears in England for some four or five centuries after the Roman invasion, but there is every probability that the cultivated varieties were introduced by the Romans, and that they were afterwards grown very much in the Monastic gardens. The early Abbots were many of them Normans, and their Institutions were intimately connected with or were simply offshoots of, similar religious houses on the Continent ; and there is doubt but that many varieties of fruit were thus introduced from time to time, as indeed is shewn by the names of the fruits themselves, and particularly of those of pears.

There is a tradition that King John was poisoned in a dish of pears—" As the devil made use of the apple for the destruction of man, so did the devil's imps use the pear to a wicked end, when the monks of Swinsted, inviting King John to a banquet, poisoned him in a dish of pears, though others write it was in a cup of ale." (*Gwillim, Display of Heraldry*). And whether this may be true or false, it certainly implies that stewed pears formed at that time

> " A dainty dish to set before a King."

A paper on early English Horticulture, by Mr. T. Hudson Turner in the *Archæological Journal*, Vol. v., contains the following interesting notice of Early Pears.—The accounts of the 4th and 20th years of Edward I. (1276 and 1292), it appears that young pear trees were produced for the royal gardens, at Westminster, of these sorts, *Kaylewell* or *Calswell*, *Rewl* or *de Regula*, and *Pesse-pucelle*. The *Kaylewell*, was the *Caillou*, or Burgundy pear, a hard, inferior sort, only fit for baking, but it seems to have been most generally grown in England, and there is extant a writ of Henry III., directing his gardener to plant it both at Westminster and in the garden at the Tower. The *Rewl* was the pear of St. Règle, a village in Touraine. It is noticed by Neckham, Abbot of Cirencester (1157–1217), who wrote an unpublished work, " *De Naturis rerum.*" (*M. S. British Museum*). The *Pesse-pucelle* may have been the variety anciently known in France as *Pucelle de Saintonge*, another variety being *Pucelle de Flandres*.

The fruiterers bills of Edward I. (1292) also enumerate the following pears : *Martins, Dreyes, Sorells, Gold-Knobs (Gold-Knopes)* and *Cheysills*. By their price, *St. Règle* and *Passe Pucelle* pears seem to have been the most esteemed. They cost from 10d. to 3s. per 100 ; *Martins* sold at 8d. and *Caillou,* at 1s. per 100 ; the rest at 2d. or 3d. per 100.

To the above list must be added one of native origin, the *Wardon* pear. It was raised by the Cistercian monks of Wardon, in Bedfordshire, a foundation of the twelfth Century. It was a baking pear of great repute, and supplied the contents of the celebrated" Wardon pies."

> " The Canon sighed, but rousing cried, I answer to thy call,
> And a " *Warden pie's*" a dainty dish to mortify withal !"
>
> *Barham—Nell Cook, "Ingoldsby Legends."*

Wardon pies were supposed at one time erroneously to be made of venison and other meats.

From the celebrity of these pears, the term "Wardons" or "Wardens" came into common use for all kinds of large baking pears, which required keeping. In an old account book of the household expenses of Henry VIII. remaining in the Exchequer, there are the following items among others of the same character.

	£	s.	d.
" For Medlars and Wardens	0	3	4
Item to a woman who guff the Kyng pears " ...	0	0	2

Lawson (*New Orchard and Garden*, 1597) says " hard winter fruit and Wardons are not fit to gather until some time after Michaelmas," and the Husbandman's Fruitful Orchard, of about the same date says, " Wardons are to be gathered, carried, packed, and laid as winter pears are." The sort now known as *Uvedale's Wardon,* or *Uvedale's St. Germain* is thought to be an improvement on this pear. The Arms of Wardon Abbey were "*ar* : three Wardon pears, *or* : two and one," but the counter seal appended to the Deed of Surrender, preserved among the "*Augmentation Records*" bears the Abbatial Arms, namely, "a demicrosier between three Wardon pears."—It is dated December 4th, 29th Henry viii. (1538).

Worcestershire has been celebrated for its pear orchards for a very long time ; and, indeed, so characteristic of the County is the pear tree, that Drayton in his poetical marshalling of the troops of Henry V. at the Battle of Agincourt, makes the Worcestershire men display it, as their standard :—

> " Worcester a pear tree laden with its fruit."
>
> *Drayton.*

An achievement on the Arms of the City of Worcester is " *Argent,* a fess between three pears sable." The story goes that when Queen Elizabeth visited Worcester in August, 1575, the City authorities caused a pear tree heavily laden with fruit to be taken from a garden and planted at the gate by which her Majesty was to enter the City. The Queen, it is said, noticed the tree with admiration, and either directed or permitted those pears to be added to the City Arms ; but why they are figured " sable" instead of " proper," does not appear. It is scarcely probable that the pears the Queen saw could have been black. It is much more likely that the present pear called " *Black Worcester,*" a large iron hearted stewing pear (*Parkinson's Warden, a Pound pear*) took its name of "black" from the sable pears on the escutcheon. The heraldic association has given it a celebrity, which except for its size, it does not deserve.

"To write of peares and apples in particular," says old Gerarde in 1597, "woulde require a particular volume : the stocke or kindred of peares are not to be numbred : eury countrey hath its peculiar fruite ; my selfe knowe some one curious in graffing and planting of fruites, who hath in one peece of ground, at the point of three score sundrie sortes of peares, and those exceeding good, not doubting but if his minde had beene to seeke after multitudes he might have gotten togither the like number of those worsse kindes ; besides the diversities of those that be wilde, experience showeth sundry sorts ; and therefore I thinke it not amisse to set downe the figures of some fewe with their seuerall titles, as well in Latine as English, and one general description, for that, that might be said of many, which to describe apart, were to send an owle to Athens, or to number those things that are without number." He then names some eight kinds of pears of which several are now well known—and goes on to say, " The tame peare trees are planted in orchards as be the apple trees, *Quorum varia insitione ex agrestibus miles ac edules fructus redditi sunt*. All those before specified and many sortes more and those most rare and good, are growing in the ground of *Master Richard Pointer*, a most cunning and curious graffer and planter of all manner of rare fruites, dwelling in a small village, neare London, called Twicknam ; and also in the ground of an excellent graffer and painfull planter, *Master Henry Banbury*, of Touthill Streete, neare unto Westminster ; and likewise in the ground of a diligent and most affectionate louer of plantes *Master Warner*, neere Horsey Downe, by London, and in diuers other grounds about London (but beware the Bag and Bottle), seeke elsewhere for good fruit faithfully delivered."

The pear is a much longer lived tree than the apple ; its wood is much more firm and less liable to decay. In Domesday Book an old pear tree is several times noted as the boundary mark to a manor or parish. In the fine deep soil of the old Abbey Garden at Lindores, on the south bank of the river Tay, in the county of Fife, there are old pear trees that still bear abundantly, though the apple trees have wholly disappeared.

On the high road between Malvern and Worcester, at Monkland farm, Newland, there is an orchard of Barland pear trees, perhaps unequalled in the world. Tradition says these trees were planted by the Monks of Malvern, and if so, they must be three hundred years old. Mr. Edwin Lees in his "*Botany of the Malvern Hills*," thus writes of them. "There are more than seventy lofty trees ; and in "a hit" as it is called, the produce has amounted to two hundred hogsheads. It has been stated of a hopyard, that in particular years the value of the produce would be equal to the fee-simple of the land occupied by the plants. Almost the same might occur with a fine perry orchard. The one in question occupies five or six acres, and the price of perry varies from sixpence to one shiling and sixpence a gallon. Now, supposing the average price of £3 per hogshead to be obtained in "the hit" a year, the perry produced would be worth £600, but "a hit" must not be expected every year. The trees are now becoming very old," p. 62.

Shakespeare mentions apples and pears, solely for the purpose of drawing some of his admirable similies from them. These are the passages :—

1.

A young man nearly full grown is

"A codling when 'tis almost an apple."

12th Night, I. 5.

2.

A close resemblance is thus marked
> " An apple cleft in two is not more twin
> Than these two creatures."
>> *12th Night,* V. 1.

3.

> " An evil soul producing holy witness,
> Is like a villain with a smiling cheek ;
> A goodly apple rotten at the heart ;
> Oh what a goodly outside falsehood hath !"
>> *Merchant of Venice,* I. 3.

4.

> " There's small choice in rotten apples."
>> *Taming of the Shrew,* I. 1.

5.

A false resemblance is shown by
> " As much as an apple doth an oyster."
>> *Taming of the Shrew,* I. 1.

6.

The English in courage are, like their bulldogs attacking Russian bears, liable to
> " Have their heads crushed like rotten apples."
>> *Henry V.,* III. 7.

7.

Trifling young men are thus hit off :
> " These are the youths that thunder at a playhouse
> And fight for bitten apples."
>> *Henry VIII.,* V. 3.

8.

Though one daughter is as like another,
> " As a crab is like an apple ;"

And again,
> " She will taste like this as a crab does to a crab."
>> *King Lear,* I. 5.

9.

A cap is ridiculed as a "custard-coffin" and a slashed sleeve is
> " Carved up and down
> Like an apple tart."
>> *Taming of the Shrew,* IV. 3.

10.

> " I am wither'd like an old Apple John."
>> *1st Henry IV.,* III. 3.

11.

> *1st Drawer.* "What the devil has thou brought there ?
> Apple John ? Thou knows't Sir John cannot endure
> An Apple John."

> *2nd Drawer.* "Mass thou sayest true: the prince once
> set a dish of Apple-Johns before him and told him,
> there were five more Sir Johns; and putting off his
> hat said, "I will now take my leave of these six dry,
> round, old, wither'd Knights."
>
> *2nd Henry IV.*, 2, &c.

12.

Apples for dessert:

> "I will make an end of my dinner; there's
> Pippins and cheese to come."
>
> *Merry Wives*, I. 2.

So Horace (Sat. 1, 3, 6).

> "Ab ovo usque ad mala,"
> from eggs to fruit, or from the first course to the dessert.

13.

Gloucestershire.   *The garden of Shallows house.*

> *Shallow.* "Nay, you shall see mine orchard,
> where in an arbour we will eat a last year's
> pippin of my own graffing, with a dish of
> carraways and so forth."
> *Davy.* "There's a dish of leathercoats for you.
>
> *2nd Henry IV.*, V. 3.

14.

> "I warrant they would whip me with their
> fine wits till I were as crest-fallen as a dried
> pear."
>
> *Merry Wives*, IV. 5.

Shakespeare calls one of his characters "Costard," and uses the word frequently in its old English meaning of "head," but it is not quite clear that he refers, except by double meaning, to the costard apple. A coffin was certainly an old term for pie crust.

> "And of the paste a coffin I will rear,
> And make two pasties of your shameful heads."
>
> *Titus Andron.* V. 2.

But whether the "custard-coffin" may mean an apple pie, is rather doubtful.

The "*Apple John*" is the "*Winter Greening*" of the "*Fruit Manual,*" which will keep for two years but gets very shrivelled—and the *Leathercoat* is an excellent culinary apple, the "*Royal Russet,*" which still grows in Gloucestershire.

There is little doubt that apples and pears were cultivated and orchards planted in England long before cider and perry were made. The national beverage of the ancient Britons was mead; that of our English forefathers was undoubtedly ale; as there is abundant evidence to shew. Mead and ale filled the flowing cups of gods and heroes in Valhalla, and of kings and warriors on earth. The Romans introduced wine and it has ever since been known in England. In succeeding

Centuries as the communications with the Continent became greater, the quantity would naturally increase. It was, however, the Norman conquest that gave the first great impetus to the wine trade with France. The quantity of wine introduced by William and his followers from Bordeaux and the neighbouring provinces, became considerable. The Vine itself, which had before been introduced by the Romans, was again carefully planted, and every effort was made by the Normans to establish it here. This is proved by the fact, that there are no less than thirty-eight entries of vineyards in Domesday Book. The quantity of French wine imported was again much increased when Henry II, married the daughter of William Duke of Aquitaine, and thus added the provinces of Anjou, Touraine, Maine, Poitou, Saintonge, Guienne, and Gascony to his dominions. From this time the consumption of wine became general, and kept increasing, until the demand for it became greater than the supply; a fact which possibly indicates also, a considerable increase of wealth amongst the middle classes of society. The price of wine was regulated by enactment, and its quality ordered to be tested twice a year. In 1396 (Edward III.) a statute has this remarkable expression in its preamble :—" The King wills of his grace and sufferance that all merchant denizens, not being artificers, shall pass into Gascoign to fetch wines thence to the intent that by this general license greater plenty may come." Still the demand increased and the price got higher until the middle of the fifteenth Century, when no wine was permitted to exceed the price of twelve pence the gallon, and a law was made that " No person, except those who could spend a hundred marks annually, or were of noble birth, should keep in his house any vessel of wine exceeding ten gallons."

When England lost the French provinces and frequent wars arose between the two countries culminating in bitterness and hatred between the people, as they did in the reigns of William III. and Anne—all commerce was necessarily restricted, and every effort was made to supply the place of the French wines. The manufacture of home-made wine of every kind was encouraged, and then it was, too, that the production of cider was pushed forward, its use generally inculcated, and its praises vaunted to the utmost by our poets :—

> " What should we wish for more ? or why in quest
> Of Foreign Vintage, insincere and mix't
> Traverse th' extremest World ? Why tempt the Rage
> Of the rough Ocean when our native Glebe
> Imparts from bounteous Womb, annual recruits
> Of wine delectable, that far surmounts
> *Gallic* or *Latin* grapes, or those that see
> The setting Sun near Calpe's ' towering height ?'
> Nor let the Rhodian nor the Lesbian Vines
> Vaunt their rich Must, nor let Tokay contend
> For Sov'ranty ; Phanæus self must bow
> To th' Ariconian Vales."—*Philips.*

It has thus been shown that the time of the origin of Herefordshire orchards and of the cider and perry made from their fruit is very uncertain. We have strong negative testimony from Bishop Swinfield's Roll, that they did not exist at the end of the thirteenth Century.—At the end of the sixteenth Century, we have again the very positive evidence of old Gerarde, not only of the existence of orchards in the fields and apple trees in the hedge-rows, but that cider was abundantly made and appreciated. The great probability, therefore, is that the peculiar adaptability of the soil in Herefordshire for the growth of apples and pears, became gradually appreciated in the course of the fourteenth and fifteenth Centuries, as orchards spread from Kent and the neighbourhood of London into these more Western regions. In the seventeenth Century, when the political circumstances of the country called for a more abundant production of cider and perry, then the orcharding was greatly extended. It was, says Evelyn, "By the plain industry of one Harris, a fruiterer to King Henry VIII. (1509-47), that the fields and environs of about thirty towns in Kent only, were planted with fruit to the universal benefit and general improvement of that country to this day :—as by the noble example of my Lord Scudamore, (c. 1630-50) and of some other publick spirited gentlemen in those parts, all Herefordshire is become in a manner but one entire orchard."

In the reign of Charles I., Parkinson published his great work (1629), "*Paradisi in sole paradisus terrestris*," or, "A Garden of all sortes of pleasant flowers, with a Kitchen Garden of all manner of herbs and roots, and an Orchard of all sortes of fruit-bearing trees." He describes fifty-eight sorts of Apples, and sixty-four kinds of Pears with numerous varieties of other fruits and plants.

Then comes John Evelyn with his "*Pomona*" in 1664, and John Philips the Herefordshire poet, with his poem on "*Cyder*," (1700) both of whose works are so full of local interest, and will be so constantly quoted hereafter, that they must not be further alluded to in this place.

The general introduction which will be published with one of the later numbers of this work—and, indeed, the descriptive account of several of the varieties of fruit illustrated, will take up the thread of history from this point. It will enter into the special adaptability of the climate and soil of Herefordshire for the growth of apples and pears, and its effects on the several varieties grown here—the manufacture of cider and perry, and all other matters of interest and use, which the inquiries instituted for this work may bring forth.

The present paper may be well concluded by a dissertation on the health giving properties of "Syder," by the Rev. Martin Johnson, M.A., of Baliol College, Oxford, who was vicar of Dilwyn, from 1651 to 1698. The fact of longevity being characteristic of the county, is also happily borne out in these days by evidence that may, perhaps, be still more satisfactory to some people.—The returns of the Registrar General make Herefordshire one of the four longest lived counties.

The following extract is taken from Dingley's "*History from Marble*," edited by Mr. Gough Nichols, for the Camden Society, (1868) who says : I received this memoriall from ye Vicar of Dilwyn (thus prefac'd) concerning his parish.

CONCERNING THE LONGEVITY OF THE SYDER-DRINKERS OF HEREFORDSHIRE.

Old age in many cases is a blessing, otherwise Abraham the father of the faithfull, had never had such a faithfull promise made him in *Gen.* xv. 55, that he should die an old man, which the Hebrew renders, " with a good hoarie head." It was denounced as an heavy judgment on Elies posterity, that there should never be any old man in his house. God himself describing the flourishing estate of his people makes the blessing of long life the best part of the descrip[con] by saying, that none should go from them or thence an infant of dayes, *Esay* lxv. 20, w[ch] *Junius* affirms, *tam ad infantulum quam ad senem referatur illud*, y[t] this belongs to infants as well as old men, for none of them shall die y[ng.] or in infants estate, neither shall any old man dep[t] until he hath fulfilled his dayes : till he hath liv'd as long as nature will p[rmit] ; neither is it a blessing onely but a beauty, soo well does a grey head become old men, *Prov.* xx. 29. Ye crown of hoary hairs which God's finger setts on their heads, carrieth more majesty and venera[con] wth it in all places y[n] a crown of gold."

" This parish, wherein Syder is plentifull, hath, and doth afford many people, that have, and do enjoy this blessing of long life : neither are the aged here bedridden or decrepit as elsewhere, but for the most p[te] lively and vigorous ; next to God, wee ascribe it to our flourishing Orchards, which are not onely the ornamet but pride of our countrey, and that in a double respect, 1st that the bloomed trees in Spring do not onely sweeten but purifey y[e] ambient air, as Mr. Beal observes in Heref. Orchards, p. 8. Next, that they yield us plenty of rich and winy liquors, w[ch] long experience hath taught do conduce very much to the constant health and long lives of our inhabitants, the Cottagers, as well as y[e] wealthier using for the most part little other liquors in their families, than restorative sider. Their ordinary course among their serv[ts] is, to breakfast, and sup with toast and cyder through the whole Lent, and the same dyet in the neighbourhood continues on Fasting dayes all the yeer after; which heightens their appetites, and creates in them durable strength to labour."

" Sider is their physick and our vessels their apothecaries shops."

The same Vicar also thus sings :—

"AN ENCOMION ON SIDER, 1677."

Of some seventy lines in length, which the following short extracts must suffice :—

* * *

" The *Hesperides* bragg of golden apples, wee
Have equal fruit, not fenc'd, but dragon free,"

* * *

" *Vin de Paris, Vin d'Orleans, Vin Sharoon,*
" With all the Gallick wines are not so boone
" As hearty Sider, y[e] strong son of wood
" In fullest tydes refines and purges blood."

* * *

" Death slowly shall life's cittadel invade,
" A draught of this bedulls his scythe and spade."

* * *

Then the following examples of longevity are given in these words :—

A. Dom. 1657—James Badam of this Parish of one hundred and five yeers of age and upwards : was so vigorous that the week before he died, he plowed most part of the day in the heat of Summer without any covering on his head.

1659—Thomas Melling a very active man of his age who was able to endure running after the doggs on foot from morning till night, neither did this nimbleness abate in him till about a quarter of an yeer before he sickned to death, at which time he was about ninety years old.

1663—Wm. Dykes above 90.

1666—Henry Seyse, an old souldier of about one hundred.

Now the diet of all these according to the best information was that which is before menconed.

1673—Avis Taylor, widow, about 100 years old. Her diet was besides Syder, bread and cheese, oatmeale and pepper, wᶜʰ she used to chew out of her pocket.

1673—Richard Tuffley, a tanner, a very laborious man now living above 100 years old.

To which may be added :

1676—Widdow Hill, of Eardesland, 111 years old.

And the Editor mentions :

1660—Joyce Andrews of Felton, died Æt. 114.

1662—Richard Wooton of Fromanton, died at Marden Æt. 104.

1756—Eliza Collier of Yarpole, Æt. 103 years.

1777—Joseph Rod of Yarpole, Æt. 104 years.

1790—Rev. Wm. Davies, Rector of Staunton on Wye, Æt. 105, who died at Hereford.

And then he quotes the memorable Morrice dance in Herefordshire, reported by Sir Walter Raleigh : and thus "reported" again by Dr. Fuller, in his "*Worthies of England*," under Herefordshire :—

. "There cannot be a more effectual evidence of the healthy air in this shire, than the vigorous vivacity of the inhabitants therein. Many aged folk, which in other counties are properties of the chimneys, or confined to their bed, are here found as able (if willing) to work. The ingenious Serjeant Hoskin gave an entertainment to King James, and provided ten aged people to dance the Morish before him, all of them making up more than 1000 years, so that what was wanting in one, was supplied in another,—a nest of Nestors not to be found in another place."

The scene of the royal entertainment has been placed at Morehampton, but James I., never visited Herefordshire. It was at the Hereford races in 1609 that this assemblage of veteran Morris-dancers actually took place, and it was recited in a contemporary tract styled : " *Old Meg of Herefordshire for a Mayd-Marian, and Hereford Towne for a Morris-daunce ; or twelve Morris-daunccrs in Herefordshire, of twelve hundred years old.*" *London*, 1609.

HENRY G. BULL, M.D.

Postscript.—The readers of this History will scarcely fail to observe the great extent, and variety of the sources from which it has been compiled. Much learning is sometimes concealed beneath a light surface, and omissions may perhaps be discovered which have been intentional ; all this is in great measure due to the information and authorities, so kindly supplied by members of our Club and other gentlemen, to wit : the President (1877) J. Griffith Morris, Esq. ; the President elect, (1878) the Rev. H. W. Phillott ; Thomas Blashill, Esq. ; Rev. James Davies ; James Davies, Esq. ; Rev. Canon Dolman ; Rev. W. D. V. Duncombe ; J. T. Owen Fowler, Esq. ; Rev. F. T. Havergal ; Rev. J. E. Jones Machen ; E. H. Jones, Esq. ; Robert Hogg, L.L.D., &c. ; Edwin Lees, Esq., F.L.S., &c. ; James Renny, Esq. ; and above, and beyond all, by the Rev. Thomas Woodhouse. This willing assistance, the writer desires most cordially to acknowledge on behalf of the Woolhope Club, in whose cause the paper has been written.—H.G.B.

"POMA QUOQUE UT PRIMUM TRUNCOS SENSERE VALENTES.
ET VIRES HABUERE SUAS AD SIDERA RAPTIM
VI PROPRIA NITUNTUR, OPISQUE HAUD INDIGA NOSTRAE.
NEC MINUS INTEREA FETU NEMUS OMNE GRAVESCIT,
SANGUINEISQUE INCULTA RUBENT AVIÁRIA BACCIS."

Virgil George II, 426-30.

"Fruit trees moreover, soon as they have known
The vigour of the stock become their own,
Push jostling upwards by their native powers,
To starry Heaven, and ask no help of ours ;
Nor less the wild grove bows its fruitful head,
And thorny bird-homes blush with berries red."

Richard Doddridge Blakemore, p. 58.

EH ! QUI SAIT QUELS SUCCES ATTENDENT VOS TRAVAUX !
COMBIEN L'ART PARMI NOUS CONQUIT DE FRUITS NOUVEAUX ;
DANS NOS CHAMPS ETONNÉS, QUE DE MÉTAMORPHOSES !

L'Homme des Champs. Delille, 1800.

"Who knows what victories await your toils,
What fruits yet new shall be your bloodless spoils,
With what amazement shall our fields behold
Changes undreamt of, marvels yet untold !"

Rev. T. Woodhouse.

THOMAS ANDREW KNIGHT, F.R.S. &c.
PRESIDENT OF THE LONDON HORTICULTURAL SOCIETY, 1811 1938.

THOMAS ANDREW KNIGHT AND HIS WORK IN THE ORCHARD.

Born August 12th, 1759.   Died May 11th, 1838.

DURING the first quarter of the present Century, Thomas Andrew Knight stood at the head of Scientific Horticulture in Great Britain. The vigour and originality with which he carried out his numerous experiments in Vegetable Physiology, and the great success which attended his efforts to introduce new varieties of fruits and vegetables by means of hybridization, proved him to be the best practical gardener of his time. The distinguished position he held for many years as President of the Royal Horticultural Society in its most palmy days, afforded him the opportunity he was so well able to use, of making known the results of his own work, and of spreading widely the influence of his example. England thenceforth took her rightful place in advancing the science of plant growth, and to Mr. Knight's original experiments is unquestionably due the great merit of beginning the work.

The only printed account of Mr. Knight's life is to be found in the introduction to a Volume of his Miscellaneous Papers, which was published in 1841, only a few years after his death. He was born at Wormsley Grange, and spent the earlier years of his life in the retired seclusion of that part of Herefordshire. For his companion he had his brother, Richard Payne Knight, nine years older than himself. Together the boys ran wild through the fields, and orchards, and beautiful woodland scenery that surround the Grange, and there doubtless they imbibed that love of nature, which distinguished them both in after life, in their respective, and very different ways.

On emerging into the world, Richard Payne Knight quickly became distinguished. His great talents found a congenial field in the study of the language and literature of ancient Greece. He became known as a refined scholar, and an elegant poet. He was the author of *"An Analytical Inquiry into the Principles of Taste,"* and of many poems and other works. His best memorial is however in the British Museum, for he bequeathed to that institution the magnificent collection of ancient bronzes and coins which he had made, and which was estimated to be worth £50,000 at that time.   Mr. R. Payne Knight represented Leominster from 1780 to 1784, when he was returned for Ludlow, and sat for that borough until 1806, when he retired from Parliament.

Thomas Andrew Knight received his early education at Ludlow, at Chiswick, and afterwards at Balliol College, Oxford.   In his youth he was less remarkable for his industry than for his natural talents, the quickness of his perception, the excellence of his memory, the ease with which he mastered any study he attempted, and the good sense which secured his great steadiness of character amidst the temptations around him.   There were no classes or honour lists in those days. Perhaps, in the case of some peculiarly constituted minds, this absence of stimulus to exertion may have been favourable to the prosecution of special studies.   Even then the University afforded some assistance to the students of natural science, though it was scanty indeed, in comparison of the ample stores of books, of specimens, and of help of all kinds, which are now so abundantly provided.   Mr. Knight was not idle, but following the bent of his own mind in his studies, he certainly laid at Oxford the foundation of that knowledge of natural philosophy, which enabled him in after life to carry out his own experiments with so much success.

On leaving the University he returned to his quiet home, amidst the fine scenery of northern Herefordshire.   Here he at once began his own special work—the study of all objects of rural interest.   He was energetic in all he did: a good sportsman, expert with rod and gun from boyhood, and ere long a zealous agriculturalist.   Nothing escaped his notice, from the very implements employed upon the land, to the objects produced by it.   He threw his mind into the work, and often with good effect.   He endeavoured, so far as his means allowed, and they were not large at this time, to improve the breed of horses, cattle, sheep, pigs and dogs; and to the end of his life ever took a deep interest in the economy of the domestic animals.

It was, however, on the study of the vegetable productions of the farm and garden, that he bestowed most thought and labour.   The mode of growth in plants; the ascent and descent of the sap in trees; the phenomena of germination; the influence of light upon foliage; the formation of roots, &c., were all tested by a series of original and ingenious experiments.   Lastly, as the result of his growth of seedlings, and of his careful practice of hybridization, many new varieties of fruits: apples, pears, plums, nectarines, cherries, strawberries, and currants: many new vegetables: potatoes, cabbages, peas, onions, &c.: and many new varieties of trees and flowers were produced; thus often making valued and important additions to the luxuries and necessities of life.

From his earliest youth Mr. Knight owed much to his brother, and to him he was also indebted for his introduction to public life.   The Board of Agriculture wished at that time to obtain certain statistics from the different districts of the country; and when Sir Joseph Banks, the President of the Royal Society, applied to Mr. R. Payne Knight, M.P. for Ludlow, he recommended his brother, Mr. T. Andrew Knight, as the best qualified man he knew to make the returns for Herefordshire. This introduction was the turning point of Mr. Knight's life.   Sir Joseph Banks quickly detected beneath a quiet reserved manner the able original mind he had to deal with.   He could fully appreciate the practical importance of the physiological experiments Mr. Knight was carrying out, and gave him the warmest encouragement to persevere with them, and to make known the result. They thus became great friends.   The evening conversaziones, in Soho Square, were thrown open to Mr. Knight, and he had the great pleasure of being introduced to many of the most distinguished men in science and literature. Sir Joseph Banks never afterwards lost an opportunity of assisting Mr. Knight, by procuring information for him, or by giving him that friendly and courteous advice, which was of still greater value.

Mr. Knight's first communication was made to the Royal Society, April 30th, 1795 ; through the influence, no doubt, of his friend the President. It was that famous paper, "*Observations on the Grafting of Trees*," in which Mr. Knight maintained the doctrine that there was no renewal of vitality in the process of grafting, but that the scion carried with it the debility of the tree from which it was taken. He advanced the same opinion in his "*Treatise on the Culture of the Apple and Pear, and of the Manufacture of Cider and Perry*," published in 1797 ; and it seemed to result, so naturally, from his observations, and from the series of ingenious experiments he carried out ; whilst, at the same time, it seemed to explain so well the cankered and diseased state of most of the trees of the old varieties of cider apples, in the orchards of Herefordshire, that it was at once generally received. It was an old theory revived, but was so well put forward that the merit of an actual discovery was awarded to Mr. Knight by common consent. It can, however, easily be shown, that the belief that a graft would not live longer than the tree from which it was taken was general from a very early period, if not from the time when grafting was first practised. This might be proved by reference to the gardening books of the age, but it is scarcely necessary to do so since they are quite obsolete.

Even Pliny contrasts the short duration of the apple with the productiveness of pears and other fruit trees when grown old. He says :—

"Celerrime vero senescit, et in senectâ deteriorem fructum gignit malus. Namque et minora poma proveniunt et vermiculis obnoxia." *Hist. Nat.* xvi, 27.

"Apple trees soon grow old, and in their old age bear inferior fruit. The apples produced are not only smaller, but apt to be grubby."

The difficulty of propagating by grafts, from old and cankered trees, was well known in all orchards. The world was prepared to believe, that "canker" would always prevent the old varieties from being prolonged in this, or in any other way ; and when Mr. Knight laid down the law so precisely, and seemingly proved it by his experiments so conclusively, that "Vegetable like Animal life has its fixed periods of duration," and that the different varieties of apple trees thus died out naturally, the opinion was universally received without hesitation.

Mr. Knight was right in attributing "canker" in the graft, to the age and debility of the individual tree, from which it was taken, but he was wrong in supposing it due to the age of the variety of apple. His theory was, that all the trees of any given variety were really nothing more than separate and isolated branches of one tree ; whereas every bud in essence is a new tree with a new life. Mr. Knight's experiments had indeed proved that "the existence of a variety of fruit trees may be protracted beyond the natural terms of the original seedling plant by grafting, or by unusually favourable circumstances of soil and situation." Had he carried out these experiments further he would have hesitated to state as he then does, "that there is a period, beyond which the debility incident to old age cannot be stimulated." This is now believed not to be correct.

"Canker" is due to debility from any cause. It may arise from the age of the tree itself ;

from direct injury; from sudden variations in temperature or climate; or, most common of all causes, from the soil in which the tree grows. Mr. Knight abundantly proved the debility due to old age to be a common cause of canker in every orchard.—Direct injury will also frequently give rise to it, whether this be produced by the accidents of wind or ladders, or by sudden alternations of temperature. The severe frosts of winter, acting on the insufficiently ripened wood, or a sudden check to growth, from a late frost in Spring, lacerate the vessels of the young wood, and give rise to "canker" in the following Summer.

The soil in which the tree grows may originally be ungenial to it; it may require drainage; or it may want the nutritive principles necessary for the tree; or lastly, it may have become too much exhausted for the continued health of the tree. "Want of food," said a good orchardist, "I have always found to be the cause of canker, and the same may also be said of the woolly Aphis. My young trees in the hedge rows became badly affected with canker, and it occurred to me that the thorns took the nutriment from their roots. I fed them with a dressing of lime, cowdung, and fresh mould, on the surface of the ground. This soon produced a good effect, and the trees recovered their luxuriance. I have never let my trees want it since, and am always rewarded by their healthy condition, and abundant crops."

Writing in 1819, Mr. Knight himself stated from his experiments in former years, "he had found that the destructive effects of canker were greatly prevented by digging up the trees once in every three or four years and applying some fresh unmanured mould of good quality to the roots." (*Hort. Soc. Trans.* Vol. iii. p. 338).

In many soils, it is well known that fruit trees can only be made to flourish by planting them on good soil, with a layer of paving stones, or a bed of concrete beneath, so as to prevent the roots reaching a soil below, that would quickly kill the tree.

It is found that "canker" will attack seedling apple trees on the one hand, and on the other it is well known that many of our old varieties of English apples have long withstood its ravages. The *Pearmain* and the *Costard* apple have been known from the 12th century: and those old varieties, *Catshead, Winter Queening, Golden Pippin, London Pippin, Leathercoat*, are still propagated with success by grafting.

The valuable cider apple, the *Foxwhelp* may be given as an example. Mr. Knight himself said of this apple, more than 60 years since : "Some attempts are still made to propagate it, but I "venture to predict they will not be successful: for the grafts necessarily partake of a life that is "two centuries old, and the young stock can give nutriment only, not new life." (*Pomona Herefordiensis.*)

This unfortunate prediction beyond doubt tended greatly to fulfil itself, for Mr. Knight's reputation as a practical orchardist was very great, and quite sufficient to put a stop to further attempts for a time. It is however beyond question, that several of our best fruit growers are now propagating successfully the Foxwhelp apple by careful grafting; and the young trees are doing, and bearing well, so that generations to come may still hope to enjoy its celebrated cider.

The notion that a graft can live no longer than the tree from which it is taken, seems to rest upon the assumption that the new wood, which proceeds from the graft, is not a new tree, but only a detached part of the parent. But this is evidently a mistake. A branch produced by a graft is

THOMAS ANDREW KNIGHT. 33

as distinctly a new and separate individual as a branch produced by a cutting. In both cases the bud is the source of new growth; and physiologically speaking a seed itself differs little from a bud, except in being more carefully protected, and in being spontaneously detached. The embryo in a seed, the bud inserted in budding, the buds in a graft, or in a cutting, differ only in their position; and each as it developes, becomes a new individual, not a mere dependent portion of the parent.

The embryo of the seed doubtless gives that mysterious rejuvenescence of life which ever dwelt so strongly in Mr. Knight's mind, and there is this great difference, that whilst the bud necessarily produces the same plant, from which it is derived, the seed even, when self fertilized, is by no means always true to the plant producing it, and thus a new and varied species may be produced; but in each case the new plant has an independent existence, a distinct and separate life, inheriting doubtless, much from the parent tree, but nevertheless capable of being largely influenced by the circumstances of its own position.

The apple tree forms no exception to the general rule in this respect. Elm trees we know have been propagated from suckers, or in these latter days by grafts, from the time of their first introduction by the Romans, and yet they are as luxuriant as ever. Individual Elms live their three or four Centuries and die from age, whilst their places are supplied by successors as strong and healthy as they were. The Common Laurel again, a tree of much shorter natural life, and which by the way bears fruit and seeds abundantly; is yet always propagated by cuttings. They grow as well and freely as the original trees; so too may it be said of *Willow* and *Poplar*, and all the free-growing trees and shrubs our nurserymen propagate so extensively by bud, or graft, or cutting.

For all these reasons it is the opinion of modern Horticultural Science, that where the soil and climate are naturally adapted to the growth of the apple tree, any variety of apple may be indefinitely prolonged with proper care and skill.

Apple trees may be said to live too long, or not long enough for the proper preservation of varieties :—so long, that short-lived, selfish man has enough for his generation, and leaves their propagation to his successors : and yet not long enough for these successors to learn to appreciate the transitory life of apple trees. Thus it comes to pass, that choice varieties die out, and are lost.

It must however be candidly admitted, that many men of great practical experience altogether demur to this opinion, and believe Mr. Knight's views to be correct. " Science may say what it likes " said a very intelligent horticulturist, whose hobby fruit-growing has been for many years, " Science may say what it likes, but it shall never make me believe that sorts don't die out, " for I know they do;" and this is still the general opinion in the orchards.

Mr. Knight fully believed that all varieties of apples died out naturally of necessity, and thus he turned his whole attention to the growth of new and improved varieties from seedlings. There seems no limit to the possible multiplication of varieties : and it is scarcely an exaggeration to say as Philips does,

" An inmate Orchat every apple boasts."

Mr. Knight had been practising hybridization for some years, and he now took up the subject with renewed energy and perseverance, and by crossing the best varieties of fruits, he endeavoured to obtain the best characters of each, with all the vitality of a new plant. From Mr. Knight's frequent reference to the views of Linnæus in describing the results of his own experiments, it would appear that it was to the writing of this distinguished naturalist, he owed the impression of the great importance of hybridization. Indeed although other observers long before him had pointed out the fact, until Linnæus had set forth and insisted on the distinction of sexes in plants, hybridization could only be mere guess-work; and as it had therefore no rational ground to rest on, there was but little inducement to attempt it. Mr. Knight was not certainly the originator of the practice, but he was the first fully to realize its practical utility so far as fruit trees are concerned; and he followed it out with so much energy and perseverance, as to make the subject his own.

Mr. Knight had already many hundred seedlings, at various periods of growth, in his beds; and in the early period of his friendship with Sir Joseph Banks, he must evidently have dwelt in one of his letters, with true Virgilian Spirit, upon the length of time his seedling fruit trees required to arrive at maturity:

> "Jam quæ seminibus jactis se sustulit arbos,
> Tarda venit, seris factura nepotibus umbram."
>
> _Georgic II. 57-8._
>
> "But slowly comes the tree which thou hast sown,
> A canopy for grandsons of thine own."
>
> _Blakemore._

for Sir Joseph writes in answer (1798) "Your experiments on apples and grapes must be very tedious, but surely the success of those on annual plants, will induce you to persevere."

Mr. Knight's after experience told him that apple-tree seedlings took from five to twelve years to come into bearing; whilst pear-tree seedlings required still longer time, and would not ear fruit until they were from twelve to eighteen years old.

Mr. Knight's chief experiments in crossing apple trees were made from the _Golden Pippin_, _Golden Harvey_, _Orange Pippin_, and _Siberian Crab_—and the results of their hybridization are shewn in the following table:—

From the Pips or Seeds of the	Fertilized by Pollen from the	He obtained the
Orange Pippin	Golden Pippin	Grange Apple ... 1802 Downton Pippin ... 1804 Red Ingestrie ... 1800 Yellow Ingestrie ... 1800
Golden Harvey	Golden Pippin	Breinton Seedling 1801
Golden Pippin	Golden Harvey	Bringewood Pippin 1800 Wormesley Pippin 1811
Siberian Crab	Orange Pippin	Yellow Siberian ... 1805 Siberian Pippin ... 1806 Foxley Apple ... 1808
Siberian Crab	Golden Harvey	Siberian Harvey ... 1807
Yellow Siberian Crab	Golden Harvey	Siberian Bittersweet

The great tendency to variety in seedlings is well shewn by the fact, that those very different apples, the *Red* and *Yellow Ingestrie* apples, were not only derived from the same parentage, as shewn in this table, but actually sprang from two pips which occupied the same cell, in the same apple; twin plants in every particular: yet the *Yellow Ingestrie*, of *Golden Pippin* colour, ripens in October; whilst the *Red Ingestrie* resembles the *Golden Reinette*, and comes to maturity in November.

Mr. Knight was one of the original members of the Herefordshire Agricultural Society, and carried into its service the same energetic activity, and the same scientific spirit, which contributed so largely to the success of the Royal Horticultural Society; and it was no doubt, owing to his influence that premiums were offered for new varieties of cider apples. These prizes were given for some years, and amongst the "*Herefordshire Tracts*," at the Permanent Library, in Hereford, is a short paper descriptive of the successful apples. It was probably written by Mr. Knight himself. It describes several of his own seedlings, and bears so great an interest in the Pomology of Herefordshire, that it must be given as an appendix in full.

Mr. Knight also edited for the Herefordshire Agricultural Society the "*Pomona Herefordiensis*," containing coloured "engravings of the old Cider and Perry fruits of Herefordshire, "with such new fruits as have been found to possess superior excellence." This was published in 1816. The letter-press was written by Mr. Knight, and the original drawings for it were made by Miss Matthews, daughter of Col. Matthews of Belmont, and by his own daughter Miss Frances Knight, under his superintendence. It is a very beautiful work and will always maintain its interest and value.

Mr. Knight took even a greater interest in the cultivation of Pears than of Apples. His very numerous experiments in crossing different varieties of plants had led him to observe—in opposition to the opinion of Linnæus—that "in the seedlings, with few exceptions, there was "always a strong prevalence of the constitution and habits of the seed-bearing plant "—and this observation was of very great service to him in raising new varieties of pears. The Pear likes a warmer climate than the apple, and in order to obtain good hardy varieties of pears to suit Herefordshire, he made a point of introducing the pollen from the rich flavoured and more delicate kinds, into the carefully prepared blossoms of those which were hardy in character—thus he used the pollen from the *St. Germain, Crasanne, Colmar, Bezi de Chaumontel*, &c., &c., with the blossoms of such hardy Autumn and Winter pears, as *Autumn Bergamot, Swans-egg, Aston Town*, &c., &c., in order that the character and habits of the new varieties might also be hardy. That he was very successful in this respect has been practically proved. His many hundred seedlings tried his patience greatly in coming to maturity, but as they did so, he sent the best of them to the gardens of the Horticultural Society in London. Here they bore fruit, and were twice officially reported upon. These reports on their character and value are most interesting, and must also be re-printed in full at the end of this paper, with Mr. Knight's own comments on the judgement; for they are all true Herefordshire varieties; they bear for the most part Herefordshire names; and have been thus officially pronounced to be most excellent pears.

Mr. Knight had the pleasant habit of presenting trees to his friends, and naming the new varieties from their residences; thus his seedling Pears have been named: *Garnons*, from the seat of Sir John Geers Cotterell, Bart.; *Foxley*, the seat of Sir Robert Price, Bart., M.P.; *Rouse Lench*, an estate near Evesham, which belonged to his son in law, Sir Wm. Rouse Boughton, Bart. ; *Belmont*, near Hereford, the seat of Col. Matthews; *Eastnor Castle*, the seat of Earl Somers, Lord Lieutenant of Herefordshire; *Whitfield*, the seat of E. Bolton Clive, Esq., M.P.; *Pengethley*, near Ross, the estate of the Rev. T. P. Symonds ; *Ross*, probably from the town of Ross ; *Moccas*, the seat of Sir George Cornewall, Bart. ; *Shobdon Court*, the seat of Lord Bateman ; *Eyewood*, near Kington, the seat of the Earl of Oxford ; *Croft Castle*, the seat of Colonel Johnes ; *Oakley Park*, near Ludlow, the seat of the Hon. Robert Henry Clive, M.P. ; *Pitfour*, the seat of Admiral Ferguson in Scotland, who was himself called, *more Scottico*, "Pitfour ;" *Broom Park*, near Canterbury, the seat of Sir Henry Oxenden, Bart., whose son was the companion and friend of Mr. Knight's only son ; and *Dunmore*, near Stirling, the seat of Lord Dunmore. These names recall to memory his personal friends, and give his hardy pears a local interest in addition to their own merits.

There were however exceptions to this rule, as the *Monarch*, *Althorp Crasanne*, *Winter Crasanne*, and the *March Bergamot*.

The following list of the new varieties of fruit raised by Mr. Knight, and considered by him worth preserving, is given as an Appendix to the memoir of his life :—

Apples.—Spring-grove Codling, Downton Lemon Pippin, Herefordshire Gilliflower, Grange Apple, &c.

Cherries.—Elton, Waterloo, Black Eagle.

Strawberries.—Elton, Downton.

A large and long keeping *Red Currant*.

Plums.—Ickworth Impératice, a large purple plum not named ; and two improved damsons.

Nectarines.—Impératrice, Ickworth, Downton, and Althorp.

Pears.—Monarch, Althorp Crasanne, Rouse Lench, Winter Crasanne, Belmont, and many others.

Many excellent and productive varieties of *Potatoes*, of which the only one named is the Downton Yam.

The Knight *Marrowfat Pea ;* and improved varieties of *Cabbage*.

If this list is as meagre with reference to other fruits, as it has been shewn to be in regard to his new varieties of Apples and Pears, it gives but a poor idea of the amount of Mr. Knight's labours.

It is not within the scope of the present sketch however to follow his experiments further, or it could be shewn that he was equally successful in producing many improved varieties of other fruits, as well as of flowers, vegetables, and trees.

A hurried glance must be taken of the great services he rendered to Scientific Horticulture, in his more public life. In 1805 he was made a Fellow of the Royal Society, an honour he had already well earned by his papers. In 1804 the Royal Horticultural Society was formed ; Mr. Knight was an original member, and in 1811 on the death of the first President, Lord

Dartmouth, he was called to the chair, and filled the office until his death in 1838.   Mr. Knight was an active member of the Society from its commencement.   When made President he infused his own practical energy into the working of the Society, and at the same time his own unselfish desire to use its organization to the utmost extent for the promotion of Scientific Horticulture.   Under his presidency, the Horticultural Society reached its highest repute and popularity, and was joined, not only by men of Science and practical gardeners, but by nearly all the rank and wealth of the Kingdom.   Its funds then became ample, and the greatest energy was shewn in procuring new plants, seeds, grafts, &c., and the best information, from all parts of the world. These were distributed amongst the members, with unsparing liberality.   " A complete revolution " was thus effected in the Science and practice of gardening, and a great public benefit was conferred " throughout the Kingdom."   (*Gardener's Chronicle,* 1877).

The complete success of the Horticultural Society was a great source of happiness to Mr. Knight himself.   As its President he was necessarily brought into communication with the most scientific men of his day on a great variety of interesting subjects.   His society was sought out, and his opinions received everywhere with the greatest deference.   Besides his great friend Sir Joseph Banks, and many other members of the London Scientific Society of the day, he became intimate with Sir Humphrey Davy, who in his work on Agricultural Chemistry adopts the results of some of his experiments.   Sir Humphrey visited Mr. Knight, and gives a pleasant description of Downton in his "*Salmonia.*"   Then there was the necessary scientific correspondence of the Society devolving upon him, English and Foreign : amongst others that with M. M. Dutrochet, Mirbel, and Decandolle, on matters relating to various points in Vegetable Physiology.

How fully Mr. Knight's services were appreciated, and their great and varied character, will be best shown by the following list of the honours given him :—In 1801 the Society of Arts awarded to Mr. Knight its Silver Medal " for a new turnip drill ;" in 1806 the Royal Society presented its Gold Medal to Mr. Knight " for his papers on Vegetable Physiology ;" in 1814 the Horticultural Society gave him its Gold Medal " for his various and important communications to the Society, not only of Papers printed in their Transactions, but of grafts and buds of his valuable new fruits."   He was also awarded the Large Silver Medal of the Society in 1815 " for the *Black Eagle Cherry* ;" in 1817 " for the *Waterloo Cherry,*" and in 1818 " for the *Elton Cherry ;* in 1822 he got the Silver Banksian Medal " for new varieties of Pears ;" in 1836 the Society presented him with the first impression of its new Large Gold Medal " for the signal services he has rendered to Horticulture by his Physiological researches ;" and, lastly, in 1835, when the Horticultural Society decided to have another medium-sized medal, it did Mr. Knight the very high honour of having his profile struck by Mr. Wyon on the die.

Since the year 1817 the Knightian Medal, in gold and in silver, has been frequently awarded ; and never, may it be added, has this Gold Medal been given more worthily than when, in 1875, it was awarded to Worthington G. Smith, Esq., F.L.S., an honorary member of the Woolhope Club, for his discovery of the true cause of the Potatoe disease, by his clear demonstration of the life history of the microscopic fungus, *Peronospora infestans.*

Distant Societies also vied with those in London, in paying tribute to Mr. Knight's talents and industry.   In 1815 the Caledonian Horticultural Society gave him its Gold Medal "in testimony

38 THOMAS ANDREW KNIGHT.

of their gratitude for his valuable discoveries—the result of patient and laborious research in Vegetable Physiology—science having been his guide." In 1826 the Massachusetts Agricultural Society sent him a Large Silver Medal "as a tribute to an eminent physiologist and a benefactor to the new world;" and in 1830 the Swedish Academy of Agriculture awarded to Mr. Knight its "Grand Silver Medal."

It must also be added, to show how widely his fame had extended, that he was made Honorary Member of many Societies not only in Great Britain, but also in Canada, and the United States, in France, Germany, Spain, Russia, Sweden, and Australia.

It would be out of place here to dwell on the more personal details of Mr. Knight's life: of his marriage and long domestic happiness: of his residences at Mary Knoll, at Elton, and at Downton Castle; all three places near each other in the wild and beautiful hilly country on the borders of Shropshire and Herefordshire; nor yet of that terrible grief of his life—the sudden death from accident of his only son at 32 years of age; an affliction that clouded with sorrow his later years; since all these facts have all been so well told in the memoir already mentioned.

Thomas Andrew Knight was a man of genius. His mind was original, clear and essentially practical. His manner was unassuming and reserved, but it concealed an ardent spirit and a warm and feeling heart. His studies were not the aimless and fruitless pursuits of a mere amateur; they were serious undertakings prosecuted with remarkable patience, industry and perseverance. He was as unselfish as he was generous, and endowed with a chivalrous sense of honour and truth. Mr. Knight was no technical botanist, but he devoted himself to the practical study of Vegetable Physiology. The one object of his life was to be useful to his generation. The great series of observations he made, were published with perfect candour, and were thus put at the service of others. However much the conclusions he drew from them have been called in question—and modern microscopic research has shewn many of his views to be untenable—the perfect truthfulness of his facts, has never been challenged. His practical papers have been the source from which Lindley, and many other writers on scientific horticulture, have borrowed largely. Much that Knight taught is now interwoven in daily practice; and many of his new varieties of fruit retain their excellence of character in the appreciation of the public. The great services he rendered to his country are not however to be measured so much by the actual work he did himself, as by the work he originated for others to follow up. His name will ever live in the annals of British Horticulture, for it certainly owes to him the beginning of its scientific character.

"Je me suis étonné souvent de notre indifférence pour la mémoire de ceux de nos ancêtres qui nous ont apporté des arbres utiles dont les fruits et les ombrages font aujourd'hui nos délices. Les noms de ces bienfaiteurs sont pour la plupart, totallement inconnus; cependant leur bienfaits se perpetuent pour nous d'âge en âge."

Bernardin St. Pierre. "Etudes de la Nature." Tome III p-270.

"I am often surprised at our forgetfulness of those ancestors, who have given us the useful trees, whose fruits and shade delight us so much. The names of these benefactors are for the most part completely forgotten, though their good deeds live from generation to generation." *(Translation.)*

HENRY G. BULL, M.D.

[The Woolhope Club is indebted now, as it has often been before, to the kindness of the editor of the *Gardener's Chronicle,* for the excellent likeness of Mr. Knight which accompanies this paper.]

A CONCISE ACCOUNT OF THE DIFFERENT APPLES WHICH HAVE BECOME
ENTITLED TO THE PREMIUMS OFFERED BY THE "AGRICULTURAL
SOCIETY OF HEREFORDSHIRE" FROM ITS INSTITUTION IN 1797,
TO JANUARY, 1809. *(Hereford, 1809.)*

THE ALBAN.—A.D., 1798.

Middle-sized ; colour splendid, and like the *Foxwhelp ;* pulp pale yellow ; juice too harsh
and austere, unless in mixture in small quantity, with that of sweet and bitter apples ; specific
gravity of the juice expressed from a perfect sample of the fruit, 1073. Trees grow freely and bear
well in some soils, but are subject to canker in others, and are only calculated for very warm
situations. Original tree on the Rev. Thomas Albans' Estate, in the parish of Leominster.

[A.D., 1799.—Premium not awarded.]

STEAD'S KERNEL.—A.D., 1800.

Middle-sized ; colour yellow with russet spots round the base of the stalk ; pulp faintly
yellow ; juice rich and well flavoured ; specific gravity of the juice at Brierly and Wormsley
Grange, 1074. Trees bear well but do not grow very rapidly, and require good soils and warm
situations. Original tree on the estate of the Rev. Thomas Alban, at Brierly. Apparently a good
cider apple and ready for the press early in November.

[This Apple is figured in the "*Pomona Herefordiensis.*" Plate XXV.]

BREINTON SEEDLING.—A.D., 1801.

Somewhat below the middle size ; colour pure yellow, and pulp faintly tinged ; specific
gravity of the juice of a moderately well-ripened sample of the fruit, 1072. Trees grow freely but
require very warm situations in which alone the fruit can ripen. Original tree at Breinton on
the estate of Dr. Symonds ; and sprang from a seed of the *Golden Harvey.*

GRANGE APPLE.—A.D., 1802.

Somewhat below the middle size ; colour similar to that of a very ripe *Golden Pippin ;* pulp
very yellow : specific gravity of the juice expressed from a well-ripened sample 1079. Trees grow
freely and bear well, but will require good soil. Fruit fit for the press the first week in November.
Original tree at Wormsley Grange, and raised from a seed of the *Orange Pippin* and the pollen of
the *Golden Pippin.*

[This Apple is figured in the "*Pomona Herefordiensis.*" Plate VII.]

MR. BAYLISS' APPLE, OR BRIERLY SEEDLING.—A.D., 1803.

Colour yellow with russet spots ; juice mild and sweet but watery : the specific gravity of the
juice expressed from a good sample of the fruit, but taken from the original tree when much
overladen with fruit 1058. Fruit fit for the press in November. Original tree vigorous and healthy
growing on Mr. Baylis's estate at Brierly.

DOWNTON PIPPIN.—A.D., 1804.

Similar in colour and flavour to a large *Golden Pippin;* pulp yellow; specific gravity of the juice of a perfect sample of the fruit in the year 1807 was 1080. Trees grow freely and are most exuberantly productive. Fruit fit for the press in the middle of November; but may be preserved till the spring. Original tree at Wormsley Grange, and sprang from the same parents as the *Grange Apple.*

[This Apple is figured in the " *Pomona Herefordiensis.*" Plate IX, with a more detailed description.]

YELLOW SIBERIAN.—A.D., 1805.

Fruit very small not exceeding the size of a common *Crab;* juice very rich, but contains much acid and astringents: the specific gravity of the juice of a perfect sample of the fruit 1085. Trees grow with great rapidity, are immensely productive, and are calculated for very high and exposed situations, where other varieties will not succeed, and for such only. Original Tree at Wormsley Grange, and sprung from a seed of the *Siberian Crab* and the pollen of the *Orange Pippin.*

SIBERIAN PIPPIN.—A.D. 1806.

Nearly similar in size to a very large *Golden Pippin ;* colour very yellow and streaked with red on one side; internally very yellow: specific gravity of the juice, which is very sweet and free from acid 1075. Trees grow very freely and bear well; but its blossoms are less hardy and its fruit inferior to either of the succeeding varieties. The original tree at Wormsley Grange, and sprung from the same parents as the preceding varieties. Fit for the press in the end of October.

THE SIBERIAN HARVEY.—A.D. 1807.

Nearly similar in size to the *Golden Pippin;* colour yellow freckled with red on one side: juice extremely sweet and rich, and its specific gravity, though expressed from another bad sample of fruit 1091 ; and of course exceeding that of any other apples of old or modern date that has been subjected to experiment. The trees grow very rapidly, and the blossoms appear to be more hardy than those of the crab tree. Original tree at Wormsley Grange, and sprung from a seed of the *Siberian Crab,* and the pollen of the *Golden Harvey.* The fruit will be fit for the press early in November. This variety will probably be found to grow and bear well in almost all soils and situations; but no grafts of it can be distributed till next year 1810. There is little doubt that the specific gravity of the juice of this fruit under favourable circumstances will be found to exceed 1100.

[This Apple is figured in the " *Pomona Herefordiensis.*" Plate XXIII.]

THE FOXLEY APPLE.—A.D., 1808.

Very small, not exceeding the common size of a crab; colour of the rind and pulp nearly that of a Seville Orange: juice very rich and free from acid, and its specific gravity 1080. The trees grow very freely, and are calculated rather for high than low situations. Fruit fit for the press early in November. Original tree at Wormsley Grange, and sprung from a seed of the *Siberian Crab* and the pollen of the *Orange Pippin.* The trees of this variety when loaded with fruit are singularly beautiful.

[This Apple is figured in the "*Pomona Herefordiensis.*" Plate XIV., with a lengthened description of its origin.]

SPECIFIC GRAVITY of the juice of some of the best Old Fruits.

That of the Golden Harvey or Brandy Apple .	. 1085
Forest Shire	. 1081
Small shrivelled Foxwhelp .	. 1080
Red Streak . . .	. 1079
Golden Pippin . . .	. 1078
Orange Pippin . .	. 1074
Red Must . .	. 1064

APPENDIX II.

A REPORT UPON SOME NEW SEEDLING PEARS RAISED BY THOS. ANDREW KNIGHT, ESQ., F.R.S., PRESIDENT. WITH REMARKS BY MR. KNIGHT (slightly condensed.)

[From Transactions of the Horticultural Society, London. Vol. I., 2nd Series, pp. 105—10. The descriptions were made in the Garden of the Society from specimens examined in the years 1827, 1828, and 1830.]

GARNON'S PEAR.

Fruit large, somewhat irregular, increasing beyond the middle, and from thence diminishing towards the eye, which is in a shallow depression or nearly level. The skin yellowish upon a wall, but remains long green if grown upon a standard. In size it resembles a Winter *Bon Chrétien,* but the quality of the latter is not by any means to be compared with it. Flesh juicy, melting and rich. A very excellent pear. Season, January.

Remarks.—I have seen this Pear, the produce of the original seedling tree, in each of the last five years; and its merits have appeared to me to be in all seasons above mediocrity. The trees grow well and bear very freely, and at an early age.

FOXLEY PEAR.

Yellowish brown, rather gritty, but extremely pleasant and rather high flavoured. Vinous like the *Galston Moor-fowl Egg.* Season, November.

Remarks.—This description corresponds very accurately with my notes. I believe, if it were ground and pressed, it would afford a finer liquor than ever has yet been obtained from the Pear or Apple.

ROUSE LENCH PEAR.—(With coloured figure.)

Skin pale green, a good deal spotted and tinged with russet. Flesh firm, rather buttery, yellowish, rather gritty, but juicy and high-flavoured. A capital pear, between breaking and buttery. Season, January.

Remarks.—I have never seen this pear in a perfect state till January. Its flesh, when it is perfectly ripe, is not, in any degree, breaking; but in every other respect the description is very accurate. The original tree has been exceedingly productive, indeed it would be more valuable, if it were less productive, for it often bears more than it can nourish properly. Some of the pears have weighed between nine and ten ounces. The fruit adheres so firmly to the tree that I have never seen one blown off by the wind; and, I think, taking its aggregate merits, that it is the best Pear for the market I possess.

DOWNTON PEAR.—(With coloured figure.)

Very like a *Passe Colmar*, Skin yellowish cinnamon colour, with very little red on one side. Flesh yellow, juicy, rich and excellent. Season, November.

The Downton Pear is very small when compared with the *Passe Colmar*. It becomes yellow early in December, but its proper season is February. The tree grows rapidly, in good form, and will bear the third year after grafting. The smallness of size is its chief defect.

WORMSLEY GRANGE PEAR.

Very cylindrical. Skin cinnamon colour with a pale speckling. Flesh yellow, melting, very juicy, sweet and rich; in perfection the end of October, and as good as a *Chaumontel*, which it is very like.

Remarks.—The merits of this Pear depend very much upon the season. It ripens early, and should be gathered early, or it will become yellow and sleepy on the tree. Gathered at the proper time and in a favourable season it has proved most excellent.

BELMONT PEAR.

Pale dull yellow, speckled with brown, the ground colour very much that of a *Blanquet*. Flesh melting, rather gritty, sweet and good. Probably an excellent variety.

Remarks.—I first saw this variety in the Autumn of 1829, when the tree produced a very heavy crop of fruit of large size; many of which proved to be in my estimation very excellent.

ALTHORP CRASANNE PEAR—(with Coloured Figure in Vol. II. 2nd Series. p. 119.)

This has much the appearance of the *Crasanne*. The Skin is of the same colour and texture; the flesh is less gritty, and the flavour good. The best description that can be given of it is, that it possessess all the richness of the *Crasanne*, with less grittiness, being perfectly melting. Season November.

Mr. Thompson, superintendent of the Royal Horticultural Society's garden, writing again of this Pear in 1836, (*Trans. Hort. Soc. Vol. II., 2nd Series, p. 119*) after it had been grown to greater perfection, says the *Althorp Crasanne* will bear competition with the first of the introduced varieties, and says it is one of the very best Pears. Specimens grown on a wall, he says, are very large and obovate; the eye is a moderately deep depression with the segments of the calyx somewhat collapsing; the stalk usually less than an inch and a half in length, or sometimes considerably shorter, thicker than that of the *Crasanne*. The skin was yellowish green with a faint brownish blush next the sun, and some russet near the stalk. The flesh was white, melting, and buttery, with very little grit, rich and excellent, but not equal to those from standards in point of high flavour.

From a standard the fruit was large, roundish-obovate, the eye was in a tolerably even-formed hollow, and open, with the segments of the calyx forming tubercles, inclining to collapse. The skin was greenish brown, interspersed with a russet-grey, not unlike the *Crasanne* in colour; but the stalk differs much being only half an inch, or an inch in length, whereas in the *Crasanne* it is one and a half, or two inches long. The flesh is buttery, rich, and very high flavoured. Season the end of October and November.

Remarks.—This variety first appeared in 1830, and fully equalled the *Crasanne* in size. It approaches it also in its globular shape. As a dessert pear, the *Althorp Crasanne* is, to my taste, the best; and its rose-water flavour will please where musk offends. Writing in 1836 Mr. Knight adds, "the original tree has borne well in all the last six years, and I believe it to be greatly more hardy than any of the Belgic varieties, and not less hardy than the *Swan's Egg.*

MONARCH PEAR.—(With coloured figure.)

Fruit middle-sized, obovate; stalk short, rather thick; eye open, placed in a shallow depression. Skin yellowish, much speckled with brown, and having a tinge of brownish red next the sun. Flesh yellowish, melting, buttery, rich and sugary; slightly musky, but not disagreeably so. An excellent pear. Season, January.

Remarks.—The *Monarch*, unlike most other pears, does not become mealy by being allowed to grow yellow on the tree; on the contrary, it is improved by being allowed to mature in this way. I named it *Monarch* under conviction that, for the climate of England, it stands without an equal; and because it first appeared in the first year of the reign of our most excellent Monarch (William IV.) The whole character of the tree is wild and uncultivated, and the young wood very thorny.

And in *Vol. II. 2nd Series.* p. 68. Mr. Knight adds, I had this year (1834) a sufficient quantity of *Monarch* pears to enable me to ascertain the specific gravity of its juices, which was 1096, that is fifteen above the *Stire* apple, and about the same which a solution of 2lbs. 6oz. of sugar would give to 8lbs of water. The taste and flavour of the juice appear to me to be very delightful.

EASTNOR CASTLE PEAR. (With coloured plate.)

An ugly brown shrivelly pear with scarcely any green. It is quite round. Stalk long and strong. Good. Season, December.

Remarks.—A very accurate description. The year 1830 is the date of the first existence of the variety.

THE WINTER CRASANNE PEAR.

Fruit large, irregularly turbinate, with a very hollow eye. Stalk thickened, and a little sunk at its insertion. Skin greenish yellow, intermixed with brown and some patches of cinnamon russet. Flesh inclining to yellowish white, melting, quite buttery, with very little grittiness, even at the core, rich, sugary, and very excellent; it has a little of the *Chaumontel* flavour. This is without doubt a very valuable sort. Season, January.

Remarks.—This variety first fruited in 1830. I retained only the smaller pears, whose merits did not seem so great as are thus represented. The large fruit of the *Crasanne* Pear, from which this description was derived, I have several times observed to be excellent in seasons when the smaller fruit was worthless. I thought the variety inferior to *Monarch*, *Althorp Crasanne*, and *Eastnor Castle*, but its weight greatly exceeds them. The growth of the original tree is enormously rapid.

THE WHITFIELD PEAR.

Colour of a ripe *Swan's Egg*, rather less russetty. Flesh melting, sweet and good, but dry. Not first-rate. Season November.

Remarks.—I first saw this variety in 1826 when I thought it very excellent. In the three following years it appeared to me to possess little merit. Last year it was better, but not so good as in 1826. A light dry soil would I think render this variety uniformly excellent.

Mr. Knight also adds the following description of another of his Seedling Pears :—

THE PITFOUR PEAR.

This is in size about twice that of a *Swan's Egg* when grown in the same soil and climate ; and in form it a good deal resembles the *Colmar Pear*. It ripens in October and November, and in the last season (1830) the first in which it was produced, its flesh was perfectly melting, and its juice rich and sweet.

What the precise female parent of each of the above-mentioned varieties was I am not prepared to say; but I believe that all with the exception of the *Foxley Pear*, and possibly the *Monarch*, sprang from the seed of the *Swan's Egg*. The *Moor-fowl's Egg* I believe to have been the female parent of the *Foxley Pear;* and I doubt whether the *Monarch* does not descend from the *Autumn Bergamot*.

APPENDIX III.

DESCRIPTIONS OF SEVERAL NEW VARIETIES OF PEARS RAISED BY THOMAS ANDREW KNIGHT, ESQ., FROM SEED, TOGETHER WITH NOTES THEREON, BY MR. KNIGHT.

[From the Transactions of the London Horticultural Society. 2nd Series. Vol. II., pp. 62—67. The descriptions were made in the Society's Gardens, by Mr. Robert Thompson, the Superintendent.]

MARCH BERGAMOT PEAR.

Fruit middle sized, in form and appearance resembling the *Autumn Bergamot*. Flesh buttery, a little gritty near the core, rich and excellent. Season March or later.

Note.—Owing to its resemblance in form to the *Autumn Bergamot*, and its ripening chiefly in March (though it may be preserved later,) I have named this sort the *March Bergamot*. This variety and the *Pengethley Pear* would probably be greatly improved if grown upon a wall.

PENGETHLEY PEAR.

Fruit middle-sized, obovate, a little curved at the stalk. Eye small and a little open ; stalk about half an inch in length. Skin yellowish brown and considerably russeted. Flesh yellowish, juicy and rich. A very good pear. Season February and March.

Note.—This Pear is larger, more juicy, and more inviting than the *March Bergamot ;* and remains in perfection quite as late in Spring. The tree is large, and its growth excessively luxuriant. The fruit first appeared in 1831, and was then very fine.

ROSS PEAR.

Fruit large, obovate. Eye open and slightly sunken. Stalk short, moderately thick. Skin yellowish green interspersed with russet. Flesh inclining to yellow, gritty near the core, but rich, juicy and sugary throughout. Season, January.

Note.—This Pear first appeared in 1832. The growth of the tree was extremely luxuriant, and the fruit all of large size.

OAKLEY PARK BERGAMOT.

Fruit middle-sized, roundish, obovate resembling a large *Swan's Egg.* Eye partly open, in a regularly-formed cavity. Stalk an inch and a half in length, rather slender, and a little sunk at its insertion. Skin greenish yellow, sprinkled with russet. Flesh buttery and melting, rich and excellent. Season, October.

Note.—Tree of free growth and bears well.

BROUGHAM PEAR.

Nearly of the middle size, obovate. Eye open in a regularly-formed depression. Stalk short. Skin, yellowish russet. Flesh yellowish white, buttery, a little gritty near the core, sugary and rich. Season, November. This sort is highly deserving of cultivation where flavour, · rather than size, is the principal object.

Note.—The Pear is as large as the *Autumn Bergamot.* I named it the *Brougham Pear* after Lord Brougham who approved of the fruit I sent him.

BRINGEWOOD PEAR.

Fruit middle-sized, pyriform. Eye open with the segments of the calyx prominent. Stalk long and rather slender. Skin yellowish brown, almost covered with russet. Flesh yellowish white, a little gritty near the core, the rest buttery, rich and very excellent, with something of the peculiar flavour of the *Monarch* Pear. Well deserving of cultivation. Season, end of October to beginning of December.

Note.—A good Pear, but variable in quality according to Season.

MOCCAS PEAR.

Fruit middle-sized, obovate, with a short stalk. Eye somewhat open and very slightly sunk. Skin brown. Flesh inclining to yellow, melting, juicy, rich and highly flavoured, resembling in this respect the *Monarch* Pear, and almost equal to that very excellent variety. Season, December.

Note.—A very fine Pear. Tree of excessively rapid growth, it blossoms and bears freely, and is hardy.

BROOM PARK PEAR.

Fruit nearly middle-sized, roundish. Eye in a moderate-sized hollow. Stalk about an inch in length, moderately thick. Skin entirely covered with cinnamon-coloured russet. Flesh yellowish, melting, juicy, with something of a Melon flavour, sugary and rich. Its very peculiar flavour may be said to partake of the Melon and Pine-apple. Season, January. A sort highly deserving of cultivation.

Not:—The singular flavour of the Pear was noticed at Downton as well as in London. The tree is fine and first bore fruit in 1830.

CROFT CASTLE PEAR.

Fruit middle-sized, oval. Eye open in a shallow depression with the segments of the calyx reclining. Stalk about an inch and a half in length, rather slender and somewhat obliquely inserted.

Skin pale yellow not glossy but rough with elevated dots and partially russeted. Flesh whiteish, a little gritty, but melting and very juicy, rich and sugary. An excellent Pear. Season, October.

Note.—A variety of dwarfish growth but very productive of fruit.

EYEWOOD PEAR.

In shape and size, very similar to an *Autumn Bergamot;* but of a deeper cinnamon-russet colour. Flesh yellowish white, melting, buttery, juicy, and very highly flavoured. It is doubtful whether it would be exceeded by *Gansel's Bergamot* in a better season for Standards than that of 1831, when the above description was made. Season, October or November.

Note.—The tree is of very free growth, hardy, and bears well in indifferent seasons.

DUNMORE PEAR.

Fruit about the size of a *Brown Beurré*, obovate, eye open, slightly depressed, stalk an inch in length, of medium thickness, rather fleshy at its junction. Skin brownish red next the sun; yellowish, with a scattering of brown where shaded. Flesh yellowish white, melting and extremely juicy, sugary and rich; a little gritty near the core, but on the whole, a most excellent pear. Season, end of September, or beginning of October.

Note.—As large and quite as good, I think, as the *Brown Beurré.* When allowed to ripen and grow yellow upon the tree, I have thought it the most melting and best pear of its early season. The birds are apt to destroy the fruit prematurely. The tree is of rapid growth, fine and hardy, well adapted to cold and late situations.

In addition to these Pears, the following varieties, raised by MR. KNIGHT, are described in Dr. Hogg's "*Fruit Manual.*"

SHOBDON COURT PEAR.

Fruit below medium size; oblate, even in its outline. Skin deep rich yellow, with a blush of red next the sun, and covered with red russetty dots. Eye very small, almost wanting, set in a small round rather deep basin. Stalk very long and slender, inserted in a small cavity. Flesh white, coarse grained, juicy, briskly acid and sweet, but not highly flavoured. A second rate pear. Ripe in January and February.

TILLINGTON PEAR.

Fruit, about medium size; short pyriform, rather uneven in its outline. Skin, smooth, greenish yellow, covered with a number of light-brown russet dots. Eye open, scarcely at all depressed. Stalk, short, fleshy, and wasted at its insertion. Flesh yellowish, tender, buttery, melting, not very juicy, but brisk and vinous, with a peculiar and fine aroma.

Note.—This is an excellent Pear, ripe in October, the fine sprightly flavour of which contrasts favourably with the luscious sweetness of the *Seckle,* which comes in just before it.

WORMSLEY GRANGE, HEREFORDSHIRE, 1877.

THE BIRTHPLACE OF THE BROTHERS

RICHARD PAYNE KNIGHT AND THOMAS ANDREW KNIGHT.

The small parish of Wormsley, with its richly-wooded hills and picturesque valleys, forms one of the most secluded districts of Herefordshire. It is mentioned in Domesday Book as one of the possessions held by Roger de Lacy of the King. The Priory, established as early as the time of King John, became a house of considerable wealth and importance. Its remains now are only to be indistinctly traced in a meadow near the Grange. At the Dissolution its site and appurtenances were granted, in 1547, to Sir Edward Fiennes, Lord Clinton; and passing through possessors named Baskerville, Scory, Vernon, Crowther, and White (who rebuilt the house), it was purchased, in 1747, by the Rev. Thomas Knight, whose gifted sons were born there, and who himself died there, in 1764. It is now a farm house in the possession of his descendant, A. R. Boughton Knight, Esq., of Downton Castle.

WYMAN AND SONS, PRINTERS,] [GREAT QUEEN STREET, LONDON, W.C.

THE

HEREFORDSHIRE POMONA,

CONTAINING

COLOURED FIGURES AND DESCRIPTIONS OF THE MOST ESTEEMED KINDS OF

APPLES AND PEARS,

EDITED BY

ROBERT HOGG, L.L.D., F.L.S.,

Honorary Member of the Woolhope Naturalists' Field Club; Secretary of the Royal Horticultural Society;
Author of 'The Fruit Manual'; 'British Pomology'; 'The Vegetable Kingdom and its Products', &c., &c.

" Hope on. Hope ever."

" Ζεφυρίη πνείουσα τὰ μὲν φύει ἄλλα δέ πέσσει.
ὄγχνη ἐπ' ὄγχνη γηράσκει, μῆλον δ' ἐπὶ μήλῳ,
αὐτὰρ ἐπὶ σταφυλῇ σταφυλή, σῦκον δ' ἐπὶ σύκῳ."
Homer Odyssey vii. 119-22.

" THE BALMY SPIRIT OF THE WESTERN GALE,
ETERNAL BREATHES ON FRUITS UNTAUGHT TO FAIL;
EACH DROPPING PEAR, A FOLLOWING PEAR SUPPLIES
ON APPLES APPLES, FIGS ON FIGS ARISE."
Pope.

LONDON: DAVID BOGUE, 3, ST. MARTIN'S PLACE, TRAFALGAR SQUARE.
HEREFORD: JAKEMAN AND CARVER, HIGH TOWN.

1878.

"INSERE, DAPHNE, PIROS; CARPENT TUA POMA NEPOTES."
Virgil Ecl. ix. 50. (40 B.C.)

. . . "Graft the tender shoot,
"Thy children's children shall enjoy the fruit."

Dryden.

"GENTLEMEN THAT HAUE LAND AND LIVING, PUT FORWARD
IN THE NAME OF GOD; GRAFFE, SET, PLANT AND NOURISH UP
TREES IN EUERY CORNER OF YOUR GROUNDS; THE LABOUR IS
SMALL, THE COST IS NOTHING, THE COMMODITIE IS GREAT, YOUR-
SELUES SHALL HAUE PLENTIE, THE POORE SHALL HAUE SOMEWHAT
IN TIME OF WANT TO RELIEUE THEIR NECESSITIE, AND GOD SHALL
REWARD YOUR GOOD MINDES AND DILIGENCE."

Gerarde. Account of the Apple. (1597).

"OF EVERY SUIT
GRAFFE DAINTY FRUIT.
GRAFFE GOOD FRUIT ALL,
OR GRAFFE NOT AT ALL."

Tusser. Marches Abstract. (1620).

MODERN APPLE LORE.

" Nor needst thou blush that such false themes engage
Thy gentle mind, of fairer stores possest ;
For not alone they touch the village breast,
But filled in elder time th' historic page.
    *   *   *   *
The native legends of thy land rehearse."
          (*Collins : " Ode on the Popular Superstitions of the Highlands."*)

" Let the World have their May-games, Wakes, Whitsun-ales ; their Dancings, and Concerts ;
their Puppet-shows, Hobby-horses, Tabors, Bagpipes, Balls, Barley-breaks, and whatever sports
and recreations please them best, provided they be followed with discretion."
          (*Burton's " Anatomy of Melancholy."*)

" Mark it, Cesario ; it is old and plain :
The spinsters and the knitters in the sun,
And the free maids that weave their thread with bones
Do use to chaunt it : it is silly, sooth,
And dallies with the innocence of love
Like the old age."
          (*Shakespeare, Twelfth Night, II., 4.*)

" If all the year were playing Holidays,
To sport would be as tedious as to work ;
But when they seldom come, they wish'd-for come,
And nothing pleaseth but rare Accidents."
          (*Shakespeare, Henry IV., I., 2.*)

The Apple has been associated with many popular customs and fancies from the earliest
period. As the favourite fruit of the people, so useful and so widely diffused, it naturally takes a
prominent place in their every-day life and their rural and domestic festivities. Apples and Apple
Trees, wherever they abound, are connected with many curious and interesting sayings and
observances. These are, for the most part, of the most fanciful and trivial kind ; sometimes they
seem to be the traditions of heathen customs, and are certainly the relics of an age when men were
more in earnest in their superstitions, and more inclined to believe in vague intimations of the
unseen than they are now. Sometimes they take the form of divination, half serious and half

jocose, the mood of a mind preoccupied but not wholly engrossed, able to mock itself a little, and smile a good humoured smile at its own hopes and fears. Some of the country proverbs which the sayings afford, are but the practical observations of rural life: or again, and by far the most frequently, the part played by the Apples is that of the natural accompaniments of rustic pleasures. In remote periods country festivities were encouraged and held on the festivals of the Church, and became thus associated with certain Saints' Days. These were always chiefly for the young, and held the place now taken by school entertainments. They were made the occasions of presenting the children with little gifts, and for the general expression of good will. As noise and merriment are ever essential to youthful enjoyment, the singing of doggrel verses became prevalent, in which the rhyme was often more thought of than reverence for the Saint, in whose honour the festival was kept. These customs and observances, whatever form they may take, are rapidly passing out of use, and they are only to be studied in the pages of "The Gentleman's Magazine," for the end of last century; Hone's "Every-day Book"; Brand's "Popular Antiquities," the Edition by Sir Henry Ellis (1813), and the Edition by W. Carew Hazlitt (1870); Dyer's "British Popular Customs"; or in the incidental notices of Local Histories, Magazines, or other Periodicals. It seemed not unfitting that the Woolhope Club should put some of them on record in its own pages, before they had passed into complete oblivion.

The importance of a fruitful year has given rise to many country sayings and omens with reference to apple-trees. In Derbyshire, and in many other counties, there is a prevalent notion, that if the sun shines through the apple trees on Christmas Day, there will be an abundant crop of apples the following year.

The danger of an early Spring, as shown by the apple tree coming into leaf too precociously, is well expressed by the rural distich:

> "March dust on an apple leaf,
> Brings all kinds of fruit to grief."

Or again, if the tree blossoms too early, it is said:

> "If the apple tree blossoms in March,
> For barrels of cider you need not *sarch*,"

because you certainly will not find them;

> "But if the apple tree blossoms in May,
> You can eat apple dumplings every day."

If, however, the apple tree should blossom when the fruit is ripe on the tree, superstition steps in, and, as this is usually prompted by fear, it is an omen of calamity, and is said to betoken a forthcoming death in the owner's family. In Northamptonshire they say:

> "A bloom upon the apple tree when the apples are ripe,
> Is a sure termination to somebody's life."
> *(Dyer's "English Folk Lore," p. 8.)*

which is so charmingly general, that it is not to be disputed. The occurrence was so very common during last year (1878) that the mortality, to support the saying properly, would indeed have been terrible.

The following is the refrain of a pretty folk song, supposed to be sung by a maiden on St. John's Eve, which speaks for itself:

> "Beau pommier, beau pommier,
> Qu'est si chargé des fleurs
> Que mon cœur d'amour.
> Il n'y faut qu'un p'tit vent
> Pour envoler ces fleurs.
> Il ne faut qu'un jeune amant
> Pour y gagner mon cœur."
>
> *(Cornhill Magazine, 1876.)*

on which Mr. Matter-of-fact might observe, that the 23rd of June is rather late for apple trees to blossom, especially in France.

In Northamptonshire, it was formerly very much the custom on St. Valentine's day for the children to go round from house to house, singing:

> "Morrow, morrow, Valentine!
> First 'tis yours, and then 'tis mine;
> So please to give me a Valentine.
> Holly and ivy tickle my toe,
> Give me red apples and let me go."

It is cruel to suggest that pricking with holly leaves, and then using those of ivy, are rural remedies for chilblains on toes; since it is more than probable here that the exigencies of the rhyme required a spirited metaphor—albeit the "red" apples might still be symbolical of the inflammation.

In the neighbourhood of Salisbury, the boys go about before Shrove-tide, singing:

> "Shrove-tide is nigh at hand,
> And I am come a shroving;
> Pray dame, something—
> An apple or a dumpling," &c., &c.

and doubtless they got something, or they would not go.

On Shrove Tuesday, apple fritters are allowed to be the best form of pancake:

> "Let glad Shrove Tuesday bring the pancake thin
> Of fritter rich with apples stored within."
>
> *(Oxford Sausage, p. 22.)*

A very harmless lottery, by means of apples, is mentioned by Collinson *(History of Somerset, III., 586)* as having obtained formerly in Congresbury and Puxton. Certain common lands, named the Dolemoors, were apportioned annually among the commoners by lot: the lots being drawn by means of apples marked in a peculiar manner, so as to correspond with similar marks cut in the turf, on the various positions of the land to be divided. Collinson says this was done on the *Saturday before Old Midsummer Day;* which suggests some speculation as to what kind of apples they were, which the worthy folk of Congresbury were able to keep so long.

The idea that the apples are christened by the showers of St. Swithin (July 15) is very general, and few country people would think of using them until this day was passed. It probably arose from the fact that they are not fit to use, before this time—indeed the "Blessing of the New Apples" was not given until St. James' Day (July 25), ten days later.

(Brand's "Popular Antiquities." Ellis, I., 346.)

St. Simon and St. Jude's Day (October 28) is generally supposed to be wet like St. Swithin's; and possibly this may explain the playfulness within doors: " A la Saint Simon et Saint Jude on envoi au temple, les gens un peu simple, demander des nèfles (medlars) afin de les attraper et faire noircir par des valets." *(Sauval Antiq. de Paris II., 617).*

Apples are by this time in full abundance, and on this day they are playfully used in love divinations. Take an apple, pare it whole, and take the paring in your right hand, stand in the middle of the room and say the following verse :—

> " St. Simon and St. Jude on you I intrude
> By this paring I hold to discover,
> Without any delay, to tell me this day,
> The first letter of my own true lover."

Turn three times round, and cast the paring over your left shoulder, and it will form the first letter of your future husband's surname. If the paring breaks into many pieces, so that no letter is discernible, alas! alas! you will never marry.

So in Gay's Hobnelia :—

> "I pare this pippin round and round again,
> My Shepherd's name to flourish on the plain ;
> I fling the unbroken paring o'er my head,
> Upon the grass a perfect " *L* " is read :
> Yet on my heart a fairer " *L* " is seen,
> Than what the paring makes upon the green."
> *(Shepherd's Week—Thursday—91.)*

It is satisfactory to remember in this case that the omen was not falsified by the event. Very soon afterwards the fair one exclaims :—

> "But hold ! our Lightfoot barks and cocks his ears,
> O'er yonder stile, see Lubberkin appears ! "

Another method of using apples to indicate the secret destinies of love, is by means of the pips. The love-sick Hobnelia tries this method also :—

> " This pippin shall another trial make :
> See, from the core two kernels brown I take,
> This on my cheek for Lubberkin is worn,
> And Booby-clod on t'other side is borne :
> But Booby-clod soon drops upon the ground,
> A certain token that his love's unsound ;
> While Lubberkin sticks firmly to the last."
> *(Shepherd's Week—Thursday—99.)*

Another method is to give to one or more apple pips the names of one or more claimants for the maiden's regard, and then to put them in the fire and watch how they burn. The charm is concise and runs as follows :—

> " If you love me bounce and fly,
> If you hate me lie and die."
> *(Edwin Lees, " Pictures of Nature," p. 300.)*

Woe betide the youth whose representative smoulders away in inglorious dulness! It is obvious how closely this resembles the nut-roasting on Hallow E'en in Scotland, immortalized by Burns.

In Lancashire it is said that the omen is consulted in another way. The enquirer squeezes a pip between finger and thumb, and moving round in a circle, utters the following charm :—

> " Pippin, pippin, paradise,
> Tell me where my true love lies,
> East, West, North, or South,
> Pulling-brig, or Cockermouth."
> *(Dyer, " English Folk Lore," p. 20.)*

Whereupon, the obedient pip starts off at once in the required direction.

The same custom is traced back by some *(Boaraldus : Suetonius—1610, col. 560)*, even to Roman times. An old Commentator on Suetonius, thus writes " Porphyrio, the Annotator of Horace, says that lovers were accustomed to squeeze apple pips between the thumb and forefinger, and watch how they fled off," as an augury of their affections.

Another method is to take the pips of the apple that has been pared, and drink them in a glass of fresh spring water, to dream of the wished-for admirer.

In Austria on St. Thomas' Night it is said an apple is cut in two, and the pips counted in each half. If the number is even, the fair damsel who makes the experiment will be married soon, and *vice versa ;* but if in the process a seed be cut in two, the course of true love will not be smooth ; and if two seeds are unhappily divided, it is a sign of speedy widowhood.

In Ireland on the Patron Day in most parishes, it is customary to dance in pairs round the " bush" and the piper. The " bush" is a cake raised on a pike, and surrounded with a garland of ever-greens, flowers, and, when in season, apples. The dancers arrange themselves in a circle, the piper plays vigorously, and the couple that can hold out longest at the exercise, wins the cake and apples.—*(Brand, by Hazlitt, II. 10.)*

Apples take their full share in the customs not only of Hallow E'en (Oct. 31), the day before All Saints', but on All Saints' Day itself, and on All Souls' Day, which follows it.

The ancient custom of providing children with apples on Hallow E'en is still observed, we are informed by Hunt, at St. Ives, in Cornwall. " Allan Day," as it is termed, is one of the chief days in all the year for children, who would think it a great misfortune if they went to bed on Allan Night without an Allan apple to hide under their pillows. So large a supply of apples is required for this purpose, that the market at which they are bought is called the Allan Market. *(Standard* Newspaper, *October, 1878.)* Hallow E'en is sometimes called " Snap-apple Night," from the game of " Snap-apple " which it was the custom to play then. A short piece of wood was suspended by a string from the centre, and balanced with the utmost care. On one end of it was stuck a fine red-cheeked apple, on the other a piece of lighted tallow candle. It was swung rapidly round, and the players, with their hands tied behind their backs, stood in a circle and tried to bite the apple. They might only get a greasy, blackened face, to the amusement of the lookers on,

> " Or catch th' elusive apple with a bound,
> As, with its taper, it flew swiftly round."
> *(Polwhele, " Old English Gentleman," p. 120.)*

" Snatching" or " Bobbing for apples " in a pail of water often varied the fun.

In Wales and some of the border counties, on the night of Oct. 31st, there is a custom still commonly practised. It is called, in the language of the country, "*Nos-cyn-gauaf*," the night before winter—or, literally rendered, meaning, everything at a standstill, nothing growing. The custom consists in making strings of apples, and hanging them to roast before the kitchen fire, with a bowl of new milk beneath each string of apples, to catch them as they fall. The dish is then partaken of by all the farm servants, and men and maidens crown the feast by dancing out the Autumn into Winter. This simplest form of "Lamb's wool" gives place to a liquor much more potent in the colder weather of the New Year's Eve, or Twelfth Night.

A much more serious and formidable kind of divination is referred to by Burns in "Hallow E'en." It must have sorely taxed the nerves of the maidens who tried it. A girl has to take a candle, and eat an apple while standing alone before a looking-glass at midnight, and is then to see in the glass the face of her future husband looking over her shoulder. "Wee Jennie" in Burns's poem, has not courage to make the experiment alone, but says :

> "Will ye go wi' me, Grannie?"

and the poet again quizzes the terror of midnight experiments, by the maiden who employs a guard :

> "She gi'es the herd a pickle nits,
> An' twa red-cheekit apples,
> To watch, while for the barn she sets,
> In hopes to see Tam Kipples,
> That vera night."

Randolph makes fairies declare a partiality for apples stolen from orchards in the night :

> "We cannot have an apple in the orchard,
> But straight some fairy longs for't."
> *(Amyntas, 1638.)*

But, in "Cataplus, a Mock Poem" (1672), the writer says of the Sybil :

> "Thou canst in orchards lay a charm
> To catch base felon by the arm."
> *(Brand, by Hazlitt, III., 273)*

So there might be danger in the theft.

On All Souls' Day (Nov. 2) there was a custom in many parts of England of "going a Souling," as it was called ; that is, going about from house to house, singing a rude song, begging for apples, as well as for "Soul cakes," (a kind of bun, made for the occasion ;) or for anything else they could get.

In Shropshire the verses ran as follows :

> "Soul ! Soul ! for a Soul cake !
> Pray, good mistress, for a soul cake :
> *   *   *   *   *
> Soul, soul, for an apple or two :
> If you've got no apples, pears will do.
> Up with your kettle, and down with your pan,
> Give me a good big one, and I'll be gone."

In Cheshire they were less exacting, for each verse of the song given in the Journal of the

Archæological Association (1850) ends :

> " *We hope you'll prove kind*, with your apples and strong beer,
> For we'll come no more a-souling, until another year."
>
> *(Dyer. " British Popular Customs," 405.)*

And another version of the song, too long to quote, ends, with admirable candour :

> " We are a pack of merry boys, all in a mind
> We have come a-souling *for what we can find* ;
> Sole, sole, sole of my shoe,
> If you've no apples, *money will do*,"

and no doubt it would.

Two other holidays in the same month, St. Clement's (Nov. 23) and St. Catherine's (Nov. 25) were observed much in the same manner. In Worcestershire, children from the cottages used to go round from house to house " Catterning," as it was called, singing :

> " Catt'n and Clement comes year by year,
> Some of your apples, and some of your beer," &c.

Always concluding with the burden :

> " Up with the ladder, and down with the can,
> Give me red apples, and I'll be gone."

The ladder alluding to the store of apples in the loft, and the can for going to the cellar where the beer was kept. In Staffordshire, the verses run variously :

> " Clemany, Clemany, Clemany mine
> A good red apple, and a pint of wine, &c., &c."

The apples were stuck thickly over with cloves, roasted on a string, and allowed to fall in hot beer, or cider.

> " Our bowl is made of the ashen tree,
> Pray good butler drink to we."

ending as usual with begging " a few red apples." *Mr. Allies, Athenæum, 1847.*

Christmastide, as a matter of course, came in for its full share of apple customs. At Harvington, in Worcestershire *(N. and Q., 1st Series VIII., 617)* the children used to go round on St. Thomas's Day (December 21) singing :

> " Wissal, Wassal, through the town :
> If you've got any apples throw them down,"

and there was also a permissable alternative, (doubtless often preferred :)

> " If you've got no apples *money will do*."

This is shown by a passage at the end of George Wither's " Juvenalia," in an old " Christmas Carroll," to be a very ancient custom :

> " Hark ! how the wagges abrode doe call
> Each other foorth to rambling ;
> Anon, you'll see them in the hall,
> For nuts and apples scrambling.
> The wenches with their wassell-bowles,
> About the streets are singing ;
> The boyes are come to catch the owles,
> The wild mare in is bringing."

An apple stuck with cloves graced the boar's head at the Christmas feast :

> " His foaming tusks let some large pippin grace."
>
> (*King's Cookery, p. 75.*)

The jovial procession of the boar's head went only from the kitchen to the hall, as it still does at Queen's College, Oxford ; but at Ripon, on Christmas Day in olden times, boys used to bring baskets of red apples. decorated with sprigs of rosemary into the minster, and present them to the congregation.—(*Gentleman's Magazine, 1790.*)

The Christmas supper table was decked as one of its chief ornaments with "Codlins and Cream," an old-fashioned dainty dish not often seen now, and a very pleasant compound it must have been. The Codlins were thoroughly roasted, arranged in a glass dish, thickly dusted over with powdered white sugar, and then covered with rich thick cream (*Mr. Edwin Lees*). The fame of this delicacy reached as far as Ireland. A poem in the 17th century written by a Dr. William King, contains the following lines :

> "Mountown ! thou sweet retreat from Dublin cares,
> Be famous for thy apples and thy pears.
>   *   *   *
> Mountown ! the Muses most delicious theme,
> Oh ! may thy Codlins ever swim in cream."

The antiquity of this " dainty dish" may also be inferred from the fact that a common kind of Willow-herb, is called " Codlins and Cream " from the odour of its blossoms.

The Eve of the New Year, and New Year's Day, again brought apples into requisition. It was the custom on New Year's Eve for young women to carry from door to door, through the villages, the " Wassail Bowl," gaily ornamented with coloured ribbons, and gilt apples, singing doggrel verses. These young women, doubtless, suggested to Ben Jonson, the personification of Wassel, which he introduces in one of his Masques. The stage direction clearly shows the allusion :

> " Enter Wassel, like a neat sempster and songster, her page having a brown bowl
> dressed with ribbons and rosemary before her."

The "Wassail Bowl" implies, the drinking of healths. " *Wass-hael*" is the Saxon toast "to your good health." The beautiful Rowena, with this salutation, presented the cup of spiced wine to King Vortigern, and secured a gracious reception for her ambitious brothers Hengist and Horsa. (*Brady—Clavis calendaria II., 320.*) Thus in ancient Saxon households, they passed the bowl round. In the " Antiquarian Repertory" (*I. 213. Edit. 1775*) is a good engraving of an old oak chimney-piece beam, on which is carved a bowl, with the inscription on one side "Wass-heile," and on the other " Drinc-heile." The bowl rests on the branches of an apple tree, in allusion to the spiced cider it was wont to contain. The " Wassail Bowl" was the *poculum caritatis* of the Abbot's table in the Monastery Refectory; and it is represented in our own time, by the "Grace Cup" of the Universities, and the " Loving Cup" of Civic festivities.

The proper compound for the " Wassail Bowl" was hot ale, or cider, flavoured with spice and sugar ; having a piece of toast in it, with roasted crabs, or apples, floating on the surface. This is Shakespeare's " Gossips bowl :"

> " And sometimes lurk I in a gossip's bowl,
> In very likeness of a roasted crab."
>
> (*Midsummer Night's Dream, II., 1.*)

When rightly made, the apples are roasted on a string above the bowl, and the soft pulp only allowed to fall into the liquor, which floating on the top gives the name of "Lamb's-wool" to the drink.

> "It welcomed with Lamb's-wool the rising year."
>
> *(Polwhele, "Old English Gentleman," 117.)*

Lamb's-wool is mentioned by Gerarde, and other writers, and in the early English ballad, "The King and the Miller," we find :

> "A cup of Lamb's-wool they drank to him there."

Herrick gives its composition :

> "Next crowne the bowle full
> With gentle Lambs' Wool ;
> Adde sugar, nutmeg, and ginger,
> With store of ale too ;
> And thus you must doe,
> To make the Wassaile a swinger."
>
> *(Hesperides, 376.)*

As an instance of the follies of antiquaries, we may mention, that Hone in his "Every-day Book" says "Lamb's-wool" is thus etymologized by Vallancy: The first day of November,(All Saints Day,) was also dedicated to the angel presiding over fruit seeds, &c., and was therefore named "*Las mas Ubhal*" that is, the day of the apple fruit, and being pronounced "Lamasool," the English have corrupted the name to "Lamb's-wool" *(c. 1416.)*

A writer in "Notes and Queries," *(Vol. V.)* say that the curious custom of "Apple howling" on New Year's Eve, was still observed in Sussex, Devonshire, and Herefordshire. Troops of men and boys would visit the different orchards, and encircling the favourite trees sing a rhyme which begins :

> "Stand fast root, bear well top," &c., &c.

with much noise, hornblowing, and merriment.

In Devonshire, a custom very similar was observed in Herrick's time on Christmas Eve. He says :

> "Wassail the trees, that they may bear
> You many a Plum, and many a Pear,
> For more, or lesse fruits they will bring
> As you do give them Wassailing."
>
> *(Hesperides, p. 311.)*

But the old customs of Christmas Eve are observed in most country places, on Old Christmas Eve, that is, the Eve of Twelfth Day. On that night in more recent times, the farmer and his men carried a large pitcher of cider to the orchard, and then standing round one of the best trees, shouted three times in full chorus :

> "Here's to thee, old apple tree,
> Whence thou mayst bud, and whence thou mayst blow !
> And whence thou mayst bear apples enow :
> Hats full, caps full,
> Bushel, bushel, sacks full,
> And my pockets full too,
> Huzza, huzza."
>
> *(Gentleman's Magazine, 1791.)*

A similar custom is observed in Somerset at the present day : and a song almost identical with this, is given in the "Glossary of the Exmoor Dialect," and said to be accompanied by throwing toast to apple trees—the toast and cider they were drinking themselves.

So in the neighbourhood of the New Forest, it is said they sang :

> "Apples and pears with right good corn,
> Come in plenty to every one ;
> Eat and drink good cake and hot ale,
> Give Earth to drink, and she'll not fail."
>   *(Christmas in the Olden Time.)*

The customs of "Apple howling" and "Wassailing" the apple trees have entirely passed away in Herefordshire. There is no trace or record to be found of them. On New Year's Eve, the curious custom of " Burning the Bush " is still practised in the central and northern districts of the county though it relates rather to the cornfields, than to the apple trees. The men light a fire in each cornfield successively, and a hawthorn bush is partly burnt in each fire, but not consumed ; for those present, carry away fragments to ensure prosperity in the ensuing year. This is done with much noise, singing the words "Old Cider" monotonously, again and again, with a most singular effect; and of course much cider drinking goes on at the same time. This custom was observed this year (1879) at Lower Lyde, and at Lyde Court, near Hereford, and at several other farms in the neighbourhood. " Our old labourers wouldn't be happy all the year if they did not burn the bush on New Year's night," said one of the gentlemen, at whose farm the custom is observed.

The labourers in Herefordshire, it is said, usually indulge in an extra glass or two on New Year's Eve. They call this, "*Burying Old Tom*," and they generally continue their uproarious festivities, till mine host makes a clearance, when they resort to the cornfields to " Burn the Bush."—
(Brand, by Hazlitt I., 12.)

On New Year's Eve on the Continent, love divinations are practised. In Silesia a maiden having bought an apple at the exact price first demanded for it, (what a fortune the apple sellers must make of it !) lays it under her pillow, and at midnight expects to see her future husband in a dream. In Swabia a widow who eats half an apple on St. Andrew's Eve, and places the other half under her pillow, expects a similar vision. The same customs with slight variations are also practised in Austria, Hesse, and Bohemia.

New Year's Day again brings forward apples. This time they are stuck with cloves, and carried round by children. An old volume of " Miscellanies " in the British Museum, of Queen Anne's time, has in its notice of New Year's Day the following quaint passage : " Children, to their inexpressible joy, will be dressed in their best bibs and aprons, (dear little oddities !) and may be seen banded, along streets, some bearing Kentish pippins, others oranges stuck with cloves, in order to crave a blessing of their godfathers and godmothers," (p. 65), which was doubtless expected to take the useful form of a shilling or a crown.

The custom of presenting decorated apples on New Year's Day is still common in some localities. The apples are mounted on three skewers, like a three legged table, ornamented with oat-grains, whited with flour, and surrounded with evergreens and berries. *(Notes and Queries, 1st Series, I., 214.)* In Herefordshire a little pyramid is formed with apples, nuts, and strings of holly berries, with holly, box, and ivy, having their leaves more or less gilt and whitened.

In Wales and one or two of the border counties, they make a far more elaborate structure (1878). A piece of board, ten or twelve inches long, by two or three inches wide, is drilled with six holes, two at each end and two in the middle; four of the holes are provided with little wooden spikes, and a handsome apple, more or less gilt, is stuck on each. In the remaining two holes sprays of box are fixed, with nuts suspended from the leaves, and the whole is completed by bunches of ivy berries whitened with flour, strings of holly berries, &c.

In Nottinghamshire much the same customs were formerly observed. (*Journal of Archæological Association, 1853.*) At Hastings, apples, nuts, oranges, as well as money, are thrown from the windows on New Year's Day to be scrambled for by the fisher boys and men.

(Dyer. " Brit. Pop. Customs", p. 11)

The eve of Twelfth Night is still celebrated in Herefordshire with the singular ceremony of the thirteen fires. This year (1879) the proceedings at Focle Farm, Upton Bishop, were carried out as follows :—At the close of the day the farmer, his labourers, friends, and visitors, assembled at the highest part of a corn-field near at hand, and twelve bonfires were made, a few yards apart from each other, in the shape of a horse-shoe. In the centre a high pole was erected and covered with straw, by twisting it round from the bottom to the top, and called " the old woman." When all was ready, the straw was lighted on the top of the pole, by the aid of a light on a second pole, so as to make it burn downwards ; meantime the twelve heaps around were also lighted, and they all blazed up together, amidst much shouting, noise, and merriment. Cider circulated freely, and healths were drunk in honour of the farmer, with wishes for good crops, and much cheering. As soon as the fires had burnt out, an adjournment was made to the farm-house, where an excellent supper of roast beef and plum-pudding had been provided. The men were all waited on by the master, mistress, and friends ; and this formed the most silent part of the evening's entertainment. After supper, the cider bowl passed round, pipes were lighted, songs were sung, and the festivities kept up to an early hour in the morning. The same ceremonies were carried out at Hill of Eaton Farm, Brampton Abbots, on the same night. Up to the year 1876, this custom was observed at many of the principal farms in the neighbourhood of Ross; but in that year the ill-feeling produced by the Labourers' Union put an end to most of the festivities, and though this ill-feeling is now happily passing away, the thirteen fires are scarcely likely to become general again.

These thirteen fires are called, " The Old Woman and her Twelve Children," or sometimes, " The Virgin Mary and the Twelve Apostles." It is believed by some antiquaries that this very curious observance is a vestige of the old heathen ceremony of the worship of the goddess Ceres— the twelve fires around the centre pole representing the twelve months of the year—and they believe that " this custom may be as successfully traced to the rule and influence of the Romans, as may be the undoubted vestiges of their arts and buildings." (*Note to " Civil War in Herefordshire," by the Rev. Jno. Webb.*) " Thirteen" is evidently a mystical number, and a very strange and unusual one it is. It seems possible, that it may have reference to the thirteen lunar months in the solar system ; a solution of the mystery that would be more satisfactory to those sceptics, who are inclined to call in question the fact, that the goddess Ceres ever was worshipped in Britain. These ceremonies were described in the " Gentleman's Magazine " so far back as 1791, when " the shouts and hallooing," it is said, " were answered from all the adjacent villages and fields, and some fifty or sixty fires could be seen all at once."

In the same article is described another old Herefordshire custom, which followed the thirteen fires on the same evening. It is thus described:—" A large cake is always provided, with a hole in the middle of it. After supper the company all attend the bailiff (or head of the oxen) to the wainhouse, where the following particulars are observed : The master, at the head of his friends, fills the cup and stands opposite the first, or finest of the oxen. He then pledges him in a curious toast : the company follow his example with all the other oxen, addressing each by his name. This being finished, the large cake is produced, and with much ceremony put on the horn of the first ox, through the hole above-mentioned. The ox is then tickled, to make him toss his head : if he throw the cake behind, then it is the mistress's (or milkmaid's) perquisite ; if before, in what is termed the " boosy," the bailiff himself claims the prize. The company then return to the house, the doors of which they find locked, nor will they be opened until some joyous songs are sung. On their gaining admittance, a scene of mirth and jollity ensues, which lasts the greatest part of the night." This custom was kept up in Herefordshire until some thirty, or perhaps twenty years ago, in some of the leading farm-steads ; but it has now wholly disappeared. Formerly there were many more men kept regularly on the farm throughout the year, when there would naturally be a greater home feeling amongst them, and a greater interest in observing old customs. Now, the permanent staff of labourers is much fewer, and the additional hands required are only employed temporarily, at the busy seasons of the year.

The old Twelfth Night rhymes are now only used by the boys on New Year's Day, who sing them from door to door, to get what they can in the shape of New Year gifts.

Some fifty or sixty years ago, apple-scoops were in general use, and were even placed on the dessert table with a dish of apples, as crackers are with nuts. Clare, the Northamptonshire poet, notices this, in his "Shepherd's Calendar " :

> "Some spent the hour in leisure's pleasant toil,
> Making their apple-scoops of bone the while."

but the fashion has changed, and it is rare now to meet with one of the old bone scoops, and still more rare to see any person scooping an apple in the old-fashioned way.

"The eating of a Quince Pear to be preparative of sweete and delightfull dayes," is noted as a wedding custom, in " The Praise of Musicke" (1586). And a present of Quinces from the husband to his bride, is noticed at an English marriage in 1725, " reminding one," says the correspondent of *Notes and Queries*, of the ancient Greek custom that the married couple should eat a Quince together." *(Brand, by Hazlitt, II., 97.)*

The "Apple Cure" was once a remedy in great vogue for the treatment of all diseases that flesh is heir to. In Heath's Account of the Islands of Scilly, mention is made of one Mr. Atwell, the Rector of St. Ives in Cornwall, who about the year 1562 attained great success in the treatment of the sick by giving to all of them apples and milk. His reputation spread into the neighbouring counties and patients flocked to him in great numbers. It was a remedy, it must be admitted, much more safe and effective than many others which have since been fashionable.

Apples are not unfrequently used as a charm to cure warts. The apple is cut in half, and each half being rubbed on the warts, they are placed together and buried in secrecy. As the apple decays away, so too should the warts disappear.

Bishop Bale writing in 1538 mentions the following charm among others:

> " For the Coughe, take Judas Eare
> With the parynge of a Peare,
> And drynke them without feare,
> If ye will have a remedy."
> *(Braud—by Hazlitt, III., 252.)*

The apple is also supposed to have other magical uses in a medical sense, but the practice and belief in them is more Continental than British, and need not be further alluded to.

The apple customs which have thus been detailed, are passing away with endless variations. These could not have been further entered into without wearisome repetition and confusion. Already it cannot be said for this account of them, as Mr. Barham somewhere says of himself in the "*Ingoldsby Legends*":

> " I am just in the order, which some folks—though why
> I am sure I can't tell you—would call Apple-pie."

but it is hoped nevertheless that the Club may like the dish that is presented to them.

The plain country life of former days is itself passing away. The habits of the people are rapidly changing with the alterations and improvements of the age. The penny-a-week school; the penny postage; the penny newspaper; and the penny-a-mile railway train; have turned the thoughts of the people, their pleasures and enjoyments, into other channels, and have almost done away with the isolation of the country villages. Children's amusements are more thoughtfully and happily provided for, by school-entertainments and summer pic-nics. The village Wake is lost in the club-feast of the district, the pleasure fair of the nearest town, and the excursions offered so cheaply by railway companies; and thus in the greater excitements of the world at large, the simple customs and observances of the country villages are rapidly passing from disuse into oblivion.

HENRY G. BULL, M.D.

NOTES TO THE
"EARLY HISTORY OF THE APPLE AND PEAR."

In the curious old Dictionary called "*Promptorium Parvulorum sive clericorum*," which seems to be a work of the fifteenth century (1440), we read : *Sydyr drynke—cisera.*

The reference here plainly is to the vulgate, which, in St. Luke I. 15, reads (speaking of St. John the Baptist), *et vinum et siceram non bibet.* So also in Judges XII. 14, it is said of Samson's mother : "*vinum et siceram non bibat.*"

This seems to show that the word *cider* was originally equivalent to σικερα, and meant any intoxicating drink, but was afterwards appropriated to that made from the juice of the apple.

In one of the earliest Law Reports known as the "Year Book," an Action which occurred at the Hereford Assizes, A.D. 1292, is reported in the following terms :

"A man and his wife named Isabel, brought a writ of waste against B. and Joan his wife, and stated that they (Defendants) held certain lands as the dower of Joan, the reversion whereof belonged to Isabel and that they had wasted Oak, Pear-trees, and Apple-trees to the value of, &c., to the Plaintiffs' damage of, &c.

The Defendants pleaded that these premises had been for ten years in the occupation of Isabel and her husband, during which term they had dug round all the apple trees and pear trees, so that they fell down of themselves." A jury was ordered for the decision of this complaint, the result of which is not recorded.

(Hereford Iter. XX., Edw. I.)

[The Woolhope Club is indebted to Mr. W. H. Cooke, Q.C., for this interesting notice, which proves that Apple Trees and Pear Trees were objects of sufficient value in Herefordshire in 1292 to go to Law about.]

"WE NOWHERE ART DO SO TRIUMPHANT SEE
AS WHEN IT GRAFTS OR BUDS THE TREE.
IN OTHER THINGS WE COUNT IT TO EXCEL,
IF IT A DOCILE SCHOLAR CAN APPEAR
TO NATURE AND BUT IMITATE HER WELL;
IT OVER-RULES AND IS HER MASTER HERE.
            *   *   *   *
WHO WOULD NOT JOY TO SEE HIS CONQUERING HAND
O'ER ALL THE VEGETABLE WORLD COMMAND;
AND THE WILD GIANTS OF THE WOOD RECEIVE,
WHAT LAW HE'S PLEASED TO GIVE!
HE BIDS THE ILL-NATURED CRAB PRODUCE
THE GENTLE APPLE'S WINY JUICE;
THE GOLDEN FRUIT THAT WORTHY IS
OF GALATEA'S PURPLE KISS;
HE DOES THE SAVAGE HAWTHORN TEACH
TO BEAR THE MEDLAR AND THE PEAR."

Cowley—" The Garden," stanza 10.

"YOU SEE, SWEET MAID, WE MARRY
A GENTLER SCION TO THE WILDEST STOCK;
AND MAKE CONCEIVE A BARK OF BASER KIND
BY BUD OF NOBLER RACE; THIS IS AN ART
WHICH DOES MEND NATURE,—CHANGE IT RATHER: BUT
THE ART ITSELF IS NATURE."

Shakespeare—(Winter Tale, IV., 3.)

JOHN FIRST LORD VISCOUNT SCUDAMORE.

SIR HENRY E. C. SCUDAMORE STANHOPE, BART.

Holme Lacy

A SKETCH OF THE LIFE OF
LORD VISCOUNT SCUDAMORE.

Born February 16th, 1600 *old style*, 1601 *new style*.   Died May 9th, 1671.

CREATED A BARONET JUNE 1, 1620,

AND MADE BARON SCUDAMORE OF DROMORE,

AND VISCOUNT SCUDAMORE OF SLIGO IN

IRELAND, BY LETTERS PATENT, JULY 2ND, 1628.

"It was by the plain *Industry* of one *Harris* (a *Fruiterer* to King *Henry* the *Eighth*) that the *Fields* and *Environs* of about thirty *Towns* in *Kent* only, were planted with *Fruit*, to the universal benefit and general improvement of that *County* to this day; as by the noble example of my *Lord Scudamor*, and of some other Public-spirited *Gentlemen* in those Parts, all *Herefordshire* is become, in a manner, but one entire *Orchard*."

(Evelyn's "Pomona," preface.)

"If men deserve to be commended for their promoting all useful Parts of Knowledge, he certainly deserves no small Commendation, who by his *great Example*, was the occasion of making his native County *The Garden* of England, for Pleasure and Delight : and of bringing no small Profit and Advantage to it." *(Gibson's Door, &c., p. 71.)*

From Norman times to the present day, the name and family of Scudamore have been distinguished. The brief touches which History affords, and the Records of early times still available, are very suggestive of interest both general and local; but this is not the place to enter into the annals of this ancient race. Its worthies must all be passed by, until we come to Sir John Scudamore Knt., the grandfather of Lord Scudamore. He was Standard Bearer to Queen Elizabeth's Honourable Band of Gentlemen Pensioners, and the holder of other honourable offices about the Court, and in his native county of Herefordshire. [1] He is alluded to here, for his great intimacy and friendship with Sir Thomas Bodley, and for his generous gifts to the Bodleian Library at Oxford. His son, Sir James Scudamore, Lord Scudamore's father, was also a very distinguished man, even in the

---

[1] Sir John Scudamore's portrait in his official costume, or uniform, is still at Holme Lacy. In Guillim's "*Heraldry*" 2nd *Edit, p. 416* (1638), it is said of him "This noble Knight hath deserved honourably of his country, by procuring, together with his worthy Lady, the building of a goodly bridge near unto Rosse over the River Wye."

brilliant age of Elizabeth. He was knighted for his valour at the seige of Cadiz, at the same time with two other Herefordshire gentlemen, Sir John Rudhall, and Sir John Scudamore, of Kentchurch, a member of a collateral branch of the Scudamores.

The fame of Sir James Scudamore however will be most enduring, as having suggested the name of "the gentle Scudamour" of Spenser's "*Faerie Queene.*" (Books III. and IV.)

His personal character for bravery, his name, and the family device "*Scutum amoris Divini,*"
"THE SHIELD OF LOVE, whose guerdon me hath graced,"
suited the poet's theme, and thus he has become immortalized :

"Blessed the man who well can use this blis :
Whose ever be the Shield, faire Amoret be his."
(*Faerie Queene, IV. 10, 71.*)

Sir James Scudamore was also fond of letters, and was too, one of the most esteemed friends of Sir Thomas Bodley, and intimate with William Laud, at that time the President of St. John's College, Oxford. These Oxford friendships of his father and grandfather, seem to have had very considerable influence on the early life of John, the most distinguished of the Scudamores.

The only history of Lord Scudamore's Life is given in a book which has now become very scarce. It is entitled "*A view of the Ancient and Present State of the Churches of Door, Home-Lacy, and Hempsted,*" by Matthew Gibson, M.A., Rector of Door, 1720, and it was written, as the author plainly states, out of "Gratitude to the memory" of Lord Scudamore. From this book and from the *Scudamore MSS.* at Holme Lacy, St. Michael's Priory, Clehonger, and the British Museum, the following particulars have been chiefly derived.

John Scudamore was born at Holme Lacy, A.D., 1600-1. He was at first educated at home under a domestic tutor. From his youth he displayed great natural ability and love of study. At the age of 16 he was sent to Oxford, after he had been already two years married, and entered at Magdalen College. Here he met Laud, who would naturally take great interest in his welfare, and who soon entertained for the young scholar an affection and respect, which ever increased in strength as years passed by. John Scudamore only remained one year in Oxford. At the early age of 17, doubtless, following the usual course of men of rank in those days, a license was obtained for him "to travel in Foreign Parts for his better Experience and Knowledge of the Languages"[1] There is no published record of his travels, but he seems to have remained abroad for more than two years, and to have written an account of them. [2] He was brought back by the illness and death of his father, Sir James Scudamore, which took place on May 18th, 1619.

The character of the young man was now well established. He had a charm and grace of manner which captivated all with whom he came in contact. Honours soon begun to fall thickly on

[1] This License to travel is now amongst the *Scudamore MSS.* at St. Michael's Priory, Clehonger. It is dated Sept. 15th, 1618, and is very curious and interesting, not only as shewing the necessity that existed in those days for such a License, but also as having the signatures of Lord Verulam ; Abbot, Archbishop of Canterbury ; E. of Worcester ; Sir Edward Coke ; Sir Robt. Naunton, and some other members of the Privy Council.

[2] Lord Scudamore's Travels in 3 volumes, MS., were sold in 1828 by Mr. Evans, of Pall Mall, among Mr. Jede's books, to Mr. Ash, of Cornhill, Bookseller (for Mr. Baring) for the sum of £2 12s. 6d. (*MS. penes Sir H. Scudamore Stanhope, Bart.*)

him. He was appointed about this time, Captain of Horse in Herefordshire, by the Earl of North-
ampton, the then Lord Lieutenant of the County. [1] The following year by a patent, dated June 1,
1620, he was created a Baronet. The circumstances of this appointment are not known, but it was
in the early days of the creation of the order, and it seems to be sufficiently explained by the fact,
that his friend Laud had, by this time, obtained very great influence at Court, and was intimately
associated with George Villiers, Duke of Buckingham, who was then the chief administrator of
patronage. The creation was aided no doubt by his own address and abilities. Certainly the
formal words of the Patent styling him " *Virum Familiâ, Patrimonio, Censu, et Morum probitate
spectatum,*" were never better merited.

It appears from the Register at Holme Lacy, that Mr. John Scudamore was married in the
Chapel thére, on March 12th, (1614–15) when only just 14 years of age, to Elizabeth, the only
daughter and heiress of Sir Arthur Porter, Knt., of Llanthony, near Gloucester, a child still
younger than himself. But it was not until six years afterwards (about 1620) that the young
couple took up their residence at Cradock, or Caradoc,[2] an estate about five miles distant from
Holme Lacy, belonging to his great uncle Mr. Rowland Scudamore. The bride brought as her
dowry, considerable property in the County of Gloucester. The marriage settlements on the Porter
side are dated 20th February, 12 Jac. I., (1614–15) and those on the Scudamore side are dated 3rd
March, 12 Jac. I. *(MSS. penes Sir H. E. C. Scudamore-Stanhope, Bart.)*

The following year, 1621, Sir John Scudamore was returned, without opposition, as member
of Parliament for Herefordshire, and during the next few years he necessarily spent a great portion of
his time in London. In 1623, April 14th, his grandfather, Sir John Scudamore, died, and he then
succeeded to the Holme Lacy estates, and moved from Cradock to the family mansion.

Neither his domestic happiness, his attention to his own private affairs, his parliamentary
duties, nor the distractions of the Court, or of London society, were able to keep him from a regular
systematic course of study, until, at last, his health began to give way, and his friend Laud, now
Bishop of St. David's—at the quiet suggestion, it may readily be imagined—of my Lady Scudamore,
had to give him the advice '" Book it not too much" *(July 10, 1624.)* " The remarkably *studious*,

---

[1] In the *Scudamore MSS.* at St. Michael's Priory, is the draught of an excellent speech by Sir John Scudamore, to show
how the County Troup of Horse could be rendered more effective ; and there is also a paper, dated May 1st, 1627, giving
" The names of mine own Company of trayned men " consisting of " Household Servants," (25) Retayners trayned," (20)
and " Retayners untrained," (20).

[2] The Manor and Mansion of Cary Craddock, or Cradock, or Caradoc, as it is variously termed, is beautifully situated on
a high wooded bank of the river Wye, in the parish of Sellack. It is the reputed residence of Caradoc Vraich-Vras, or Strong
Arm, a British Chieftain, and one of the Knights of King Arthur's Table. He is the hero of the well known ballad of
" The Boy and the Mantle " *(Percy's Reliques)* and there is to be found near it the traces of a small encampment. But to
come to authentic history:—The estate was purchased from Mr. Roger Mynors in 1594 by Mr. Rowland Scudamore, who
made it his chief residence. He probably rebuilt the house, which is one of the finest specimens of the Elizabethan style of
Architecture in the County. It remained in the family, and appears to have been generally occupied either as a jointure
house, or by some junior member of the family. In the year 1864 it was purchased from the Trustees of the late Earl Digby
by Mr. Elisah Caddick, who is the present proprietor. This gentleman has much enlarged, and in a measure re-modelled the
house, but the grand oak-panelled hall and its other old features, have been preserved, so that it is not unworthy of its beautiful site, and of the associations with which its past history is linked.

(" Manor Houses of Herefordshire," by Rev. C. J. Robinson, M.A.)

and *sober*, and *pious* and *hospitable* life he led " says Gibson " made him esteemed and respected by all good men." His intimacy with Laud was great and sincere; "true Friends they were and heartily disposed to serve each other, in all the Vicissitudes of their Lives. The Bishop often visited him at Holme Lacy, and preached at the church there on Nov. 20th, 1625, as shown in the " *Diary of the Life of Archbishop Laud* " *(f. 21.)*

The Duke of Buckingham did not fail to notice the brilliant qualities of Sir John Scudamore. During his attendance in Parliament they became warmly attached to each other. Party spirit was already running very high, and it may well be believed, that the Duke would be glad to secure the active services of a man of so much worth. It was to him Sir John Scudamore owed the next promotion that befel him. He was created Baron Scudamore of Dromore, and Viscount Scudamore of Sligo in Ireland. The patent is dated July 2nd, 1628, and the very next month we find Lord Scudamore at Portsmouth, engaged as a volunteer, to take part in the projected expedition for the relief of Rochelle. The Duke, who was a proud, overbearing, unscrupulous man, had become very unpopular. He was stabbed at Portsmouth, at the " Spotted Dog " inn, by Lieut. Felton, an officer who felt himself deeply aggrieved by the conduct of the Duke in his former expedition. This act, shocking and desperate as it was, met with much public approval. It probably took place in the actual presence of Lord Scudamore. It certainly affected him very deeply. [1] He however boldly braved the popular feeling against the Duke, and stayed at Portsmouth to attend the removal of his remains, " An act of such *Generosity* and *Honour*, as gain'd him a great Character and Reputation " says Gibson "even of the Duke's Enemies as well as Friends " *(Gibson's " Dore," &c., p. 70.)*

This assassination put an end to the expedition, and " Lord *Scudamore* retired to his *Country* Course of *Life* again ; diverting himself sometimes with Planting and Grafting of *Apple-trees*, and making Experience of their several sorts of Fruit " (p. 70.)

Lord Scudamore seems to have remained at Holme Lacy for the next seven years, and beyond

---

[1] Lord Scudamore wrote immediately to Laud, then Bishop of London, to give him an account of the assassination, and received the following answer, given by Gibson (p. 68), which his own after-fate renders the more interesting :

" My very good Lord,

I received by your last Letter the saddest News that ever I heard in my Life. Yet I must and do heartily thank you for writing so lovingly to me. For if you had not written as you did, I had been left to the Wildness of the many Reports which spread about the City. And I know your pen writt those Letters with a Heart full of Sorrow, and in special for the barbarous and damnable Manner of it. I purpose not to write then, either to declaim in his Commendations, which so few would believe ; or to express my Grief, which as few would pity : but only to let your Lordship know, that though I have passed a great deal of Heaviness, yet I have cause to expect more to come. And the Benefit of this will be, that I shall for ever less esteem what the Malice of the World can lay upon me. Under which, if any fall (as much is threatened) I thank you heartily for your second Letters, that you will appear, what I have ever hoped, a Friend in the time when Friends fall off. And I hope you think I shall in my way be ready ever to serve your Lordship.

*   *   *   *

I was not with his Majesty since this execrable Fact was committed, till now he came to *Windsor*, but stayed in *London* to give the best Comfort I could to the Lady Duchess, who, good Lady, hath been in great Extremity. Now the Court seems new to me, and I mean to turn over a new Leaf in it, for all those Things that are changeable. For the rest I must be the same I was, and patiently both expect and abide what God shall be pleased to lay upon me. To whose Gracious Providence I leave you and myself and shall ever rest,

Your Lordships very loving

Windsor,   *Friend and Servant*

Sept. 12, 1628.   GUIL. LONDON.

all doubt entered into matters of rural industry for the improvement of his estate, with all the intelli-
gence and practical energy which distinguished him. It was about this time tradition tells—and
history can scarcely be expected to notice the small matters of rural economy—that he met with the
seedling apple, which was named the " Redstreak," or " Red-strake." He grafted it extensively
himself, and encouraged its propagation with all his influence, until, as Evelyn states in his " *Pomona*,"
the famous " Redstrake " of Herefordshire obtained the pre-eminence of all other " Cider Fruit."
He speaks of it, as " a pure Wilding, and within the memory of some now living, surnamed the
Scudamore Crab." Later on Philips in his poem on " *Cider*" takes up the subject and after
speaking of the virtues of the " Musk " apple says :

> " Yet let her to the *Redstreak* yield, that once
> Was of the *Sylvan* kind, unciviliz'd,
> Of no regard, till *Scudamore's* skilful hand
> Improv'd her, and by courtly discipline
> Taught her the savage nature to forget :
> Hence call'd the *Scudamorean* plant, whose wine
> Whoever tastes, let him with grateful heart
> Respect that *Ancient, Loyal* House." [1]

Tradition also states, that Lord Scudamore introduced several of the varieties of the Norman
Cider-apples, which are at this day so popular in our orchards, though it is nowhere supported by any
written authority. These varieties are found to succeed so well in Herefordshire ; they are so hardy
in character ; so prolific in bearing ; and some of them possess such excellent characteristics for
cider making ; that they have become more and more appreciated as time has passed on. It is said
that there has scarcely ever been a complete failure of the apple crop since they have been introduced; and so it has come to pass, that in all fresh plantations in the orchards, a full allowance is always
made for Norman apples.

In the short first and third Parliaments of Charles I, (1627-1628) Lord Scudamore was
Member for the City of Hereford, and this political connection with the City rendered him still more
intimate with the citizens. He was appointed High Steward of the City (1631) and took the most
active interest in everything relating to its welfare—and more especially in all matters relating to
the Cathedral or the Clergy. In later life he was appointed (Aug. 29, 1660) High Steward of the
Lordships and Manors of the Dean and Chapter of Hereford.

After the death of Rowland Scudamore, his great uncle, which took place on January 8th,
1630-1, the Mansion and Estate of Cradock passed to Lord Scudamore, and he seems to have kept
that house in his own hands, and gone there for change from time to time. One of his sons was
baptized in Sellack Church. The entry in the Parish Register is as follows :

" Rowland, the sonn of John Viscount Scudamore and Madam Elizabeth his wife, was baptized
the 22nd of May, 1631." This child died the following year.

The same Register also bears curious testimony to the delicacy of the family health :

" 1632, ye 16th of November.

M^{dum,} that upon the day and year above written, a license was granted by Richard Pritchard,

---

[1] It may be observed that the Poet makes "Scudamore" a word of two syllables, as if spelt " Scudmore," or "Skydmore,"
and there is reason to believe, that this was the original form of the word.

Vicar of Sellack, unto John Viscount Scudamore, his Lady, and their Sonn, in respect of manifest sickness and infirmityes, to eate flesh upon dayes prohibited during the time of their sickness and infirmityes. Registered in the sight of Walter Collowe, one of the Church Wardens, the 22nd of November, 1632."—This son was James, then eight years old.

From his earliest years, Lord Scudamore had been embued with a deep religious feeling, which strengthened more and more as life went on. It was this that gave rise to his great friendship with Archbishop Laud, and it was itself doubtless much heightened by that great prelate's influence. At the age of 26, Lord Scudamore had become so strongly impressed with the impropriety of receiving the great tithes which his ancestors had acquired in several parishes that he consults Laud, then Bishop of Bath and Wells, on the subject. The Bishop's long letter is given by Gibson in full, but the essence of the advice contained in it, is, that he should "restore back to the Church what had been bought from those, who had no right to sell." Lord Scudamore immediately resolved to follow this advice, but it was no easy matter to make the restoration. It was not until six years after that he obtained from King Charles I a *License in Mortmain* to restore and endow with the impropriated tithes, all the parishes in which he held them, viz: Dore, Holme Lacy, Boulstone, Bosbury, and Bredwardine. The License is dated December 6th 1632.[1]

Lord Scudamore at once set about the restoration of Dore Abbey Church.[2] He "rebuilt the Quire and Cross Isle," repaired and roofed the whole church; and it was eventually re-consecrated and dedicated on March 22, 1834, the anniversary of his own baptism. This restoration of the Church of Abbeydore was almost the first, if not quite the first, which had taken place since the dissolution of monasteries. It was therefore much commented upon, and the ceremonies used at the consecration, which were doubtless arranged by Laud, were much called in question.[3]

Lord Scudamore afterwards re-built the rectory and out buildings and endowed it with all the tithes, "as well *great*, as *small*, *Predial*, *Personal*, and mixt . . . . . . I find" says Gibson " by the

---

[1] It affords a curious example of the troubled state of the times, and of Lord Scudamore's caution, that many years after (Anno 13-14 Car. II., c. 44,) he should have thought it necessary to obtain a Special Act of Parliament to legalise the acts done under the authority of this License.

[2] "This venerable Place" says Gibson "had been reduced to a condition so *Ruinous* and *Mean*, that one, who well remembered the Rebuilding of the Church of Door, saith, *Mr. John Gyles, otherwise called Sir Gyles, Curate here, before the present Church was rebuilt, read Prayers under an Arch of the old demolished Church, to preserve his Prayer-Book from wett in rainy weather*' (Deposition of Hugh Powell in the case of Tithes inter Watts *and* Watkins, &c., Curia Scac. 4, W. & M.)."
Gibson compares its condition to the desolation of Jerusalem ".As the *Altar* there had been prophan'd, so the *Communion-Table* here had been *pulled down*, and buried in the Ruins of the Church ; till carrying a great deal of Stone away for *Common-uses*, it was dug up, among the rest ; and appropriated (if by way of abuse I may be allowed to call it so, tho' I tremble at it) to the *salting of Meat, and making of Cheese upon*. Thus it continued for a while, till it was very strangely (tho' without a Miracle) discovered where it was. Whereupon the *Lord Scudamore*, when he rebuilt this Church, with great Awfulness ordered it to be restored, and set upon three Pilasters of Stone : where now it stands, the most remarkable *Communion-Table* of any in these parts, being one entire Stone, 12 Foot long, 4 Foot broad, and 3 Inches thick " (p. 41.)

[3] The form and order of the service used on this occasion is in the Library of the British Museum *(Add. MSS., No. 15,645)* and another MS. copy is also preserved in the Archiepiscopal Library, at Lambeth. It was reprinted by the Rev. John Fuller Russell, in 1874 *(Pickering, 196, Piccadilly,)* to prove the church usage of the XVII. Century.

Book of Pensions and Annuities, for several years successively, that sometimes 48s. and sometimes 50s. sterling was paid to *John Phelyps,* under the style of *Wages* for serving the cure of *Dore" (p. 27.)* Thus by Lord Scudamore's munificence "a most *scandalously poor, precarious stipendiary Cure,* became one of the best Livings in the Deanery of *Webley ;* but inferiour to few in the Diocese of *Hereford" (p. 45.)* [1]

Under the power granted by the License Lord Scudamore gave up also all his impropriated tithes in the parishes of Boulstone, Little Birch, Bosbury, and Bredwardine. He also repaired the Church of Holme Lacy, built the vicarage house, and completely restored the tithes to the parish. " He would not hold a Foot of that Land, nor retain that to himself, which should not pay Tythes to the Minister " *(p. 125.)* When only 18 years old he had commissioned Laud to purchase a service of Communion plate for Holme Lacy Church ; and for some reason or other eight years afterwards, he had again to supply the plate. This time he purchased a very handsome and costly service, and provided an iron chest with three locks to it, for its safe preservation. [2]

The parish of Hempstead, near Gloucester, had lost all its church income. " There had been no Minister for many years nor any maintenance for one." Its property had been entirely appropriated to the Priory of Llanthony, near Gloucester, and when this monastery was suppressed it lapsed to the Crown and was sold. Lord Scudamore purchased the church lands and advowson, and endowed it with "all Tythes, Oblations, and Obventions whatever." He built a very ample and handsome house, as a vicarage, and the second rector, Mr. Gregory, afterwards made Archdeacon, caused these lines to be written in gold letters on the front of it :

"WHO'ERE DOTH DWELL WITHIN THIS DOOR
THANK GOD FOR VISCOUNT SCUDAMORE."

This estate of Llanthony was settled on his wife with reversion to their son, and Lord Scudamore could not therefore restore them to the church "as he most heartily desired to do," but " he *charged himself with the Arrears* of all these Tythes from the very first time of their coming into his Possession " *(Gibson, p. 126.)* and used the money for charitable purposes.

On June 9th, 1635, Lord Scudamore was appointed by the King, Ambassador to the Court of Louis XIII., at Paris. His acquaintance with modern languages, and his repute as a man of letters,

---

1 Dore was a Cistercian Abbey, richly endowed ; and at the suppression of the Monasteries, its site and demesne lands were granted by Henry VIII., to John Scudamore, Esq., of Holme Lacy, who afterwards purchased the great tithes ; and thus they descended to Lord Scudamore. The Chapter house, cloisters, and other buildings, with a considerable part of the church itself were pulled down soon after its suppression, and the materials sold. The present church, as restored by Lord Scudamore, has a spacious nave, chancel, north and south aisles and porch ; with a well proportioned and massive tower. " He re-edified this place at his own proper cost and charges, and hath caused it to be furnished with a chancell and seates ; a bellfry and bells ; a churchyard ; and all things also requisite to a parish church." *(Scudamore MSS., British Museum.)*

A story was current at one time that the stone slab of the altar was conveyed by Mr. Scudamore to Holme Lacy, and used there in the dairy ; from whence it was afterwards restored to its proper place and use.

2 A curious fate befel this service. At the time of the Civil War it was buried secretly in the park to the knowledge of the two church wardens only ; one of them proved treacherous, stole the plate and fled to France, "where having consum'd his execrable theft he died in extreme misery and want." *(Gibson, p. 128.)* Lord Scudamore left the sum of £45 in his will to supply it a third time.

together with other personal qualifications for the post, explain the appointment; the great dignity and courtesy with which he fulfilled the duties of the office, amply justified it. Lord Scudamore forthwith repaired to Paris and remained there nearly four years. He then tendered his resignation, which was accepted by the King with regret, but with much kindness, "only for those important reasons" says Mr. Principal Secretary Coke, in his letter of Revocation dated November 27th, 1638, "which you allege concerning your own particular Occasions, which now call you thence by your own Desire. Your four years Employment, and the Memory of your good Service, will honour you and yours." (*Scudamore MS.*, British Museum.) It must be added also that he left Paris to the regret of King Louis XIII., who presented him with his own portrait and with that of his Queen, Anne of Austria. [1]

When Lord Scudamore first arrived in Paris it was a remarkable circumstance, that he should at once discountenance the Huguenots. The English people had always felt the deepest interest in their welfare; and it is a fact, that the Duke of Buckingham was never popular, except when in answer to the national will, he made the expedition to the Isle of Rhé to assist them. It had always been customary hitherto for the English Ambassador at Paris to attend the service at the Huguenot Chapel, but Lord Scudamore never did so.

"Whether" says Gibson "by the Inclinations of his own Nature, or by Advice from others, he not only declined going to *Charenton*, but furnished his own Chappel in his House, with such Ornaments, (as Candles upon the Communion-Table and the like) as gave great Offence and Umbrage to those of the Reformation there, who had not seen the like." (*p. 74.*) This conduct on his part was clearly pre-arranged; and there can be little doubt that Lord Scudamore acted under the influence of Laud in this matter, or certainly with his full sanction.

During his residence in Paris, in addition to the ordinary duties of his office, Lord Scudamore, having become very friendly with the celebrated Hugo Grotius, acted as the mediator between him and Bishop Laud in the vain endeavour to bring about a union of the Protestant Churches in the North of Europe. He lost no opportunity and spared no expense to procure important MSS. for the Bodleian Library at Oxford. [2] He ever welcomed men of learning when the opportunity occurred to him of doing so. He was very civil to the philosopher and politician, Thomas Hobbes, when he was at Paris, travelling as tutor with the young Earl of Devonshire. In 1638 Milton visited Paris,

---

[1] These pictures are original paintings of great value. Their description in the catalogue at Holme Lacy, is as follows :—
"Louis 13th standing, truncheon resting on floor in right hand : Rich dress mantle, 13 sémi-fleurs-de-lis, open shoes with rich roses; inscribed "Louis 13, æt 38. Beaubrun fecit." Whole length collar of Holy Ghost."

"Anne of Austria, his Queen, whole length standing before a chair of state in dress of times; black gown trimmed with gold and jewels; rich embroidered petticoat, right hand on her body, left hanging down, inscribed :

"Anne de Autriche, Reyne de France, agée 37 ans, grosse de 3 mois, fait par Bobrun, en l'an 1638."

There was also a third fine original painting in the collection, which is thus described :—" A Lady ¾ sitting, in yellow dress of the time, going to give the breast to an infant, with an order appendant to a blue ribbon round his neck—a fine piece, an original of Louis 14th when young, presented at his birth to Viscount Scudamore, Ambassador."

(*MS. penes Sir H. Scudamore Stanhope, Bart.*)

[2] "No good Manuscript indeed shall escape me loosely" he writes to Laud (Oct. 2, 1637) and he speaks of being unfortunate in being unable to procure one on the *Basilic Constitutions*, from the Library of *Peter Faber*, for which he had offered £500, but Cardinal *Richelieu* heard of it, and refused to allow it to be taken out of France.

and he gives this account of his reception: "The noble Thomas *(sic)* Scudamore, King Charles' Ambassador, to whom I carried letters of commendation, received me most courteously at Paris. His Lordship gave me a card of introduction to the learned Hugo Grotius, at that time Ambassador for the Queen of Sweden, to the French Court, whose acquaintance I anxiously desired, and to whose house I was accompanied by some of his Lordship's friends."

The large and interesting collection of *Scudamore MSS.* in the British Museum contain many papers relating to this period of Lord Scudamore's, Life:—Instructions for himself (dated June 9th, 1635,) with the autograph of Charles I [1]: letters of introduction: draughts of Despatches received and sent: his correspondence with Laud and Grotius: letters of advice from Lord Talbot, his correspondent at Venice: papers of negociation about the Prince Elector Palatine and the Restoration of the Palatinate: copy of the Treaty of Peace with the Emperor: message from the French Government to the Emperor: Pass for his Secretary, John Browne, to England, and letters from him: copy of letter of Charles I. to Louis XIII., upon the recall of Lord Scudamore, with the official notification to himself from Mr. Secretary Coke: and also "Passages at my taking leave of the French Court" which is curious as containing the messages sent by the King, Queen, Princesses, Lords and Ladies of the French Court to the King and Queen of England, with the expression of good will to himself personally by all these parties, by Cardinal Richelieu and the Prince de Condé. [2] The limits of this paper do not admit of any more extended notice of these papers, or of his public proceedings at Paris.

The urgent private reasons which induced Lord Scudamore to retire from the Embassy are not stated ; but on carefully considering all the circumstances, aided by details scattered here and there among his papers, the inference becomes very strong, that the real reasons were of a pecuniary nature. His income did not equal his expenditure, and he was compelled therefore to raise money on his Estates. The expenses attendant on an embassy were much greater relatively in the XVII century, than in our own times : and Lord Scudamore was not the man to spare any expense that might be required to maintain the dignity of his office, as the representative of England at the French Court. During the four years of his Ambassadorship, the King, his master, was ruling without a parliament, and was so pressed for money, that it seems highly probable, that he left his Ambassador very much to his own resources in this respect. Lord Scudamore was a good man of business, and, as may be seen from his papers, extremely careful in money matters.

Among the *Scudamore MSS.*, at Holme Lacy, and at St. Michael's Priory, is an account current rendered to Mr. Edward Ashe, of London, for the year 1638 (his last year of office) in which the expenditure amounts to the very considerable sum of £6,422 3s., and it is remarkable, that to meet this heavy expenditure, there is but one official payment given, viz., "Nov 12, (1638,) Received out of the Exchequer £521," and this was less "£5 19s.," charges thereupon, as appears on the other side of

---

[1] " Your weekly advertisements (not collected from Gazetts, or vulgar rumour, but faithful discoveries) must be sent to our Secretarie ; that we thereby may be informed in all that may occurre and thereby take notice of your good services, which we will graciously accept " *(Instructions for Lord Viscount Scudamore.)*

[2] The Prince de Condé sends a message to Laud whom he greatly praises, and recommends the King " neither to give way at all to the Puritans, nor to yield to the Scotch, but to cut off three or four heads and so make peace."

the sheet ; almost all the other remittances come from his own Agent, Mr. Mansfield, with a balance still deficient of £3,698 17s.   And on the expenditure side also appears the significant entry, of the payment of interest for money previously borrowed.   This sum has been stated as the whole of his expenditure for the four years at Paris.   There is nothing on the papers to show that it was so, although the first four payments, amounting to £2,600, are bracketted together and may refer to previous expenditure.

At St. Michael's Priory, is also a copy of the bill for his " Transportation into France," signed by Lord Scudamore *(See Appendix I.,)* which amounted to £852, and from a draught in the British Museum of the first portion of these expenses, £553 13s. 9d., it appears, that there was a gain (on exchange) by paying in English "peeces," £67 3s. 8d., which is duly debited to the credit account.   Lord Scudamore's expenses of preparation must also have been considerable.   Besides his suite of servants and their handsome liveries, his Lordship took over 33 horses, with carriages and carts for luggage, &c.

Lord Scudamore was now at the zenith of his fame, and on his return home " he was received in Herefordshire with a hearty welcome from all classes ; his friends and tenantry met him on horse-back, and congratulatory addresses poured in upon him from all sides." *("The Civil War in Here-fordshire," by the late Rev. John Webb, p. 19.)*   A long congratulatory letter on his return home, in Latin, is preserved in the British Museum, from " Rt. Tetloe, Incumbent of Much Dewchurch."   The Christmas of the year 1639, by special proclamation of King Charles, was ordered to be kept by all Royalists with great festivity, as a counter demonstration to the rigid austerity which the Puritans inculcated.   Lord Scudamore obeyed the king's command "with his own openhearted concurrence," says Gibson.   He kept open house at Holme Lacy in antique and magnificent style "from the 23rd of December to the 11th of Januaire followinge, beinge 2 weeks and 5 days " *(Scudamore MS., British Museum.)*   " So wide was his popularity, that presents and contributions according to feudal fashion, were received not only from Hereford, and Ross, and about fifty of the neighbouring places, but also from Dore, Clifford, Leominster, Worcester, and Glou-cester" *(News Letter Dec. 3rd, 1639, and Scudamore MSS. as quoted by Mr. Webb.)*   It seems very probable however that these presents were made and received, not simply as a tribute to his personal popularity, but also with the full knowledge of the very great pecuniary liabilities he had incurred during his Ambassadorship at Paris.   *(See Appendix II.)*

It is worthy of notice here, that though there are accounts of Apples and Cider at Holme Lacy and at Cradock, and he sometimes sent cider up to London, yet at these Christmas festivities only twelve hogsheads of it were consumed, against fifty-four of beer, and one of ale.   Its value however per hogshead was 17s. whilst beer was only set at 10s. and ale at 15s. the hogshead.

(Scudamore MSS., British Museum—See Appendix II.)

The " Gennet Moyle " apple is most particularly mentioned.   This was the favourite cider apple through the county until it was supplanted by the Redstreak.

"Gathering grafts for London 8d. per d^m. 7 days.	A cooper 1s. per d^m.
Mending the Cider Mill 1s. per d^m.	A common labourer 6d. per d^m.
Apple gathering 6d. per d^m.	Making a Hogshead 5s."

And in the Winter " Apples and cider sent to London weighed 2,600 at 6s. the 100, £5 16s. Pears carried to London, 6 bottles of Cider ditto."

1641.

" A common labourer per diem 8d.

13 Hogsheads of cider made in 1639, 12 of them spent in the Xmas, one sent to my Lady at Bowe, worth 12s. each, making each 5s. per hogshead. (There is some uncertainty as to whether this last 5s. was for cider-making, or hogshead making, but it means probably here, the expense of making the cider.)

Gathering fruit at Cradock, Oct. 21, 1641.

1 man 3 days ; 3 maids, 2 days ; 1 maid 2.	May 12, a ffirkin of Cider sent to London.
2 boys 2 days ; Higgins and his wife (1 day) 1s.	Nov. 25, beer vessals repaired for cider (probably) 1642.
Jane Meend 1 day.	"July, John Cook grafting 3 days, 1s. per day.
Nov. 4, Corn and Cidercarried to Llanthony (Gloucester.)	May 5, a Butt and ffirkin of Cider, 2 Hampers, Apples, &c." [1]

(Scudamore MSS. at Holme Lacy ; extracted by the Rev. John Webb.)

For the next four years and upwards Lord Scudamore lived in strict retirement at Holme Lacy. It was in the troublous times of the struggles between the King and the Parliament, the outbreak of civil war. When Lord Scudamore saw that open war was inevitable, he prudently and cautiously resolved to set his own affairs in order. On the 20th of June, 1642, he executed a Lease and Deed of Trust of all his Herefordshire property for 99 years to a Mr. John Whittington, Mr. Mansfield and another, in trust to pay off the debt £5,000 still due upon it; to raise £7,000 for his daughter's portion when it might be required ; and to secure her jointure to Lady Scudamore, and the estate for his only son. Two months afterwards the King raised his standard at Nottingham, (Aug. 22, 1642,) and the Civil War began. Lord Scudamore remained all this time quietly at Holme Lacy ; and, so far as is known, neither gave help to the King, nor was he expected to do so. It may be that he dreaded to incur further pecuniary liabilities in a cause he judged hopeless : and it may be also, that, like Lord Falkland, he lamented the excesses on both sides and for that reason remained comparatively neutral. He made no concealment of his views however, which were always those of loyalty to the King and strong attachment to the Church ; " He was the only nobleman of any rank at that time resident in the county on his estates " *(The Rev. Jno. Webb)* and the influence of his example was no doubt very great. His quiet life of retrenchment did not save him, and the time soon arrived when he was to suffer for his loyalty.

On April 25th, 1643, Lord Scudamore, and his only son, were in the City of Hereford when it was attacked by Sir William Waller, and taken the same day. Lord Scudamore was made prisoner, but was treated with great courtesy. He was ordered to keep to his lodgings, as was the case with all the other leading men who were taken ; and a few days afterwards it was intimated to him that he must appear before the Parliament in London. He was however allowed on *parole*, to

---

[1] At the beginning of the XVII. century, orchards were more numerous in Monmouthshire than in Herefordshire. In Mr. Webb's copy of a "Booke of ould remembrances " kept by Walter Powell, a yeoman and retainer of Raglan during the Civil War, is this entry :—   1612.

"This was the greatest yeare of ffruite that ever I saw. I made 50 Hogsheads of sider of the tieth of both parishes."

One of these parishes was Llantilio (or Llandelio) Cressenry, near Raglan. The other may have been Tre'rgaer, Llanarth, or Prurliôs. At the time in question Mr. Powell occupied the Vicarage at Llantilio. The tithe was evidently taken in kind, and the entry shows the provincial use of the word "fruit."—*(Rev. T. W. Webb.)*

present himself there at his own convenience, and the following Pass was made out for his journey :

"These are to will and require all souldiers to permit the Lord Scudamore with his traine, to passe their guards ; and for theire soe doing this shall be theire warrant. Given under my hand, at Hereford, this 29th April, 1643.

WILLIAM WALLER."

The original Pass is now amongst the *Scudamore MSS.*, at St. Michael's Priory. There also is the following letter written by Lord Scudamore to Sir Robert Pye immediately after his arrival in London. The letter pays a high tribute to the courtesy of Sir William Waller, aud gives Lord Scudamore's own view of the circumstances :

"For my noble friende Sir Robert Pye, Knt., at his house in Westminster."

"Sir,

It was my fortune to bee in Hereford when Sir William Waller took it; and being a person that abounds in civilitie, he did mee the honour to come to mee to the place where I lodged, and after some passages of noble respect, hee desired mee to consider him as the Centurion's servant, who was to doe as he was commanded; that hee was governed by instructions, and according unto them was to intreat mee to apply myself, in what convenience I might, to goe to London to the Parliament. The answer I made him was, that I should be readie to obey his commandments. And speaking to him concerning the way, and time, of my going up, hee was pleased to leave mee to myself to goe how I would, and to take my word that I would be heere by last night. After this I waited upon him at his Quarters, and carrying with mee a copy of the Treaty upon which the town was rendered, I observed to him, by the sixt article of the Treaty, I conceaved both my person and goods to bee free for anything past, as being a citizen, and having had the happiness long since to serve under that quality, as an unworthy member of the honourable house of Commons. And, therefore, I desired that I might enjoy the justice and benefit of that article, and if there was no other reason than my being in Hereford when the town was rendered up, I presumed it was in his power to excuse mee from the journey to London. He made answere, that hee had already written up how that he had found mee in Hereford, and that I would be shortly in London, and that he had taken my word for it ; and that besides he had represented how much I had suffered, and how little I had acted, and that having gone thus farre, it would not be decent for him to thinke upon freeing mee in the country, but that he did not doubt but I should find so easy a passage in the parliament as would be even beyond my expectation. Whereupon I continued myself to my former engagement, and am accordingly come hither. It was yesterday in the evening before I arrived heere, so that I cannot make offer of myself to the parliament till to-morrow morning that they sitt. But then my request, Sir, to you is, that you will favour mee so much as to acquaint the honourable house of Commons, both with the contents of this my letter, and that I am arrived, and doe attend heere with all humbleness, to receive and submit unto their pleasure and commands.

"Being come, I finde the house I lived in lately, in Petty France, and the goods in it, are newly sequestered upon a general Ordinance of Parliament. I trust and humbly move, that since I have brought my person hither, the house will be pleased to give order that this sequestration may be taken off, or otherwise to refer it to a committee. And I hope further, that when a thorough search shall have been made of mee, it will be found that neither bitterness of mind against persons, nor greedy desire of any worldly thing, have moved mee to or fro in the carriage of my self amidst these dismall distractions and divine judgments, upon my deare mother England, but that I have desired and laboured to keep a good conscience, according to the best of my understanding ; and though it should prove to bee an erring conscience, yet it had been sinne in mee to goe against it, being mine.

So, Sir, I take leave and rest

Your affectionate friende

To serve you

SCUDAMORE.

Petty France, May 14th, 1643.

The 6th Article of the Treaty on which his Lordship relied for his freedom was as follows :

"That the Mayor, Aldermen, and Cittizens, shall be freed from plunder and their persons left at liberty for anything past."

but he soon found that he was too rich a prize to be let off so easily.

A few days after he wrote the following letter to Sir Robert Pye, of which the draught is amongst the *Scudamore MSS.* of the Longworth Library, at St. Michael's Priory :

"I humbly desire to redeeme my liberty with a sume of money, and doe conceive that the honourable house of Commons will not onely incline to suffer mee to receive this favour which they grant to many, but will also use me without rigour in the proportion of the mulct, the rather in regard my sufferings have been already great, and my doings onely such as have expressed conscientiousness of duty according to my understanding, without bitterness of mind towards persons, or sinister designes upon things." He then details the injury to his property in Herefordshire and Gloucestershire—the hardships his wife had to go through, and his own sufferings, as being "above the proportion of his desert . . . . for this action of Hereford, wherein I was but a Volunteer and had no command, and being heere casually and a sworn citizen and steward of the town, I knew not how in honor to run away from it just then when a force appear'd before it."

Endorsed by Lord Scudamore "Draught of Letter to Sir R. Pye, 1643, May."

Lord Scudamore thus found in London, that he was not to be so fortunate as he had been led to expect, either with regard to his personal liberty, or the possession of his property. He was delivered to the custody of the Serjeant at Arms, and having already been marked as a Delinquent, his house and property at Petty France, Westminster, had been sequestrated, in short he was a prisoner without other means than such as he had brought with him. His house had been seized, and his goods, even to the wardrobe and personal effects of Lady Scudamore, were ordered by the Parliament to be sold :

"Die Lunæ, 29 Maii, 1643.

It is this day ordered by the Commons House of Parliament, that ye goods seized of the Lord Scudamore bee forthwith solde and ye Proceeds thereof employed for ye service of ye fforces under ye command of Sir William Waller.

H. Elsynge, Cler. Parl. Dom. Com," *(Scudamore MSS.)*

Endorsed by Lord Scudamore.

"Copy of the order for sale of my goods in Petty France."

The goods were seized and removed to the Guildhall for sale. There is an Inventory among the *Scudamore MSS.* at the British Museum, dated May 11th, 1643, and also another at St. Michael's Priory, made by "Norris a Joyner, Addle St., and Stone a Joyner, in Barbican." They seem to have been sold at a great sacrifice since the whole contents of the mansion only realized £176 15s., whilst the valuation put upon them was £700, in the list of Lord Scudamore's losses.

On the 21st of June, 1643, Lord Scudamore had to give a Bond for £1,000, with two sureties to John Hunt, Serjeant at Arms, to attend the Orders of Parliament. On July 10th, 1643, he was assessed on his property in London, that is on the house and tenements of Petty France, by the Parliamentary Assessors (sitting at Haberdashers Hall, London,) for the sum of £1,500, which was ordered to be paid in ten days.

It would seem that the removal of the goods from Petty France had stopped the income derived from it, for at the end of the year the Committee of Sequestration issue this order :

Dec. 14, 1643.

"Foreasmuch as it appears to this committee that the Rents issuing out of Lord Scudamore's House and Tenements at Petty ffrance doe amount to more than can be made of them, whereby noe benefit can come to the State by them, it is therefore ordered that the sequestration of the said House and Tenements be taken off and discharged."

John Jackson, Solicitor.

This removal of sequestration and the sale of all the household contents, did not prevent another assessment being issued on the same property the following year (November 1st, 1644,) for the sum of £2,000, in an equally peremptory form, which shews that they expected to get the money from some indirect source. *(Scudamore MSS., St. Michael's Priory.)*

Lady Scudamore remained in Herefordshire, and occupied the mansion at Holme Lacy. The property at Llanthony and Hampstead, in the suburbs of Gloucester was seized in the same spring, and the devastation committed upon it by the Parliamentary Army was soon reported at Holme Lacy. Lady Scudamore immediately wrote to Sir William Waller, then in command there, and received the following answer, which recalls the saying that " Sir William Waller took Hereford by his courtesy."

The original letter is with the *Scudamore MSS.*, at St. Michael's Priory :

" NOBLE LADY,

    I shall ever take itt as a great honour to receive your comands, and I shall, with a ready obedience, entertaine them. In obedience to your Ladyshipp's letter, I sent for Alderman Pury, and questioned with him what wast had been comitted either upon your Ladyshipp's house or grounds. I finde some Trees have been felled, and have given order there shall be no more touched, but I am assured nothing about the house hath been defaced, only a Tower of an ould chappell adjoyning thereunto was pulled down in regard itt might have been some annoyance to the workes. For your Ladyshipp's rents, I have given order the sequestration should not be executed ; so that, Madam, they are still att your comand. If there be anything else wherein I may advance your Ladyshipp's service, I humbly beg the favour to be comanded, that I may have opportunity to give some demonstration with what passion

I am, Madam,

Your devoted, humble servant,

Gloucester, 4th June, 1643."   WALLER.

Directed :

" For the Right Honourable the Lady Scudamore, att

Home Lacy,

Humbly present these."

    Other rulers however came to Gloucester, who were not so courteous and considerate. Among the *MSS.*, at St. Michael's, is one by Lady Scudamore, entitled " My true sufferings to be presented to Collonell Massie." It has no date, but was probably written at the end of 1643, or in the beginning of 1644. Her Ladyship complains :

    First, it is well known I have not received one penny of rent, being £1443 1s. 10d., that was due to me as my inheritance and present maintenance, but have patientlie submitted without doing anything, notwithstanding an Order of Parliament, from the Lords and Commons was sent downe to stay the ruine and waste committed both of houses and trees, which order was delivered to Mr. Wise, Maior of Glocester ; now let any indifferent man judge what I have suffered in both without any end, untill all be ended ; it being the uncharitable disposition of some to build out of the materialls, which have been left unto the owners even by enemies. Beside the taking of my best goods even to all my wearing apparell, in Pettie ffrance in West-minster, to a verrie great value ; beside the house and garden of £150 per annum, it being all the house I have by joynture.

    These sufferings I had hopt might have favoured me, without the losse of the presence of my Lord, which considerations being rightlie weighed, I shall not doubt but that his person being under Parliament, they will be pleased to procure a live-lihood for him ; which cannot be if the example of Mr. James Kirle should goe on, for his estate being siezed on by the Governor of Hereford and others for supposed delinquencies ; he proposeth to do the like to my tenants, as he hath alredje on James Collins and Richard Mecke, so that I must lose my rent allotted unto me and my child," &c., &c.

Her Ladyship no doubt got her rents, and equally without doubt, had to send the money for his Lordship's maintenance in London. Some short time after however, the order obtained by Sir William Waller was set aside, and the Gloucester estate was sequestrated.

There is another letter written by her Ladyship to Mr. Sheppard, a Lawyer in Gloucester, which is too curious to omit. It shows that it was sometimes necessary for her to take care of herself and that she was equal to the occasion. It is as follows :

"MR. SHEPPARD,

I am verrie sensable of your efforts, which shall not be forgotten by me, and now I shall present you with a passage from the Governor of Hereford, that hee was pleased to send a petition of four gentlemen, Mr. Herbert of Coldbrooke, Mr. Baker, and Mr. Morgan and one other, unto me with a letter which his Lieutenant would not suffer me to keepe; with an order from the Governor under this petition of the aforesaid gentlemen that I must urgentlie lay down £60 which my brother the late Governor of Hereford borrowed of them, as they say, withóut any hand for it. My nature was never to dispute with gentlemen, but to believe them, that I should give satisfaction unto them; which is the manner I did to Captaine Baskett, that I did not know why I should be answerable for the debtes of my Lord's brother; who had but an annuitie of £50 a year during his life, which he had taken at Michallmas last; yet, to render the Governor a sevill answer, I did ashower Mr. Baker, one of the aforenamed gentlemen, with the same Captaine Haskett, that I considered myself to be under the honble Committee of Glocester and Hereford; and begged that he would be pleased to let the said demands rest untill the Committee of the said parties come hither, and what they please upon a right examination I will submitt unto. Yet notwithstanding these sevillities on my part and my patience, I was the onlie woman that had goodes in towne not to be compounded for by any meanes as I could make. Yet this day I received a message from him, if hee did not receive the money, hee must fall foul upon me. I shall beseeche the Committee to take it into their consideration, to take their due and favour

Hom Lacie,   Your affectionat ffriend,
  this 26th of January, 1645.   ELIZABETH SCUDAMORE.

And then comes the postscript which shews that my Lady was a little afraid of refusing the demand, and would rather pay it unjustly than give any offence.

"I doe desire you that little which I have heare might be reserved for the bennefite of those gentlemen who have power over it, and an order to forward it for them."

The Mayor of Hereford for that year, was William Cater, and it is right his name should be given, since it is extremely doubtful if he got the money. He might have had the opportunity however of getting it legitimately later on in the year.

The real siege of Hereford took place the same autumn. The Earl of Leven, with 8,000 foot and 4,000 horse besieged the city, from July 31st, to September 2nd, 1645, when it was successfully defended by Col. Barnabas Scudamore, Lord Scudamore's only brother, and Col. Coningsby.

Lord Leven's army was badly paid, and badly fed. The men were compelled to eat the yet green corn from the fields, and the unripe apples and pears from the surrounding orchards, and were in consequence much affected by illness, from which many of them died. Dr. Beale[1] states it in this way : "The Scottish soldiery made themselves ill with eating apples, to them an unwonted luxury."

The King himself visited Hereford on September 4th, 1645, but the Scotch had notice of his approach, and withdrew from the city two days before his arrival. He knighted Col. Scudamore, and

---

[1] The author of "Herefordshire Orchards a Pattern for all England" (1724) in his "Letters" to Samuel Hartlib, Esq.

78   LIFE OF LORD SCUDAMORE.

two of his companions, for their gallant defence of the city. The King remained in Herefordshire, marching and counter-marching for several days, as is shewn concisely in the *Iter Carolinum*. Coming from Oxford through Worcester, he visited Bromyard, Wednesday, September 3rd : Hereford, September 4th, dining at the Bishop's Palace : Leominster and Weobley, on Friday the 5th : back to Hereford, Saturday the 6th, dining again at the palace : He went to Ragland Castle, Abergavenny, and Monmouth, from the 7th to the 14th, when he returned to Hereford : Monday, September 15th, it says " we marched half way to Bromyard, but there was *leo in itinere*, and so back to Hariford again " : Wednesday 17th, to *Rendezvous* at Arthurs Stone[1] and there dined : to Hom Lacy, supper, (Lord Scudamore's) slept one night : Thursday 18th, to *Rendezvous*, five miles from Hom Lacy, with intention for Worcester, but the Parliamentary officers, Poynz and Rosieter barred the way, so they turned through Hereford, Leominster, and Weobley, to Presteign, by a march lasting from six in the morning to midnight. So for Chirk Castle, &c.

Dr. Beale, who wrote in the time of Charles II., says that "when the King (of blessed memory) came to Hereford in his distress, and such of the *Gentry* of *Worcestershire* as were brought thither as *Prisoners* ; both *King*, and *Nobility*, and *Gentry*, did prefer *Cider* before the best *Wines* those parts afforded." *(Evelyn's " Pomona." 4 Edit., p. 94.)*

Some little time seems to have elapsed before the committees of sequestration appointed through the country, got into working order. The Committee appointed for Gloucestershire and Herefordshire sat at Gloucester, and the first notice relating to Lord Scudamore is the following :

" January 27, 1644.

Upon information then given, John Viscount, or Lord Scudamore, was (with others) voted a Delinquent, for that he, and they, and every of them, had been in actuall arms on the King's side against the Parliament within the space of two years, then last past."

                                   A Sworn Coppie,   STEPHEN HALFORD.

They forthwith proceed to enquire into his estates, for the purpose of sequestration, when they are met by the Lease, which Lord Scudamore had had the forethought to execute just two months before the Civil War broke out. The Committee seems to have regarded it at first with all due credit ; but as time went on, they became more sceptical ; though it was not until the next year, that they resolved to make a special enquiry with regard to it. The following order was then made at the Committee by Ordinance of Parliament for Gloucester and Hereford :

" 9th February, 1645.

Ordered that Lord Scudamore's Lease, or Deed of Trust, made to one Mr. Whittington and others, be brought before the Committee with such witnesses as shall be thought good to be produced as conducing to the proof thereof, upon the xxiii day of this instant February ; and doe further desire that Col. Byrch, Governor of Hereford, doe permitt the agent of the Lord Scudamore to take from amongst the rest of his Lordship's writings the said Lease, and any other writings concerning this matter to be produced before the Committee the day afforesaide, whereby upon that day in examination upon the case, the Committee may give their resolution thereon : and further that in the meane space, the lands of the Lord Scudamore shall not be sett[2] to any person or persons living within the bounderie of the Citie of Gloucester.

                         (Signed)   J. BROMWICH,
                                          GYLES OVERBURY, HEN. JONES.

───────────

[1] A cromlech on the ridge of Meerbach Hill, not far from Dorstone, and about fourteen miles from Hereford.
[2] This use of the word " set " in the sense of " let " is a Herefordshire provincialism not yet extinct.

Lord Scudamore's friends and agents used every exertion to comply fully with the order of the Committee. Mr. Thomas Edmonds, the Attorney at law, who witnessed the sealing and delivering of the Lease at the time it was dated, makes an affidavit, as to its *bonâ fide* character, a copy of which is with the *Scudamore MSS.*, at St. Michael's Priory. Mr. Whittington also makes an attestation to the same effect, which is with the other *MSS.* in the British Museum. Mr. Whittington states that "he knew Lord Scudamore was truly indebted to divers persons (he being surety for some,) and that he made this Lease to Mr. Mansfield and himself to secure their debts; and to raise £7,000 as a portion for his daughter; and for the maintenance of his Lady:" and adds that "he hath paid divers sums according to the lymittation of the said trusst and intendeth to go on in payment of the same if he may be permitted." Mr. Mansfield seems to have been dead at this time. The other witnesses to the Lease, Robert Edwards, Michael Bayard, and John Shephard, probably appeared in person before the Committee. Lord Scudamore was represented by counsel also.

The "Instructions for counsell" in Lord Scudamore's hand writing are at St. Michael's Priory. Some of its passages are interesting. It begins :

"It is not unknown to your honourable Committee what my Lord hath already suffered in his person and his estate both reall and personall for his Delinquency, which was neyther *early begunne, long continued*, nor *much prejudiciall* to ye service of ye Parliament. His losses past being irrecoverable; his person and personall estate beinge still in your power, it is now humbly desired, they would not extend theyr severity to others that are guilty of this only crime, of being deare unto him, and otherwise have not offended at all."

He then goes into the objects and terms of the release, and his Counsel is to argue that :

"There is no cause in Lawe, or Equity, to disalow this Lease. The truth of ye execution of it, at ye time it beareth date, will be proved upon oath, and that will satisfie the Lawe. The valuableness of ye consideration for which it was made will make it good in Equity. . . . . . My Lady is now willing to compounde for my Lord's personall estate, for as much of it as is in her power, and for ye rest when it shall be restored by Colloneil Birch. . . . . . . If my Lord could . . . . forfeit as much as his estate is worth, yet former debts must first be discharged. . . . . . Lastly, my Land is charged with an annual Portion of Corne to ye Poore, which must be allowed however. . . . . . . If these gentlemen say that it is beyond their power to take off ye Sequestration; it must then be desired that ye whole estate may be protected and preserved from any injury till my Lady in reasonable time may acquaint ye Parliament."

These arguments for the most part were pleaded at Gloucester before the Committee with the good effect, at any rate, that the great charge and trouble of carrying witnesses to London was thus avoided.

The Committee at Gloucester came to no decision and transferred the matter to London with all the evidence taken. These papers are now among the *Scudamore MSS.*, at the British Museum, and with them is the Report of Serjeant Bradshaw (the regicide), upon the whole case, in 1646, but they are all much injured and mutilated by damp.

The London Committee after much delay were obliged to admit the validity of the Lease; but they retained the personalty at the Mansion, as a matter of course : a certain demise to one Lane of Hampton Bishop, and one Parsons of Little Birch, of the Messuages and Farm Tythes of Hom Lacy and other properties at a rent of £204 17s. 9d. : and also an annuity of £50, on some of the above properties payable to Sir Barnabas Scudamore, Knight, "a voter and active delinquent against the Parliament."

In the British Museum there is a very complimentary letter to Lord Scudamore, though somewhat fulsome, dated October 5th, 1646, from Mr. J. Bromwich, the Chairman of the Gloucester Committee of Sequestration, who it seems was a cousin of his Lordship's, in which he speaks of himself, as "a much honorer of your parts and personall abilities" and offers to send up a certificate in his favour, and do all in his power to assist him.

Lord Scudamore used every effort he could make, and all the influence he possessed to regain his liberty. His friends however for the most part were not in favour with the Parliament, and all his efforts for a long time were in vain. Gibson gives the secret of his want of success. " I find that after all his *Losses* and *Sufferings*, he was very hardly admitted to compound ; having stood out so long and still standing out against the *Negative-Oath*, which he would never take : Because *he conceived himself bound* not to withdraw his obedience from the King." *(Gibson's " Dore, &c,"* *p. 110.)*

There is no mention made in the *MSS.* of the prison in which Lord Scudamore was confined ; nor is the date or order for his release to be found there. The order from the Gloucester Committee to take off the Sequestration, " in regard that he hath compounded at Goldsmith's Hall for his Estate" is in the British Museum, and is dated November 24th, 1646, but it would seem that his refusal to take the Negative Oath still barred the way. Gibson states that "he suffered three Year's and ten Months' Imprisonment, which occasioned his Lordship so great Sickness, as he very hardly survived" *(p. 109.)* This would make the date of his release to be in February, 1647.

The following rough draft of the " Catalogue of Losses" is from the *Scudamore MSS.*, at the British Museum, and is meant to represent the whole of his Losses by the Civil War :

LLANTHONY, NEAR GLOUCESTER.

	£
Household furniture	600
Trees	1,200
Besides sheds and mounts, that tenants quarrell *(sic)*	,,
Pay and rent two Mansion Houses, with extrordinary outhouses ruined for ye Defence of Glocester, which will not be got into the condition they were in (for less than)	10,000
Much of the materials being carried into Glocester to make upp their owne buildings there.	
PETTY FRANCE.	
Household Furniture, Clothes, &c.	700
HOME LACEY.	
Household Furniture, Lynin, Books, &c.	1,200
Horses	500
Sequestrating my Estate in Glostershire, 4 years	,,
Sequestration of my Estate in Herefordshire, 1 yeare, and	,,
My sonne hath out of my Estate settled upon him pound £500 10s.—at 15 yeare's purchase is	7,500
Debts	,,
Fine at Goldsmith's Hall (besides the Charges of Sequestration	2,790

These figures amount only to £24,490, but Gibson says " by a fair estimate which I have seen *he suffered in his Estate to the value of thirty seven thousand, six hundred, and ninety pounds* " *(" Dore, &c." p. 109)* so that the figures of the preceding blanks were no doubt filled in.

In a MS. book at Holme Lacy, containing the names of persons, who compounded for their Estates during Cromwell's Usurpation, in the following entry :

"Scudamore, Lord Viscount John, and James, his son   ...   ...   £1690"

which may very possibly have been the composition fee without the charges of Sequestration.

On Lord Scudamore's return into Herefordshire, the condition of public affairs had greatly changed. The King had just been surrendered by the Scotch into the hands of the Presbyterians, and was a prisoner at Holmby House in Northamptonshire: the Presbyterians ruled the Parliament, which was in high dispute with the army: Laud had been executed ; the Bishops were set aside ; and the clergy everywhere oppressed. In the city of Hereford and in his own neighbourhood the Cathedral and the Churches had been defaced and mutilated, and all the most faithful Ministers of the Church had been ejected from their livings to make way for the " Preaching Ministers " appointed by an Ordinance of the Commonwealth for the City and County of Hereford, March 28th, 1646. He thus found many of those friends he esteemed most highly, in great distress, and as far as his own circumstances would allow, was most kind and charitable to a great many of them. His house at Holme Lacy was always open to them ; and he gave " Annual Pensions to some of the *Canons*, *Prebendaries*, and *Vicars Choral* of the *Church*, and divers of the *Parochial Clergy* of the *Diocese of Hereford :* viz., to some of them *ten*, to others *eight*, to others *six Pounds* a Year."

(Gibson's " Dore," &c., p. 111.)

In later years he sent considerable sums for distribution to sequestered Ministers of the Church in other parts of the kingdom. [1] Fuller in his " Worthies of England " speaking of his Lordship says " During the Tyranny of the Protectorian Times he kept his secret loyalty to his Sovereign " *(Fol. 47) :* he speaks of his kindness to loyalists in prison : and then gives the following examples of his Lordship's " Charity to the distressed Clergy, whom he bountifully relieved." Among them were some of the highest Order and Dignity in the Church. " To Dr. Wrenn, Bishop of Ely, his Lordship gave sometimes £50, sometimes £100 at once " : to Dr. Bramhall, Bishop of Derry, he also gave liberally : " to Dr. Stern (afterwards Archbishop of York,) Chaplain to his friend Archbishop Laud, and attendant upon him at his *Martyrdom*, he always allowed *Twenty pounds a year.*" Dean Fuller, Dean King, and the names of upwards of twenty others, are also given by Gibson, as receiving contributions from him. " The arrears of the Llanthony Tythes formed part of the fund out of which the distressed Clergy were so bountifully relieved during the tyranny of the Protectorian Times. In 1652 his Lordship charged himself with £1200 from this Source. It was spent every year in gifts and pensions to poor ejected Bishops and inferior Clergy." *(Gibson's " Dore," &c., p. 168.)* Thus Bishops, Deans, Canons, and Parochial Clergy in their distress were indebted almost for bare sustenance to the kind munificence of Lord Scudamore. " I am but a Steward" he would say, if any one alluded to his generosity.

To the Cathedral at Hereford, and to the Cathedral Clergy, he was always generous. He gave £100 at one time : timber for the roof of the buildings : and many valuable books to the Cathedral Library, after it had been ransacked by the rebels in the Civil War. [2] To several of the Cathedral

---

[1] Among the *Scudamore MSS.* at the British Museum, is a letter of thanks from Dr. Hammond, for £100 sent by Lord Scudamore to him, to relieve the distressed Clergy.

[2] Among the same *MSS.* at the British Museum, is also an Acquittance of John Clarke, Custos of the College of Hereford, for an Act Book of the Dean and Chapter, preserved, and restored by Lord Scudamore after the Usurpation.

Clergy he allowed annual stipends, so long as it was necessary. In "Collin's Baronettage of England" it is said that "during the dismal confused usurpation he sent money privately to the exiled king, and relieved all his fellow sufferers to the utmost of his power ; particularly he gave yearly pensions to several deprived orthodox ministers, and entertained so many others in his house, that he was justly styled *a Nursing Father to the Church.*" *(Vol. II., p. 157.)* " Bishop Kennet estimated Lord Scudamore's benefactions to the Church at fifty thousand pounds." *(Rev. J. C. Robinson.)*

It is a remarkable circumstance, that after the Restoration, when the Non-conformists were ejected (1662) Lord Scudamore befriended at least one of the most active of them, viz., Mr Tombes [1] an act as charming in character as it was extraordinary for the time at which it was done.

Lord Scudamore felt very much, as time went on, the restrictions on his movements which was made a condition of his release. What this was is not stated ; it is probable that he was not allowed to pass the boundary of his own Estate or Parish ; whatever it was he became very desirous to have greater freedom.

The following draught of a letter is amongst the *Scudamore MSS.*, at St. Michael's Priory, and it shows the efforts he made in this direction :

" 5 December, 1655.

Endorsed " Copy of my letter to Mr. Edward Rawlins."

" SIR,

By ye encouragm^t of some good friends of yours and myne, I addresse mysealf to you. I doe understand y^t my Lord Protector hath vouchsawffed to free some persons of honor who fall under his highness's declaration and orders, by having been on ye late king's partie ; ye inclosed sertifficat from ye governor of heerford and other commissioners testifieing my peaceabell demeanour since my composition, makes hope in me, y^t I maie be capabell of ye like favour, and be free by your management and promotion of this business to bring it to efect. If you please to accept, you will both doe y^t w^h will be welcom to some friends of yours heer ; and allsoe laie upon me surely an obligation as will make me blush to meet you if I should not deserve it, so far as maie be in ye power of your affectionate friend and servant.

hom lacie, ye 5th of December, 1655."

The copy of the certificate is as follows :

"Wee whose names are heer underwritten at ye request of John, Lord Viscount Scudamore, of Homlacie, within ye County of Hereford, doe certifie all whom it maie concern, that he hath since his Composition for his delinquency lived verrie peaceable and quiet at his habitation here without offering ye least disturbance to ye publick peace of this Commonwealth, though we believe he hath been much attempted hereunto by evill mynded men, and hath alwaies, as far as we could discover, endeavoured ye peace of ye Nation, and showed himself redie to doe all office of kindness to ye friends of this present government, and notwithstanding ye severall atempts which hath been made by ye enemies of ye peace of the Commonwealth in this countie and elsewhere, we have not the lenst cause of suspicion of any dainger by him In testimony whereof we have subscribed our names, this last of November, 1655.

AR. ROGERS
THO. RAWLINS
JO. CHOLMELEY.   MILES HILL   FRANC. PEMBER

(Endorsed) 30th November, 1655.

Copy of the certificatt from the Governor and Commissioners."

---

[1] Mr. Tombes obtained considerable eminence during the Protectorate. " He was frequently called upon to preach before the House of Commons, and in 1653 was nominated one of the Triers for the appointment of Ministers. This high office brought him into connection with the chief religious movements of the day, and gave him a voice in the disposal of vacant Benefices. Like his associates in office, he did not neglect his own interests, but was presented to, and held at the same time with the Vicarage of Leominster, the Rectory of Ross, the Mastership of the Hospital at Ledbury, and the Vicarage of Bewdley." He is described as a little, neat-limbed man, with a quick eye. He was a learned, eloquent and energetic preacher, but was still more celebrated as a keen and vehement controversialist." *(Fyler Townsend—History of Leominster, p. 117).*

The result of this application is not recorded. The very next month (December, 1655,) a return is made to the Committee in London shewing the annual income derived from Lord Scudamore's Estate in Herefordshire to be £1,062 19s. 4d., but the date of this return was only a coincidence, since it is endorsed " My particulars for the extraordinary tax for Herefordshire." *(Scudamore MSS., at Holme Lacy, and St. Michael's Priory.)*

So far as his own freedom went the application could not have been altogether successful ; since some fifteen months later, he was obliged to obtain a special order to visit London for his health. This order is also at St. Michael's Priory, as follows :

> "Saturday, 6th of March, 1657,
> At the Councell at Whitehall.
> Ordered by his Highness the Lord Protector and the Councell, that John, Lord Viscount Scudamore, have liberty to remayne in and about the citye of London and Westminster for recovery of his health, the late proclamation to the contrary notwithstanding.
> HEN. SCOBELL, Clerk
> of ye Councell.

Lord Scudamore seems to have lived at Holme Lacy for the remainder of his days, and to have solaced himself by the exercise of good deeds, and by his attention to all matters of rural interest. Dr. Beale, in a letter, dated May 3rd, 1656, writing of the "admirable contrivers for the publick good " in Herefordshire, say " Lord Scudamore may well begin to us : a rare example for the well ordering of all his family : a great preserver of Woods against the day of *England's* need : maintaining laudable hospitablity, bounded with due sobriety : and always keeping able servants to promote the best expediences of all kinds of agriculture." [1]

Lord Scudamore was very attentive to the animals on his Estate, and took great pains to improve them. " His flock of sheep on one occasion amounted to six hundred, a very great number in those days ; and in the old house at Holme Lacy, as was customary in large country houses of that period, there was a large room set apart as a wool chamber." *(Rev. John Webb.)*

It was during the later years of his life that Lord Scudamore imported the live stock from the Continent from which our celebrated breed of Herefordshire Cattle have been derived. There is no evidence that any special care was taken as to the breeding of cattle before this time : indeed there is a passage in a letter from John Ellyott to John Scudamore, Esq., grandfather of Lord Scudamore, dated November 11th, 1564, which shows, that in that reign no particular attention was given to their herds. It is this : "Althoughe we be no breders in Herefordshire, yet we be accompted to be ffeders of oxen." *(Scudamore MS., quoted by Rev. Jno. Webb.)*

The Tradition is, that in the reign of Charles II Lord Scudamore imported seven cows with white faces and red bodies from the Low Countries. It is now near a century since Mr. Thos. Andrew Knight put this tradition on record, and expressed his belief in its correctness. He noticed the curious fact, that the cattle represented in the pictures of Cuyp and other Flemish painters, often show the familiar colours of our own meadows.

The importation of these cows must have been a very difficult as well as a very expensive

---

[1] Dr. Beale also names other orchard planters of that time, " Sir Henry Lingen : our learned Mr. Bennet Hoskins of Harewood ; Mr. Reed, of Lugwardine ; Mr. Smith, of the Weir ; Mr. Freeman, of Brockhampton : Mr. Charles Morgan, of Blakemere ; and many others," as men, who " metamorphise the Wilderness to be like the Orchards of Alcinus."

proceeding in those days; and another local tradition which has been preserved by Mr. W. H. Cooke, Q.C., and most kindly communicated for this paper, has the merit of showing how the difficulty was overcome. The tradition comes through the family of the Lanes, long resident in the parish of Hampton Bishop, and themselves famous as cattle breeders. Mr. Cooke heard from Mr. James Lane, of Hampton Bishop, who died in 1827, *æt* 70; (the father dying in 1809, *æt* 90; and the grandfather in 1754) that the family of the Herefords, of Sufton Court, in the adjoining parish, helped Lord Scudamore in the importation of the cattle.

On reference to the Sufton pedigree, it appears, that the second and third sons of Roger Hereford, who died in 1659, emigrated to Dunkirk, in which port they established themselves as merchants. Sir Edward Harley, M.P., for Herefordshire, was made Governor of Dunkirk, in 1660, and he would naturally aid Lord Scudamore in any plan for the benefit of their common county. Dunkirk was taken by Cromwell in the year 1658, and sold by Charles II. to the King of France in 1662. It is most probable, therefore, that Lord Scudamore obtained the cows from Dunkirk, through the assistance of Sir Edward Harley, and the sons of his neighbour at Sufton. They would not be Dutch, but would be obtained from the rich residents of the Belgium of our day.

We have only to suppose that the cattle were imported in this way about the year 1661, and all the local traditions with reference to the origin of the Herefordshire Breed of Cattle are fully borne out. The importation of the cows might possibly have taken place some six or eight later, since Lord Scudamore lived until 1671, and the Messrs. Hereford remained at Dunkirk after it was sold to the French King; but it is most probable that they were brought over at the time suggested. [1]

There is no record of any special festivities being held at Holme Lacy on the Restoration. The rejoicing was so general throughout the kingdom, that it scarcely seems to have been called for. Lord Scudamore was invited to become a candidate to represent the City of Hereford in the 1st Parliament of Charles II; since in the British Museum, there is a Copy of his Letter to the Mayor and Council of Hereford declining to be proposed. He did take a part in the Election however, for there are copies of two speeches of Lord Scudamore at this Election among the *MSS.* at the British Museum.

Lord Scudamore especially took delight in the planting and grafting his Orchards. "This part of Husbandry," says Gibson, "was the main diversion his Lordship had," and the influence of a man so highly esteemed and beloved, could not fail to be as great, as Evelyn represents it to have been, in extending the orchards all around him. The Norman apples which he was the first to introduce are more planted for cider all through the County than any other kinds. It is singular also that that the modern system of growing pears, the "Cordon System" as it is called, should also have been first introduced at Holme Lacy, to spread as it deserves, and probably will spread, from the same centre through the County.

The last letter of Lord Scudamore, preserved among the MSS. at the British Museum, is dated October 8th, 1667. It is one written to his grandson, at Christchurch, Oxford, as follows :

"JACK,

Your Father hath delivered your letter of the 26th September. Your readiness and profession to attempt whatsoever

---

[1] It is stated in the " History of the Hereford Breed of Cattle," by Mr. Thos. Duckham, that " there are red-with-white-face breeders who advance, that they can trace them as being the breed of their Ancestors for the past two hundred years." (p. 5.).

studey I shall please, calls me to give my advise, which I doe heartily for no other end but your good. When you have finished the last book of the Phisicks, I wish you would goe through all the rest of Aristotle's Naturall Philosophy, and then to reade over his Metaphisicks; and so from thence to proceede to his Morall Philosophy. Constant I am to what I have often declared, that I would have no study to take you off from Aristotle untill you have read him All over. Hasten likewise to finish Euclide. Justinian's Institutes of the Civill law I would exhort you to reade. But in all action the right Timing of a thing is a circumstance of great weight. The time I could wish for Justinian is when you resolve to reade the grounds of the Comon Law of England, which will be a study that will make you that are an Englishman fitt for employment. And Justinian will furnish with Notions which will prepare the Understanding and make it more susceptible of the maxims of the Comon Law. For your afternoon study I doe much approve of Mr. Old's motion, that is, the pleasant and usefull study of the Globes. To which purpose your Tasker hath undertaken to informe himself in London what are now the best Globes, the best Mapps, and who it is that hath lately written in the low countreys in Mercator's way, and is to be preferred (as hee saith) before Mercator. Hereof hee will write mee an account to the ende that upon my purse your desire may be promoted. This letter I would have you communicate to Mr. Old.

I heare that the last yeare's cider proves dead. This bearer, who is my houschold servant, I have sent with a remedy, and in case the Vice-chancellor should bee gone up to London to the Convocation before his coming to Oxford, I wish you to intreat Mistress Benson, with the remembrance of my service to her, to admit this bearer to perform his part which hee is sent to doe. I presume I need not use many words to move you to assist in all things, that this bearer may doe handsomely and effectually that which my love to you is the motive of. God bless you with his Grace, and I rest

"Your loving Grand-Tasker,

"Home lacie, Oct. 8, 1667." "JOHN SCUDAMORE.

Lord Scudamore has been accredited by some late writers with the establishment of the terraced walks and the neatly clipped hedges of Yew-tree, which now form so striking a characteristic of the Holme Lacy gardens, but this was not the case. The times were far too troublous for the exercise of the "topiary art" as it was called, even if the idea had been in unison with Lord Scudamore's character, which it certainly was not. This Dutch system of gardening required a far more peaceful time for its introduction—a proper time for the practice of

"Retired leisure
That in trim gardens takes its pleasure."

This "topiary art" did not become fashionable until the reigns of William III., and Anne. [1]

---

[1] The Mansion of Holme Lacy was erected about the year 1545, on a site previously occupied by buildings that, tradition asserts, formed the dwelling of Walter de Lacy and his heirs. It has twice undergone great alterations since Lord Scudamore's time. He spent nothing on it himself, but he is thought, during his residence in France, to have procured the designs for rebuilding it, which were afterwards carried out in part by his grandson, the second Viscount. It was in the style of a French Chateau, built of reddish stone, with high pitched roof, and projecting carved oak cornices. It was strikingly similar in design to Clarendon House in London, which is so frequently mentioned by Evelyn in his Diary The approach to the house was from the North, through a lofty arch which led into a quadrangle, formed by the steward's house, laundry, dairy, stables, cider house, &c. Passing through the quadrangle under a second and deeper archway, the broad gravelled terrace was crossed to reach the house door. The steep roof of the quadrangle, its projecting cornices, with the bell tower on the one side and the clock tower on the other, while strictly in character with those of the Mansion, gave to both a singularly picturesque appearance.

"In keeping with such a house were the gardens, in which the topiary artist reigned supreme, and alley answered alley, with a regularity to modern eyes monotonous. Yet the trim yew hedges and straight walks help us to recall the smoothly-polished lines in which Pope, a frequent visitor, sang the praises of the 'Man of Ross;' and we may fancy Gay as wandering through the maze, gaining there a fresh experience in 'The art of walking the streets,' although the poet's friendship was rather with Lady Scudamore (Frances, daughter of Simon, Lord Digby) than with her husband, yet it serves to complete the chain of poetic associations, with which Spenser was the first to link this noble family." ("*Manor Houses of Herefordshire*," *by the Rev. J. C. Robinson, M.A.*)

The second grand alteration was not so happy. Sir Edwyn Scudamore Stanhope (c. 1825) swept away the quadrangle with

In deeds of charity, in the exercise of benevolence, and in the encouragement of industry, Lord Scudamore spent all the time he spared from his books. He educated and employed the poor and provided for the infirm and aged, and gained a degree of esteem that falls to the lot of few men to enjoy, or as Gibson hath it, "established him such an Interest and Respect in his native County, as without Detraction from or Disparagement of others, hardly any before him had, or hardly will ever have again." *(p. 111.)*

Lord Scudamore had his full share of domestic affliction. His four eldest sons he lost in their early infancy ; one other son and a daughter also died in their first year : his fifth son James, and his last child, a daughter, Mary, were the only two who reached maturity. Lady Scudamore died in 1631. His brother Sir Barnabas Scudamore died in 1652. His son-in-law, Mr. Russell, the son of Sir William Russell, of Strensham, Worcestershire, died in 1657; and his only son died in 1668, leaving however a son, who succeeded his grandfather, and became the 2nd Viscount Scudamore. *(MSS. penes Sir H. E. C. Scudamore Stanhope, Bart).*

Lady Scudamore had delicate health and received great benefit from the Bath waters. At her death she left directions that £200 should be paid to the Mayor, Aldermen and Citizens of Bath, to purchase a rent charge of £10 yearly ; of which 40s. was to be paid to the Mayor for a dinner for himself and the Aldermen ; and £8 to the Physician they were to appoint annually, to give advice free of charge to the sick poor. In conformity with this settlement, the Corporation regularly pay £8 a year to a physician, whom they annually appoint to attend the poor in Bellot's Hospital.

(Report of Charity Commissioners, Somerset, p. 297, 1820.)

The publick notification of this benefaction is copied in Dingly's "History from Marble" *(I. xlix.)* as follows :

"Ascending on the right hand to the Pump House, read this upon a brass monument, which is of such publick concerne that it ought to be printed and distributed for ye good of the diseas'd poor of the Nation :

ALL POOR PERSONS NOT BEING CONVENIENTLY ABLE TO MAINTAINE THEMSELVS AND RESORTING TO THE BATH FOR CURE OF DISEASES OR INFIRMITIES, may take notice that there ought to be a Physitian yeerly nominated and appoynted by the Mayor and Aldermen of Bath, who is to give his best advice from time to time to the sayd poor persons without any reward from them, there being a salary provided for that by the charitable guift of Dame ELIZABETH VISCOUNTESS SCUDAMORE.

MDCLIII.

This plate is now attached to the wall on the left hand side of the entrance to the King and Queen's Private Bath, from Stall Street, and is still in good preservation (1879). [1]

all its great conveniences, its fine archways and its picturesque effects : pulled down the chapel, and the remaining part of the old house, to build the west wing ; dwarfed the roof by raising a stone ballustrade around it ; put Bath stone architraves to the windows, and added a large classic portico ; thus quite changing the character of the house. The approach is still very handsome with a broad terrace drive of considerable length ; and the situation of the mansion with the park surrounding it, is as varied and picturesque as a fine site, and magnificent timber can make them.

The intérior of the house is but little altered. The main plan of the rooms is unchanged, and it thus retains all the interest its old associations call forth. The beautiful carvings by Grinling Gibbons remain over the fireplaces ; the birds, flowers and fruit they so faithfully represent forming characteristic and appropriate frameworks to the flower pieces by Baptiste and other distinguished artists.

[1] In consequence of this notice, and of Acts of Parliament passed to encourage the poor to go to Bath, and authorising them to demand assistance from the parishes through which they passed, the place became so infected with vagrants that it was soon found necessary to restrict the gratuitous advice to the inmates of the Hospital established by Sir Thomas Bellot in 1609 for the reception of twelve of the poorest strangers who should come to Bath for the benefit of the waters ; and this is done at the present time.

Lord Scudamore lived through a period of great anxiety and danger, but he reached a good old age, and passed away in the quiet of his own home, and in the peaceful possession of his Estates. He died May 9th, 1671, aged 70 years, 2 months, 23 days. He was buried on Thursday, June 8th,[1] in the family vault at Holme Lacy, "where, on its being opened August 22nd and 23rd, 1822, for the examination of the family coffin plates for legal purposes, Lord Scudamore's coffin was found curiously constructed in lead, so as nearly to fit the body, with a long inscription on a brass plate, recording his virtues and services." *(MSS. penes, Sir H. E. C. Scudamore Stanhope, Bart.)*

By a codicil to his Will, Lord Scudamore bequeathed £400 to the Lord Bishop of Hereford for the time being, in trust, to be settled as perpetual stock, to set to work the poor people of the City of Hereford. In 1840 this sum had increased to £6,035 6s. 5d. in 3 per cent. annuities, when by an Order of the Court of Chancery it was applied for building "The Scudamore Schools" in Hereford. These schools have been in full operation for nearly 40 years, and are well attended. Thus, at this time, hundreds of children are chiefly indebted for their Education to the beneficence of Lord Scudamore. So prosper the deeds of good men.

No good Thing can be said of any Man, which may not justly be said of him; "who lived so rare an Example of *Piety* towards God, *Loyalty* to his Prince, *Love* to his Country, *Hospitality* to his Friends, *Æconomy* in his Family, *Bounty* to the Clergy, *Charity* to the Poor, and *Munificence* to the Church, upon which 'tis known·he bestow'd above *ten thousand Pounds*. He dy'd universally lamented, A.D. 1671." *(Collins's Baronettage of England, Vol. II., p. 156-7, London, 1720.)*

The learned author of "*The Civil War in Herefordshire*," (1879) the late Rev. John Webb, so often quoted before, thus sums up his character: "There can be no hazard in asserting, that to the eye of a dispassionate observer—so far as an estimate can now be formed of his character—for natural and acquired ability, piety, integrity, charity, even to munificence, and most of the virtues that adorn a publick, or private life, few persons of that age in this or any other country could be found to have surpassed Lord Scudamore" *(p. 20.)*

In the words of Archdeacon Gregory, of Hampsted, in Gibson's Appendix:

"SIT MEMORIA EJUS IN OMNE TEMPUS BENEDICTA,
ET EXEMPLUM EJUS SEQUENTES
SUB SCUTO AMORIS DIVINI PROTEGANTUR
OMNES IN POSTERUM ECCLESIO PATRONI.

AMEN."

HENRY G. BULL, M.D.

---

[1] As appears from a letter of Sir Wm. Gregory to Sir E. Harley, in the Appendix to Lady Brilliana's Letters, published by the Camden Society.

APPENDIX I.

The Scudamore MSS. after the death of the Duke of Norfolk passed into the possession of Mr. Howard, of Corby, who gave them, with a curious antique chest in which they were deposited, to the late Thos. Bird, Esq., of Hereford. At the sale of his effects in 1835 they were divided into several lots, and became distributed. A small portion of them returned to Holme Lacy: another and larger share was purchased by Mr. R. Biddulph Phillipps, of Longworth, who gave them with his fine library to St. Michael's Priory, near Hereford: and the remainder found their way through the hands of Mr. W. H. Vale of Hereford, and the late Rev. John Webb, of Tretire, to the British Museum.

The *Scudamore MSS.* at the British Museum contain many other papers besides those already alluded to here. There are Summonses from the Irish Parliament, Proxy Papers, Receipts for Fees, for Subsidies, and for Rent: many Papers relating to Dore Abbey, viz., the Deed of Endowment; Contract with John Abell for re-building the Church, dated March 22nd, 1632; contract with Francis Stretton for Ceiling the Church, dated August 22nd, 1634; Agreement of Wages for Parish Clerk; Two Letters from Bishop Godwin; Papers and Drawings, &c. An Autograph Letter of Thomas Hobbes, of Malmesbury, giving reasons for absenting himself from England on the approach of the Civil War: a curious Letter from a Servant, J. Porter, respecting his services in Lord Scudamore's premises in London, 1649: many papers relating to his personal troubles besides those already noticed; Notes of the opinion of Mr Whittington in respect to the condition of the Lease on his Herefordshire property, in which this passage occurs, " The said Viscount Scudamore had in marriage with his said Lady, a fine Estate of the inheritance of the yearly value of £600 and better, which would have reverted, in case of no issue, to her family." Letter from Lord Scudamore to Sir Barnabas Scudamore upon the hardship of his sequestration and advising him to apply for a mitigation, dated Nov. 22nd, 1651. A letter from James Scudamore to his father on his going clandestinely abroad: Lord Scudamore's Address to the County of Hereford for the restoration of the Cathedral after the Civil Troubles: with many private Letters from Jeremy Stevens, Richard Cornewall, George Andrews, &c., &c., &c.

The *Scudamore MSS.* at St. Michael's Priory also contain many curious papers not mentioned here. There is an original Letter dated 9th of August, 1626, from the Lords of His Majesty's Privy Council, to Sir John Scudamore, directing him " to arrest and lodge in gaol two priests, Geo. Berington, O.S.B., and — Hanmer, who are lurking about the County of Hereford, and doing much mischief." It has the autograph signatures of no less than eleven Lords of the Council. This Letter is also supported by one from the Bishop of Hereford (Dr. Francis Godwin) dated Whitbourne, 30th September, 1626, addressed to " The Right Worthy, my very worthy friend, Sir John Scudamore, Knt. and Baronet." Of Geo. Berington the Bishop says, " I have had layed all the gynnes I can think of, and as soon as I can learn where he is, I will, by God's grace, travel speedily to the place and make search for him. To do anything sooner were but to give him warning to hyde himself. The other (Hanmer) converseth for the most part (as I am told) in your neighbourhood, at my Lady Bodenham's, or els at her son's.   If you can catch him there, you may do a very good deed, and that for which I would account myself much beholden unto you."   There are also Letters from Sir J. Brydges, from John Warner, Bishop of Rochester, Mr. Serjeant Wyldes, &c., Particulars of Lord Scudamore's Estates, and many other papers of great local interest.

[The cordial thanks of the Woolhope Club are due to the following gentlemen, who have so kindly assisted the writer of this paper in examining the *Scudamore MSS.* ; to Sir Henry E. C. Scudamore Stanhope, Bart., for allowing the books and papers at Holme Lacy to be searched ; to the Very Rev. Prior Raynal, at St. Michael's Priory, near Hereford, who not only permitted the *MSS.* to be copied, but who most kindly rendered great assistance in reading the most difficult passages : and lastly to James Renny, Esq., who spent several days in examining the *MSS.* at the British Museum, and who has himself supplied all the information derived from that source.]

APPENDIX II.

(Endorsed) 1635.

" Copy of my bill for my Transportation into France," from the Longworth *Scudamore MSS.*, at St. Michael's Priory, Clehonger.

John Viscount Scudamore, his Majties Ambassr with the French Kynge, Lewis 13th, humbly craveth to receive allowance according to his Majties Privy Seale, bearinge date the 23rd Aprill, in the eleventh year of his Majties Raigne, for extraordinary disbursements concerninge my transportation, &c., made by me, in the particulars followinge.   For

		£		£
Myselfe	Myselfe and wife at two severall times from Holmlacy to London	68		353
	Myselfe and company from Greenwich to St. Denis by the way of Calais, being 12 nights, including gifts to gards, drummers, &c., in towns	285		
Goods and Horses	Two carts for apparell, &c., necessary to be carried with me, being uncertaine at the partinge of my Barke from Greenwich where and when my first audience would be, with horses of relay for part of my family	61		244
	Horses transported in three barkes and shaloopes for my company and goods, shippinge and unshippinge	77		
	Goods with part of my family by long sea from Greenwich to Paris	90		
	Goods from Homelacy to London	16		
Gifts	Gifts to Captaine, Lieutenant, and others in the ship in which I passed myselfe			80
Expresses	Two expresses, th' one to provide my house when it was thought I should have been sent away before my stuffe and goods could arrive at Paris. Th' other to give notice of my landinge, and to return to me on the way			15
My reception and audience	My attendance at St. Denis for my reception and goinge to Meaux for audience, being seven nights, as alsoe to gifts to this Kinges gards, drummers, trumpets, and musike, &c.			160

Summa totalis   £852

(Signed) J. SCUDAMORE.

APPENDIX III.

AN ABSTRACT FROM THE STEWARDS ACCOUNT OF THE EXPENSES INCURRED AT HOLME LACY AT CHRISTMAS 1639, IN OBEDIENCE TO THE KING'S COMMAND. *(Scudamore MSS.,* British Museum.)

"Hom lacie. Spent through Christmas in 1639, from the 23rd of December to the 11th of Januarie followinge, beinge 2 weeks and 5 days."

The presents sent for the festivities are first enumerated, analyzed, and set down as being worth altogether, £131 3s. 2½d. They come from a great number of persons, whose names are all given, with a full detail of their several gifts ; and are catalogued as coming from 59 localities arranged in alphabetical order, from Acornburie, Bullingham, &c., to Worcester, Woolhope and Yatton. The following summary is given of their value :

	£	s.	d.			£	s.	d.			£	s.	d.
Larder	...	46	13	6	ffish	...	1	4	8	ffruits and Spices	...	23	13	4
Acates[1] of Store	...	28	11	9	Wine	...	22	0	9	Storehouse	...	2	15	6
Dayrie	...	1	5	2½	Bakehouse	...	1	10	0	Stable	...	3	8	6

£131 3 2

Then follow the exact details under each heading, for example

Larder.

	£	s.	d.			£	s.	d.			£	s.	d.
Oxen 1	...	7	0	0	Muttons 24	11	4	0	Bacon cheynes 1	...	0	2	0
Cowes 1	...	6	0	0	Bacon 3 fliscbes	2	0	0	Sparibs 1 ...	...	0	1	6
Veales 11 and 5 joynts	13	13	0	Brawnes 2 ...	4	0	0	Pigs 32 ...	...	2	8	0
				Beef cheynes 1	0	5	0					

£46 13 6

and so on through the list to show the exact value of the presents.

The Steward's Account of the general expenses is equally precise, and contains a greater variety of articles. The two hundred and forty years that have since passed, give an interest not only to the dishes set before the guests, but also to the prices of the articles which are so exactly given.

---

1 ACA'TES [Old Fr. *acat, achat,* purchase ; *acheter,* pronounced *acater* in Picardy and Languedoc, to purchase. Ital. *accattare* to beg or borrow.] Provisions ; victuals ; viands ; in more modern language *cates.* This is a frequent word in our elder writers.   *Johnson.*

"The kitchen clerk, that hight Digestion,
Did order all th' *acates* in seemly wise."   *Spenser F. Q. II. ix. 31.*

Shakespeare, punning on the word "Kate," uses the modern form of the word :—

"My super-dainty Kate,
For dainties are all *cates.*"   *Taming of Shrew, II. 1.*

The quantity and cost price of the chief articles purchased under the several departments, is as follows :

	£	s.	d.				£	s.	d				£	s.	d.
Kitchinge ...	83	5	0	Cellar	...	...	32	17	0	Storehouse ...		...	4	18	0
Acates of Store	41	8	8	Wine	...	...	44	14	1	Stable	...	...	16	11	10
Dayrie ...	12	7	11½	Buttrye	...	...	11	7	6	Hurll	...	...	27	18	3
ffish ...	8	10	0	Bakehouse ...		...	6	18	6	Gifts	...	...	27	3	0
				ffruits and Spices		...	20	6	5						
												£233	8	6

Kitchinge.

Beefe 2,904 po.	...	att 2d. po.	Neates tongues 13	...	...	att 8d. ye tongue
Muttons 33 ...	...	att 11s. ye mutton	Calves feete 13 sett	...	...	att 4d. ye sett
Veales 11 ...	...	att 1 £ and 7d. ye veale	Calves heads 11	...	...	att 8d. ye heade
Does 21 ...	...	att 13s. 4d. ye Doe	Udders 8 ...	...	...	att 8d. ye udder
Porks 4 ...	...	att 8s. ye pork	Tripes dishes 20	...	...	att 4d. ye dishe
Bacon flisches 2	...	att 13s. 4d. ye flisch	Calves races 11	...	...	att 1s. ye race
Brawne collers 12	...	att 3s. 4d. ye coller	Calves Haggases 11	...	...	att 4d. ye haggas

Then there were Souse [1] dishes ; Sparibs ; Bacon cheynes ; Milk and Crayme ; Garnish ; Gold and Colors for ye Pasterie ; Civet ; Ambergrease ; Muske and many other more ordinary trifles.

Acats of Store.

Turkyes 53 ...	...	att 2s. ye turkye	Partriges 150	...	...	att 6d. ye partridge
Geese 54 ...	...	1s. ye goose	Phesaunts 29	...	...	3s. 4d. ye phesaunte
Capons 249 ...	...	1s. ye capon	Woodcocks 69	...	...	4d. ye woodcock
Hennes 70 ...	...	8d. ye henne	Heathcocks 4	...	...	1s. ye cock
Pullets 26 ...	...	6d. ye pullet	Grouses 4	...	...	1s. ye grouse
Chicks 36 ...	...	3d. ye chick	Peacocks 3	...	...	5s. ye cock
Ducks 47 ...	...	8d. ye duck	Blackbirds 2 dozen	...		1s. ye dozen
Rabbits, couples 94	...	1s. ye couple				

Dayrie.

Cheeses 6 ...	...	att 2s. ye cheese	Candles 359	...	att 5d. ye po.
Butter 195½ po.	...	4d. ye po.	Eggs 1,017	...	4d. ye id.

ffish.

Olde Lynge, couples 3	...	att 10s. ye couple	Eeles 30 ...	...	...	att 2d. ye eele
Haberdine, couples 3 ...	...	4s. ye couple	Samons 4	...	...	9s. 5½d. ye samon
Carps 100 ...	...	4d. ye couple	Codde, couples 7 ...	...	...	3d. ye couple
Oysters 300 ...	...	1s. ye 100	Pickelde oysters, 3 barrels	...		2s. ye barrel

With trouts and other fresh fish.

Cellar.

Beere, hog^{shd} 54	..	att 10s. ye hog^{shd}	Aile, hogshead 1	...	att 15s. ye hog^{shd}
Cider, 17 hog^{shd}	...	17s. ye hog^{shd}			

---

[1] SOUSE [*souse*, salt, Dutch.] Pickle made of salt, or anything kept parboiled in salt pickle. *Johnson.*

"And he that can rear up a pig in his house,
Hath cheaper his bacon and sweeter his *souse*. *Tusser.*

"They were seething of puddings and *souse*."
Old Ballad.—" King and Miller of Mansfield."

Wine.

The Wines were Claret, Sacke, Muscadine, and White Wine. *Butterie* and *Bakehouse Meale* for ye Coocks. Wheate in corn or meal is charged 4s. the bushel, and Rye, for ye hall and poore, 2s. 6d. a bushel.

ffruits and Spices.

Sugar loafe po. 94	...	... att 1s. 8d. ye po.	Raysons sonne po. 80	...	... att 5d. ye po.
Sugar powder po. 40 ...	...	1s. 6d. po.	Raysons Maligo po. 40	...	... 4d. po.
Currans po. 80	...	... 5d. po.	Pruens po. 40 ...	...	... 2d. po.

There are also Rice, Pepper, Almonds, Dats, Nutmegs, Ginger, Cloves, Mace, and Sinamonde with prices attached.

Storehouse.

Here is a great variety of articles, the chief of which are Olives, Capers, Sampire, Sallet oyle, Rose Water, Oynions, Lemons (24), Oringes (9), Viniger, Beniamin, Perfuminge Cloves, Rushes, Barme, Wardens (350) price torn off, and many other small articles.

Stable.

The Hay is only priced in mass. Oats (102½ bushels) and Peas are both charged 1s 6d. buss.

Hurll.

Under this head, Wood 110 loads is charged 3s. 4d, ye load, then there is "haling and cuttinge," "Smythe making charre coales," &c.

Gifts.

	£	s.	d.		£	s.	d.
To those that brought presents...	... 8	3	0	To the Singinge boys.. ...	... 0	5	0
Musicke of Hereforde ...	... 4	5	0	To Mils ye Taborer	... 0	5	0
Welsh Harper	... 0	10	0	To ye Coocks and others in ye kitchinge ...	11	10	0
Blinde Harper	... 0	10	0	To two servinge women ...	... 0	8	0
To the Organiste	... 1	0	0	Danes helpinge ye beere ...	... 0	7	0
				Total	£27	3	0

These Stewards' Accounts were originally very precise and complete, but the *MS.* has been so much injured at the edges, by damp and time, that the short abstract here given presents, perhaps, as full an account as would be interesting.

"VERTUMNUS AND POMONA BRING THEIR STORES,
FRUITAGE AND FLOWERS OF EV'RY BLUSH AND SCENT,
EACH VARIED SEASON YIELDS."

Mason's "English Garden."

"AND TASTE REVIVED,
THE BREATH OF ORCHARD BIG WITH BENDING FRUIT
OBEDIENT TO THE BREEZE AND BEATEN RAY,
FROM THE DEEP LOADED BOUGH A MELLOW SHOWER
INCESSANT MELTS AWAY. THE JUICY PEAR
LIES, IN SOFT PROFUSION, SCATTERED ROUND.
A VARIOUS SWEETNESS SWELLS THE GENTLE RACE
BY NATURE'S ALL REFINING HAND PREPARED,
OF TEMPERED SUN AND WATER, EARTH AND AIR,
IN EVER CHANGING COMPOSITION MIXED."

Thomson.

"PROUD OF HIS WELL-SPREAD WALLS, HE VIEWS HIS TREES,
THAT MEET (NO BARREN INTERVAL BETWEEN)
WITH PLEASURE MORE THAN EVEN THEIR FRUITS AFFORD,
WHICH, SAVE HIMSELF WHO TRAINS THEM, NONE CAN FEEL,
THESE THEREFORE ARE HIS OWN PECULIAR CHARGE;
NO MEANER HAND MAY DISCIPLINE THE SHOOTS,
NONE BUT HIS STEEL APPROACH THEM."

Cowper.—The Task. Book III.

I Edward III. (1326.)
    "Lord Berkeley sent a dish of pears from
Berkeley, to Ludlow, to his mother-in-law, Lady
Mortimer, *pro novitate fructûs.*"

Fosbroke's Berkeley Family, p. 133.

THE CORDON PEAR WALL AT HOLME LACY.

PLANTED IN 1861, AND 1865, BY

SIR HENRY E. C. SCUDAMORE STANHOPE, BART.

(From a Photograph.)

THE CORDON SYSTEM OF GROWING PEARS AT HOLME LACY.

[The Pear trees at Holme Lacy, grown upon the Cordon System, have become so noted for the regularity with which they bear and for the size and flavour of their fruit, that the Pomona Committee of the Woolhope Club gladly welcome the following paper, which gives their history and the practical details of their management. The Committee are much indebted to Sir Henry E. C. Scudamore Stanhope, Bart., for writing it at their request.

The Cordon System of growing pears is as well adapted for small gardens, as for those of a larger size, and that it may be the more clearly understood, a full page wood cut of the Holme Lacy trees, copied from a photograph, is also given.]

It seems to be very generally admitted that of the fruits adapted for out-door cultivation in our variable climate, none surpasses the Pear in its merits as a dessert fruit ; whether in regard to the number and excellence of the varieties produced ; or in regard to the length of time it is in season ; for with good management, Pears may be had from the latter end of July to the end of March.

The successful experiments of Thomas Andrew Knight in the production of new varieties of fruits by means of hybridization, which were published at the beginning of the present century in the "Transactions of the London Horticultural Society," gave that great stimulus to fruit culture, which has enriched our tables so much during the past fifty years. This progress has been nowhere more marked than it has been with Pears, for many of our most esteemed varieties are of recent origin. In France, Knight's example was most perseveringly followed out, and with the best results. The climate is better adapted for Pears than our own, and the number of new kinds produced has been much greater. In addition to these new varieties, we owe also to France, the improved method of growing Pears, which is called the "Cordon System."

There is much waste by the ordinary method of growing Pears, for when grafted on the pear stock and trained on the fan, or other old forms of culture the trees require from 12 to 15 years to cover the space allotted to them, and are then, except in skilled hands, apt to bear only on the ends of the branches, to the loss, as well of time, as of wall space. To obviate these defects, Monsieur Du Breuil, the eminent professor of fruit culture in France, in the year 1852 introduced the method of training Pear trees which he calls "the Simple Cordon Oblique." In his excellent little book "*De la conduite des Arbres Fruitiers*" he gives the following directions for their growth *(pp. 90-99,)* which are here somewhat briefly translated.

"Choose Pear trees (grafted on the Quince stock,) one year old, healthy and strong, having only a single shoot. Plant them 18 inches apart, inclining at an angle of 60° each in the same direction, and prune back about a third of the length to a strong bud. During the next year, favour the free growth of the leader as much as possible by stopping all side shoots. The second pruning commences by pinching back the lateral branches through the summer to aid the formation of fruit buds, and again cutting the leader back to one third, or even back to the second year's wood if there is no bud sufficiently vigorous on the last season's growth. When the time arrives for the third pruning, the stem of the young tree has usually attained two thirds of its full growth; it should then be inclined still lower, that is, from the angle of 60° to one of 45°. If it had been inclined so much before, it would have favoured the growth of the strong buds at the bottom of the stem, to the weakening of the leading bud.

The trees are treated in this way until they reach the top of the wall. They must then still be cut back, year after year, below the top of the wall; so as to encourage a strong leading shoot, and at the same time, to cause the sap to circulate freely throughout the whole length of the stem.

On walls built from East to West, the trees may incline in either direction; but on those built from North to South, they should incline towards the South, and on walls built on a slope, they must incline towards the highest part of the wall, and since they all incline in the same way, the wall will be equally and regularly filled. The wall should not be less than from eight to twelve feet high to grow Pears on the Cordon oblique method; if the wall is higher than twelve feet, the upright, or vertical Cordon is best: but if the wall is below eight feet in height, the old forms of growth are preferable.

By this method the wall will be covered in 5 years, a gain of 10 or 12 years over all other methods of training; and the trees will begin to bear the 4th year and arrive at full bearing on the 6th year, instead of having to wait until the 15th or 20th year by other methods. Many varieties (in this way) may be planted in place of a single tree, on the ordinary method, and thus fruit may be obtained for a longer period. If a tree dies moreover, with a little care another tree may readily be put in its place, and but little wall space be lost meanwhile, and this only for a short time. [1]

My attention was drawn to this method of training Pear trees during a residence in France from 1859 to 1862. I saw it in operation in the central and southern Departments, and it succeeded so well there, that I was anxious to try it in Herefordshire, a county where the love of Pomology in its various branches has so long prevailed among all classes. In the year 1861 I sent over a number of maiden trees from France, and since the price is an important consideration where a large number of trees is required, it may be mentioned that their cost was only four pence each, exclusive of transport, which by *petite vitesse* was not heavy. At that time Pear trees grafted on the quince stock were not easily to be obtained from the English Nursery Gardens, though they can be had now almost as cheaply here as in France.

The Cordon Wall at Holme Lacy has a south aspect. It is thirty seven yards in length and eleven feet high. In this space eighty three pear trees are trained, comprising forty three varieties. One half of the wall was planted in 1861 and the trees bore fruit in 1864; and the other half was planted in March, 1865, and bore fruit in 1868. It has been essentially an experimental wall, both in regard to the system of training the trees, and to the varieties of Pears grown.

In the year 1874 the following account of the progress of the experiment was published in the "Garden." From the time these trees began fruit-bearing up to the present date they have been perfectly healthy and have yielded large crops of fruit every year, with the exception of perhaps some

---

[1] The Cordon System of training Pear trees is fully described and illustrated with M. Du Breuil's own drawings, in Mr. Robinson's interesting work on "The Parks, Promenades, and Gardens of Paris" (p.359-363) and it is now becoming very generally adopted.

two or three trees which may have missed bearing. As to the superiority of the fruit in size, appearance, and flavour, over fruit grown in the same garden, on pyramids and espaliers, there can be no question. In hot summers it has been necessary to mulch and even to water the trees, for the quince stock roots nearer to the surface and has more small fibres than the pear stock. As to pruning I certainly think that summer pinching back cannot be carried out so successfully in our climate as it can in France. My late gardener (Mr. Wells) who planted the trees and managed them for some years writes as follows: "My experience in pinching has never been what I was led to expect: never but in one solitary instance have I found the fruit bud to be the result of that practise, and that one was so far from home, so to speak, that it had to be cut off to keep the spur short. Nor was this all, for I have found that what was once a decided fruit bud, would lengthen and grow into wood before the growing season closed." The result of summer pinching on the Cordon Wall has certainly been to produce too many wood shoots, which may have been caused by a richer soil or a more moist climate.

Five summers have passed since this report was given; summers by no means favourable to the production of fine fruit; and yet the Cordon trees have far surpassed those trained as pyramids, bushes, espaliers, or on walls of various aspects : they have been more uniformly productive, and have produced fruit much finer in quality. The further experience moreover has caused the practice of summer pinching to be altogether abandoned. The fruit buds, not ripening so quickly as in a warmer climate, burst into wood shoots. The spurs are now shortened back twice during the summer and autumn, once after the Midsummer shoot and again in September. Under this treatment the trees are abundantly furnished with fruit buds.

As a rule it is not advisable to depart from the single Cordon. In the garden of a friend, triple Cordons have failed to produce as fine fruit as that grown on the single Cordon ; this is attributed to an unequal distribution of sap.

The Cordon Wall has thus far been a great success ; but with reference to the varieties of Pears that can be grown on it, there have been some failures. I was anxious to ascertain whether several sorts of Pears which are of excellent quality in France would ripen equally well in Herefordshire; but some of those planted have failed to do so. *Suzette de Bavay*, for example, though it bears excellent crops, does not mature its fruit, and with *Madame Millet, Doyenné d'Alençon*, and some others, which have also here proved inferior as dessert fruit, will have to be replaced. All the popular sorts succeed well ; and some also that seem to have failed in other places. It has often been said, as for instance by a clever writer on Pears, in the "Journal of Horticulture" last autumn, that "*Doyenné Boussoch* and *Beurré Bachelier* are very large and handsome but utterly worthless." From my Cordon Wall, both these kinds are excellent, though the *Doyenné Boussoch* has the fault of many of the early sorts of being liable to decay at the core, almost as soon as it is fit to eat.

It would be difficult to give a list of Pears best adapted for growing on the Cordon Oblique System ; for not only do tastes differ, but differences of soil, of climate, and of situation account for some Pears being highly esteemed in one county and considered worthless in another. There is abundant choice for experiments, since in the last edition of the "Fruit Manual," the best work of the kind in our language, Dr. Hogg gives no less than six hundred varieties of Pears. Without pretending then to name those best adapted to all places, a short list may be given of those which have proved to be of first rate quality, in a succession of seasons at Holme Lacy, viz.:

Beurré Giffard.	*Marie Louise.*
Beurré d'Amanlis.	*Doyenné du Comice.*
William's Bon Chrétien.	*Thompson's Pear*
Fondante d'Automne.	*Beurré de Jonghe.*
Seckle.	*Winter Nelis.*
Beurré Hardy.	*Beurré d'Aremberg.*
Beurré Superfin.	*Josephine de Malines.*
Louise Bonne of Jersey.	*Glou Morceau.*
Beurré Bosc.	*Monarch.*
Maréchal de Cour.	*Bergamotte Espéren.*

These varieties are named in their usual order of ripening but in this respect Pears are capricious and should be carefully watched, especially as so many first-rate October and November varieties may become ripe at the same time. Many other excellent sorts could be named, but this list of twenty varieties should give a succession of fruit in maturity from the first week in August to the end of March.

It may be interesting to add the weight of some of the pears grown on the Cordon Wall at Holme Lacy, and which have been produced moreover by trees bearing a full crop of fruit:

Duchesse d'hiver (a cooking pear)	...	...	18½ oz.		
Easter Beurré ...	...	16 oz.	*Glou Morceau* ...	...	14 oz.
Beurré Diel ...	...	15 oz.	*Doyenné d'Alençon*	...	13½ oz.
Van Mons Leon Leclerc	...	15 oz.	*Beurré Bosc* ...	...	12⅝ oz.
Beurré Superfin ...	...	14 oz.	*Zéphirin Grégoire* ...	...	11½ oz.
Maréchal de Cour ...	...	14 oz.	*Bergamotte Espéren*	...	11½ oz.
Triomphe de Jodoigne	...	14 oz.	*Josephine de Malines*	...	11 oz.

and fruit from *Doyenné Boussoch, Beurré Hardy, William's Bon Chrétien, Madame Millet, Figue d'Alençon, Beurré Sterckmans, Duchesse d'Angoulême* and several others have given similar weights.

The Cordon System, established by Monsieur Du Breuil, simply means, as Mr. Robinson has well defined it, "a tree confined to a single stem"; and the great advantages derived from growing Pears by this method are: (1) The wall space is quickly covered: (2) An early return of fruit is obtained from the trees: (3) The fruit is large, free from spots, and of fine flavour: (4) Many varieties may be grown on a limited wall space: (5) And therefore fruit is obtained in season for a much longer period: (6) A greater quantity of fruit is obtained in proportion to the wall space occupied: (7) The equal distribution of the sap through the tree: (8) Simplicity of management: and (9) if one tree dies from any cause, only a small amount of wall space is lost, and, with a little care, its place may be quickly supplied.

Many of the best Autumn Pears do not keep long, and it seems undesirable, in small gardens, to grow them in large quantities, for example, in the place of three trees on the pear stock trained in the ordinary horizontal way, thirty Cordon trees might readily be grown, and they would furnish an ample supply of ten different varieties to ripen in succession.

The real difficulty with the Cordon System consists in making the trees grow equally to the full height of the wall. Some varieties will not grow so strong on the quince stock as others, the *Easter Beurré* for example. They will form a leading fruit bud, instead of a leading wood bud, and hence they have to be shortened further back to a strong wood bud. Thus they will not reach the top of the wall so soon as those kinds which form proper terminal wood buds. Sometimes a tree fails altogether, and when it does so, or when an unsatisfactory variety has to be replaced, M. Du Breuil recommends the insertion of a broad piece of board edgeways on both sides of the fresh planted tree, so as to preserve its growing space from the neighbouring trees.

The objections urged against the Cordon System of growing Pears are (1) that the original cost of planting so many trees is greater; this is true, but it is well met by the fact that they come into full bearing in 5 or 6 years, instead of in 15 or 20 years under the ordinary system : (2) that with one single stem the tree would be too vigorous and make wood instead of blossom. This is not found to be the case, for the reason that if the size of the tree is thus limited so also is the root space. Moreover the natural tendency of the Pear grafted on the quince stock is to form fruit buds more freely than wood buds : (3) It is said that the trees will not live so long under this method of growth. This may be so, but the Pear is a very long lived tree, and sufficient time has not yet elapsed to prove it. The Cordon trees at Holme Lacy have been in bearing for fourteen years without shewing any signs of canker or decay. They are seemingly as healthy and vigorous now as at the time they were planted : and (4) It is said, that the Cordon System is not adapted for supplying fruit in sufficient quantities for the ordinary market. This it does not pretend to do. For this purpose, standard trees must be had recourse to ; but how many of our choicest Pears refuse to ripen their fruit as standards !

I have written these notes as an *Amateur* with much diffidence, but the experiment has answered so admirably at Holme Lacy, that I could not refuse the pressing request of the Pomona Committee, to place the results on record, and thus to shew what great advantages the Cordon Oblique method of growing Pears affords to all who have the necessary wall space at their command.

I will take this opportunity of noticing two instances of remarkable Pear trees in this neighbourhood :

Worlidge mentions "a Pear tree growing near Ross in Herefordshire, in 1675, that was as wide in circumference as three men could encompass with their extended arms, and of so large a head, that the fruit of it yielded seven hogsheads of perry in one year."

The other is a still more remarkable tree. It grows in the Vicarage garden and adjoining the Glebe land, in the parish of Holme Lacy, and has been celebrated for upwards of a century for its peculiarity of growth and its abundance of fruit. In the Parish Register of Holme Lacy, for the year 1776, is the following entry :

> "*Mem.* It is likewise inserted as a great natural curiosity, that the great Pear Tree upon the Glebe adjoining to the Vicarage house produc'd this year fourteen hogsheads of Perry, each hogshead containing one hundred gallons."

Since this time, tradition states that it has yielded fifteen, sixteen, and upon one occasion, as much as twenty hogsheads of perry in a single season.

It has been well described by Duncumb in his Agriculture (1813) in these terms :

> "An extraordinary (pear) tree growing on the glebe land of the parish of Hom-Lacey, has more than once filled fifteen hogsheads in the same year. When the branches of this tree in its original state became long and heavy, their extreme ends necessarily fell to the ground, and taking fresh root at the several points where they touched it, each branch became a new tree, and in its turn produced others in the same way. Nearly half an acre of land remains covered at the present time. Some of the branches have fallen over the hedge into an adjacent meadow, and little difficulty would be found in extending its progress " *(p. 90.)*

The tree is very curious and interesting at the present time. In the vicarage garden the group of stems is very picturesque, and several of the trees rise to a height of from thirty to forty feet. There are still nine stems on the lawn, one in the hedge, and seven in the adjoining meadow. Some creeping prostrate stems, and a few upright ones have been removed for lawn improvement. The fruit trees in the garden have a circumference of from 7 ft. to 9 ft. 6 inches in the bole, and added together, the trees give a total circumference of no less than 94 feet.

In another meadow at a little distance, is a complete grove of the same kind of Pear tree. They all seem to have sprung from one original tree creeping along the ground in the same way as those in the garden have done. Its trunk lies prostrate and perishing whenever a new tree is not springing from it. The trees grow closely together and form a cluster nearly forty-four yards in diameter. This group grows upon land which formed part of the glebe, until the year 1875 when it was exchanged for other land belonging to the domain, more convenient for the Vicarage.

H. SCUDAMORE STANHOPE.

THE

HEREFORDSHIRE POMONA,

CONTAINING

COLOURED FIGURES AND DESCRIPTIONS OF THE MOST ESTEEMED KINDS OF

APPLES AND PEARS,

CULTIVATED IN GREAT BRITAIN,

EDITED BY

ROBERT HOGG, L.L.D., F.L.S.,

Honorary Member of the Woolhope Naturalists' Field Club ; Secretary of the Royal Horticultural Society ;
Author of ' The Fruit Manual' ; ' British Pomology' ; ' The Vegetable Kingdom and its Products', &c., &c.

" Hope on.   Hope ever."

" Ζεφυρίη πνείουσα τὰ μὲν φύει ἄλλα δέ πέσσει.
ὄγχνη ἐπ' ὄγχνῃ γηράσκει, μῆλον δ' ἐπὶ μήλῳ,
αὐτὰρ ἐπὶ σταφυλῇ σταφυλή, σῦκον δ'ἐπὶ σύκῳ."
                                                Homer Odyssey vii. 119-22.

"THE BALMY SPIRIT OF THE WESTERN GALE,
ETERNAL BREATHES ON FRUITS UNTAUGHT TO FAIL ;
EACH DROPPING PEAR, A FOLLOWING PEAR SUPPLIES,
ON APPLES APPLES, FIGS ON FIGS ARISE."
                                                Homer, Odyssey vii.—Pope.

LONDON: DAVID BOGUE, 3, ST. MARTIN'S PLACE, TRAFALGAR SQUARE.
HEREFORD: JAKEMAN AND CARVER, HIGH TOWN

1880.

"Nec longum tempus et ingens
Exiit ad cælum ramis filicibus arbos,
Miraturque novas frondes, et non sua poma."

Virgil. Georg. II, 80.

" And in short space the laden boughs arise,
With happy fruit advancing to the skies.
The mother plant admires the leaves unknown
Of Alien trees, and apples not her own."

Dryden.

"Next will I cause my hopeful lad,
If a wild apple can be had,
          To crown the hearth."

Herrick to his friend John Wickes.

THE CRAB,

ITS CHARACTERISTICS AND ASSOCIATIONS.

The " Early History of the Apple and Pear" has already been given, but the Crab, also, may well deserve special attention, since it has been generally considered that most of our cultivated Apples are derivable from an improvement of the native Crab. The pips vegetate readily, but they grow slowly, and it requires from eight to ten years for them to bear. The majority of the seedlings will be like the parent plant, but a few will produce a finer or more eatable fruit, and from these again still better apples could be produced. The process is tedious and thus the young Crab trees that are produced are now used only for grafting. " Certainly," Phillips has observed in his "Companion to the Orchard,"—"it is on this stock that most of our valuable Apples have been grafted and raised by the ingenuity of the gardeners, who have by saving the seeds and studying the soil, so improved and multiplied the varieties of this most excellent fruit, that it has now become of great national importance, affording an agreeable and wholesome diet in a thousand shapes to all classes of society " (p. 33). This statement however has reference to the educated and improved Crab, and we must first contemplate it in its natural wild state.

The Crab, or Wild Apple, designated *Pyrus Malus* by Linnæus, is now to be found growing in most British woods, as far north as Morayshire, in Scotland. Modern botanists have made two varieties of the Crab, *Pyrus acerba*, having smooth leaves and small sour fruit ; and *Pyrus mitis*, having the leaves, pedicels, under side of the leaves and young branches, woolly. It is believed, however, that this last is a derivative from seeds of the cultivated Apples accidentally dispersed. *Pyrus acerba* is the indigenous Crab, and to this our remarks will apply. Blossoming as it does in company with the Cowslip and other May flowers exciting to poetry, it presents where located by the side of rural winding brooks, one of the loveliest sights of Spring :—

> "The jay's red breast
> Peeps over her nest,
> In the midst of the crab-blossoms blushing."—*Howitt.*

Indeed as far as beauty is conferred, the "wildings of nature " may have a more charming aspect than cultivated varieties, and so Dr. Withering remarks in his British Botany, that " the delicately blended pink and white of the Crab blossom renders it still more exquisitely beautiful either collectively or individually than that of the apple." Clare, the Northamptonshire poet, has thus feelingly alluded to the blossoms of the Crab :—

> "Cares have claimed me many an evil day,
>   And chilled the relish that I had for joy ;
> Yet when crab blossoms blush among the May,
>   As erst in years gone by, I scramble now,
> Up 'mid the brambles for my old esteems,
>   Filling my hands with many a blooming bough ;
> Till the heart-stirring past a present seems,
>   Save the bright sunshine of these fairy dreams."

Even the philosophical Lucretius could notice beauty in the appearance of Crab trees when in flower :—

> "Quæ pomis intersita dulcibus ornant." (L. 5.)
> "Such places which wild Apple trees throughout adorn."

thus showing that in Roman times there were parts of Italy where the Crab or wild Apple made an appreciable show.

That some kind of wild apple has existed both in Europe and Asia *ab origine* is clear from various authorities, and its fruit must have been utilised from a very early period. The prophet Joel when he declares the destruction of the fruits of the earth by a long drought, mentions "the Apple tree," among the "trees of the field" that are "withered because joy is withered away from the sons of men." (*Joel* I. 12.) So Solomon in the Canticles, derives a simile from the Apple tree "among the trees of the wood ;" though the kind of apple referred to may not be specifically the same as our Crab, yet it would seem to be one growing spontaneously in woody places. Wherever the wild apple was first brought into culture in aid of the wants of the human family, as it must have been "time out of mind," it was in all probability in Eastern Countries. However it is clear from what Pliny has written in his great work on Natural History, that in the first century of our æra, the apple tree, as then cultivated, was considered of high value among fruit trees with the Romans. "There are Apples," he says, "that have ennobled the countries from whence they came, and many apples have immortalized their first founders and inventors." He here alludes to the process of grafting, which even at that time seems to have been well understood and experimented upon with various trees. Virgil in his Georgics, inculcates grafting, and on Pliny's authority, no less than twenty-nine kinds of apples had been thus produced in Italy in his time. But it would appear that in warmer climates than England, apples are produced fit for eating without the trees on which they grow requiring any artificial process to improve them. Thus Thornton in his History of Turkey, says—"Apples are among the most common fruits of Wallachia, and one variety appears natural to the climate, as it bears without culture a fruit called *domniasca*, which is perhaps the finest in Europe both for size, colour and flavour."

But to return to the history of the truly native Crab of Britain. Although the Druids are reported to have had orchards in the vicinity of their sacred oak groves, this rests on no certain authority, for though they would be familiar with the Crab, it is very dubious whether the ancient painted Britons in pre-historic times knew anything of grafting. As, however, Crabs when roasted, "hissed in the bowl" at a much later period, so the Britons as well as their honoured instructors the Druids, may have enjoyed a luxury that was easily within their reach, though their orchards, if they had any, did not produce fruit fit for converting into cider. The "Horan Apel-tree" is sometimes mentioned in Saxon charters as a convenient conspicuous boundary, but it was probably nothing but an old Crab-tree, whose bark had been *whitened*—hoary was a favourite term with the Saxons for any white stock or stone—by being encrusted with the continuous growth of lichens, which is often the case with old trees in the present day. William of Malmesbury, writing of Saxon times, says, that once in the year 973, King Edgar lay down to sleep "under a *wild* Apple tree." The expression of wild, which evidently means the Crab, also implies that there were cultivated Apples in the country. But after the Romans had accomplished the conquest of the southern parts of Britain, and their military and civil officers formed villas in the neighbourhood of their cities and stations, they then no doubt introduced into their gardens all they had been accustomed to see in Italy, and the Apple among the rest. They would also probably direct the grafting of the native Crab, so extensively practised in Italy at this period, following the dictum of Virgil—

> " Poma quoque ut primum truncos sensere valentes,
> Et vires habuere suas ad sidera raptim
> Vi propria nituntur."
> > *Virg. Georg. II., 426.*
>
> " Apple trees grafted with the tender shoot,
> Quickly gain strength and bear superior fruit."

That the Romans during their occupation of Britain did exercise a powerful influence upon the horticulture of the country they had colonized, is clear from the following statement of a learned writer :—"When the Saxons got possession of Britain, they found it not such as Julius Cæsar describes it, but cultivated and improved by all the Romans knew of agriculture and gardening ;"* and he goes on to mention a number of plants that were unknown in Britain previous to the Roman invasion, and were afterwards commonly cultivated both for ornament and use.

Although our native Crab may be praised for the beauty of its blossoms when called forth by vernal influence ; and its fruit in autumn,

> "Sun-reddened with a tempting cheek,"

(as Clare the rural poet writes), may allure rustics and farmers' boys, it is but an ugly-looking tree at the best, and often gets terribly deformed with knobs and excrescences. Thus its deformities with attendant hardness, have become proverbial, not only colloquially as designating an ill-natured hard-hearted "crabbed" fellow, but various poets, and especially Shakspeare, have brandished the crab-stick unmercifully, as the following quotations will prove :—

> " Fetch me a dozen crab-tree staves, and strong ones, these
> are but switches to 'em—I'll scratch your heads."
> > *(Henry VIII., Sc. 3.)*
>
> " We have some old crab trees here at home, that will not
> Be grafted to your relish."
> > *(Coriolanus II., Sc. 1.)*

* Chronicles and Memoirs of Great Britain during the Middle Ages. Published by authority of Her Majesty's Treasury, under the direction of the Master of the Rolls, 1864.

"Thy mother took into her blameful bed
Some stern untutor'd churl, and noble stock
Was graft with *crab-tree slip ;* whose fruit thou art,
And never of the Neville's noble race."

> *(2 Henry VI., III., Sc. 2.)*

" That was when
Three *crabbed* months had sour'd themselves to death
Ere I could make thee open thy white hand."

> *(Winter's Tale, I., Sc. 2.)*

In " Divers Crab-tree Lectures," &c. (1639), there is a cut representing a woman beating her husband with a ladle, inscribed " Skinnington and her Husband ;" and the " irrand Scole" is made to say among other exciting words in rhyme—

" But all shall not serve thee,
For have at thy pate,
My ladle of the Crabb-tree
Shall teach thee to cogge and prate."

> *(Brand—British Antiquities by Hazlitt, vol. ii., p. 129—30.)*

Southey has in one of his Poems introduced the Crab-stick as an instrument of domestic discipline, which ladies in the present day would not much approve of, though unless merely uplifted as a show of what *might* be inflicted, the law has now given a remedy to a castigated wife, if she carries her case before a justice of the peace :—

" Richard Penlake a scolding would take
'Till patience avail'd no longer ;
Then Richard Penlake his Crab stick would take,
And show her that he was the stronger !"

> *("St. Michael's Chair.")*

This Crab-stick discipline in households " of the baser sort," would appear to be an old story or an old practice, going back most likely into savage times when the weaker sex found severe masters. Beaumont and Fletcher make one of their characters say—

" Get you to bed, drab,
Or I'll so *crab* your shoulders."

> *(" Monsieur Thomas.")*

Another simile has been adduced from the *sourness* of the fruit of the native Crab, which has passed into a proverb. Excellent Vinegar is made from the liquor of Crabs squeezed to a pulp, which is called Verjuice, and was formerly used to cure sprains and scalds, and was often kept by good housewives in the country for that purpose. It was also used to give an agreeable acidity to the rustic delicacy called syllabub, only now to be met with in farm houses where old-fashioned observances are kept up. Good Isaac Walton in his "Complete Angler," thus offers to treat a friend. He says—" When next you come this way, if you will but speak the word, I will make you a good syllabub of *new Verjuice*, and then you may sit down on a hay cock and eat it." No doubt palatable enough on a hot day.

Pliny in his great work on Natural History, says but little in favour of the Crab, he remarks— " As to Wildings and Crabs, little they be all the sort of them, in comparison. Their taste is well enough liked, and they carie with them a quick and sharpe smell: how be it this gift they have for their harsh sournesse, that they have many a foule word and shrewd curse given them, and that they are able to dull the edge of any knife that shall cut them." *(Trans. Holland, p. 438.)*

In the following passage sourness of temper is suggested as the result of looking at a person stigmatised as a Crab :—

> " *Petruchio.*—Nay, come, Kate, come, you must not look so sour.
> *Katherine.*—It is my fashion when I see a Crab.
> *Petruchio.*—Why, here's no Crab, and therefore look not sour.
> *Katherine.*—There is, there is."
>
> *(Taming of the Shrew, II, Sc. I.)*

A crabbed unfeeling temper may appear in animals as well as in the human race—

> " I think Crab, my dog, be the *sourest-natured* dog that lives :—
> My mother weeping, my father wailing, my sister crying, our maid howling, our cat wringing her hands, and all our house in a great perplexity, yet did not this cruel-hearted cur shed one tear,"
>
> *(Two Gentlemen of Verona, II, Sc. 3.)*

> ————————— " O she is
> Ten times more gentle than her father's crabbed ;
> And he's composed of harshness."
>
> *(Tempest, III, Sc. I.)*

> "She will taste as like this, as a crab does to a crab."
> "Thy other daughter will use thee kindly ; for though she's as like this as a crab is like an apple, yet I can tell what I can tell."
>
> *(King Lear, I, Sc. 5.)*

> " Something too crabbed, that way, friar."
> .
> *(Measure for Measure, III, Sc. 2.)*

So Milton has used the simile in reference to what may be thought uninteresting and distasteful to the mind :—

> " How charming is divine philosophy,
> Not harsh and *crabbed*, as dull fools suppose."
>
> *(Milton—Comus, 476.)*

Southey in depicting Winter, takes a simile from the stubborn harshness appertaining to Crab Trees :—

> " A wrinkled *crabbed* man they picture thee
> Old winter, with a ragged beard as gray
> As the long moss upon the Apple tree."

But what the poet calls a "long moss," is in reality a Lichen called *Usnea barbata*, and if any old Apple tree in an orchard is neglected for a few years, it is sure to get its branches covered with the gray beardlike *Usnea*.

There are several references to crabbedness—meaning a surly temper—in the Plays of Beaumont and Fletcher:

> "It does me good to think how I shall conjure
> And crucify his crabbedness."
>
> *(The Pilgrim.)*
>
> "Hast thou forgot the ballad, *Crabbed Age*?"
>
> *(The Woman's Prize.)*

Age is often assumed though not always justly, to be "crabbed" with a verjuice face, and thus Beaumont and Fletcher in their "Four Plays," mention "Old Crabbed Saturn!" and so rare Ben Jonson in his "Silent Woman," stigmatises "A crabbed coxcomb." But though this notion of moroseness taken from the sour taste of the expressed juice of the Crab is often used both in writing and colloquially, yet the "crab sauce" of a knotted stick as an "*argumentum baculinum*," is the most favourite idea, as thus introduced by Beaumont and Fletcher in their Dramas :—

> "*Petronius.*—Give her a crab-tree cudgel !
> *Petruchio.*—So I will ;
> And after it a flock bed for her bones."
>
> *(The Woman's Prize.)*
>
> "*Malicorn.*—Ay there's the point ; we would expect good eating ;
> *La Poop.*—I know we would, but we may find good beating.
> *Laverdine.*—You say true, gentlemen, and by my soul,
> Tho' I love meat as well as any man,
> Such *Crab-sauce* to my meat will turn my palate.
>   *   *   *   *   *   *   *
> ——————— If cudgell'd,
> I hope I shall outlive it : I am sure
> 'Tis not the hundreth time I have been serv'd so."
>
> *(The Honest Man's Fortune.)*

Crab sticks made from the often curiously knotted branches of the old neglected Crab trees that in obscure places become ugly and distorted, were formerly much in demand, but are now out of fashion ; and Crab trees are at present only valued as stocks for grafting various sorts of Apples upon, being well adapted for that purpose.

> "Art bids th' illnatured Crab produce
> The gentle Apple's winy juice."
>
> *(Cowly.)*

and this is very often done when a farmer finds a strong Crab-tree rising up in one of his hedges, and so converts it to a useful tree bearing good fruit :—

> "As fruits ungrateful to the planter's care,
> On savage stocks inserted learn to bear."
>
> *(Pope.)*

The fruit of the Crab, small as it is, taken when ripe, is yet utilised in secluded rural districts, on account of the quantity of Verjuice it yields, which is converted into vinegar :—

> " Oft from the forest, wildings he did bring,
> Whose sides empurpled were with smiling red."
> *(Spenser.)*

In the present day, however, the large manufactories of malt vinegar advertised by show bills in every grocer's window, have supplanted Crab vinegar, now not easily obtainable except in old-fashioned farm houses. Perhaps, however, even Crabs might be made to produce a palatable beverage adapted to rustic palates, not much caring for the sharpness of drink, and ready to swallow any liquid that is not too watery. "Crabbs make a mordicant Cyder, which doth well please our Day Labourers" says Dr. Beale in his "Herefordshire Orchards," and they are not unfrequently used with pears to give piquancy of flavour to the Perry. The following anecdote was communicated to me by a friend familiar with the habits of country people :—

A gentleman farmer who resided in a part of Worcestershire, in the vicinity of the river Teme, where Crab-trees were rather plentiful, always caused a hogshead of Verjuice to be made every year, and thus had a good supply of capital home-made vinegar. This was kept in an out-house of the farm, and used for various purposes as occasion required. When he died, my friend who was an executor, had to look over the effects on the farm, and arrange for their sale by auction. He knew there was this hogshead of Crab vinegar in the barn, but he reserved it as he fondly thought for himself, kept it out of the catalogue of sale, and left it to the last, not imagining such a sour liquid would be furtively tapped. But when he came to draw off the vinegar, the hogshead was found to be empty, the labourers about the place while it was untenanted, having made free with the contents, and drained them to their entire satisfaction. If they had found " a body " in the drink, as is often said of sharp cider, they had carried it off to my friend's great chagrin.

Crabs in the good old times were wont to be used in a more agreeable way, making a rural dish when scalded or roasted, and floated in semi-solid cream, the latter at any rate being good, however the Crabs might taste, which would of course require some sugar to soften their acidity. Ben Jonson in his play of " The Sad Shepherd," mentions " Crabs and Cream," as if it was a dainty to be thought well of, and so one of his characters presents Marian

> " With a choice dish of Wildings here, to scald
> And mingle with your cream."

Wilding is a term often given by the poets to the Crab, or wild Apple, and thus Clare, a rural poet makes rustics—

> " Hunting the hedges in their reveries,
> For wilding fruit that shines upon the trees."

Another "choice dish" in which Crabs made a show, consisted of spiced ale into which roasted Crabs were thrown, and of course as Shakspeare intimates, made a hissing noise, exhilarating to the ears of the roisterers, who doubtless liked the ale that was ladled out to them better than the garnishing Crabs. This was a jovial winter dish, as—

> " When roasted Crabs hiss in the bowl,
> Then nightly sings the staring owl."
> *(Love's Labour Lost, V. Song.)*

Puck, in the " Midsummer-Night's Dream," is made to boast, that—

> " Sometimes lurk I in a gossip's bowl
> In very likeness of a roasted Crab ;
> And, when she drinks, against her lips I bob,
> And on her wither'd dew-lap pour the ale."

So from this it would appear that the fun was to induce persons to stoop to drink out of the flowing bowl, and then push their heads into it, and no doubt the immortal bard had himself often seen this done ! Mr. Tom Burgess in his " Legends and Traditions of Warwickshire," says that " Warwick-shire school boys still know where Crabs grow, and can dig pig-nuts, and in remote districts, particularly to the east of the county you may still find the ' roasted Crab' lurking in the gossip's bowl."

Southey mentions how school-boys in his youthful days were accustomed to gather Crabs, and alludes to

> " The Crab tree where we hid the secret hoard,
> With roasted Crabs to deck the wintry board."
> *(The Retrospect.)*

There is not very much of Folk-lore attached to the Crab, since it appertains more to the Apple, but a few things belong especially to the Crab. Thus Brand mentions the following lines as in common use in Suffolk at Michaelmas :—

> " At Michaelmas time or a little before
> Half an apple goes to the core ;
> At Christmas time, or a little after,
> A Crab in the hedge, and thanks to the grafter."

At the Michaelmas season, village maids in the West of England, go up and down the hedges gathering Crab apples, which they carry home, putting them in a loft, and forming with them the initials of their supposed suitors' names. The initials which are found on examination to be most perfect on Old Michaelmas Day are considered to represent the strongest attachments, and the best for the choice of husband.

A very curious and extraordinary custom was formerly practised at St. Kenelm's, near Hales-Owen, in the northern part of Worcestershire, called " Crabbing the Parson," but its significancy, and how it arose, is entirely unknown. It is thus alluded to in the *Gentleman's Magazine,* for September, 1797 :—

" At the wake here called St. Kenelm's Wake, alias Crab Wake, the inhabitants have a singular custom of pelting each other with Crabs; and even the Clergyman seldom escapes, as he goes and comes from the Chapel."

Brand in his "Popular Antiquities" goes into the particulars of this "Crabbing" affair more fully, and apparently on the authority of an eye-witness. The following is the account given by him :—

"On the feast of St. Kenelm (July 17th), a fair or wake was wont to be held, and the Sunday after the fair it was the annual practice to *Crab the Parson*. The last person but one who was subject to this process, was a somewhat eccentric gentleman named Lee. He had been chaplain to a man-of-war, and was a jovial old fellow in his way, who could enter into the spirit of the thing. My informant well recollects the worthy divine after partaking of dinner at the solitary house near the church, quietly quitting the table when the time for performing the service drew nigh, and reconnoitring the angles of the building, and each buttress and 'coign of vantage' behind which it was reasonable to suppose the enemy would be posted, and watching for a favourable opportunity, he would start forth at a fair walking pace (for he scorned to run) to reach the chapel. Around him thick and fast fell from ready hands a shower of Crabs, not a few telling with fearful impetus onhis burly person amidst the intense merriment of the rustic assailants. But the distance is small; he safely reaches the old Saxon porch, and the storm is over." *(Vol. I., p. 32-44.)* A later incumbent, the Rev. John Todd, frequently ran this gauntlet, and on one occasion there were two sacks of Crabs, each containing at least three bushels, emptied in the church field in readiness for throwing at the poor parson. All things in time get abused, and it becomes necessary to alter them, or entirely to abolish what has become a nuisance. So in the present case, rude fellows began to use missiles of a more unpleasant nature than Crabs, and the practice was interdicted. It has been abandoned for some years, and not a solitary Crab is now thrown.

A few years since there existed by the road-side a mile from Bidford, in Warwickshire, five miles from Stratford-on-Avon, and some few hundred paces from the river, an old Crab-tree, whose gnarled trunk and giant size bespoke the growth of centuries, and for many years this tree had been associated with the name of Shakspeare, and it was known in the vicinity as "Shakspeare's Crab-tree." The story connected with this Crab-tree was, that the poet went with some boon companions on a particular occasion, to a drinking bout at Bidford, a customary thing in those rude days, and returning late and rather overcome with the liquor imbibed, he and his companions lay down under this tree and passed the night there. The tale though a mere tradition, has always been believed, and the Crab-tree thus became celebrated, and was always visited accordingly by admirers of the immortal bard. A quarto volume has been published illustrating this legend, and describing the villages mentioned by Shakspeare in some off-hand rhymes he is said to have uttered on awaking in the morning. The author of this book* gives an engraving of the tree as it appeared in 1823, and says "My earliest recollection of the Crab-tree was about the year 1814, at which time it was frequently called "Shakspeare's Canopy"; it was regarded with almost superstitious veneration by the peasantry of the neighbourhood, and was then rich in foliage and fruit. The autumnal winds of 1816 blew off several of its stalwart boughs, and year after year it suffered equally from the effects of time and the depredations of unthinking visitors." From these untoward circumstances it became

*Shakspeare's Crab-tree with its Legend, &c., by Charles Frederick Green, 4to.

at last a leafless rotten trunk, and its remains were carefully removed to Bidford Grange at the end of 1824. The engraving of the Crab-tree in the work alluded to, shows it to have had an hollow trunk of great size, and considering that it was probably a large and old tree when Shakspeare and his companions slept beneath its shade, it had very likely existed full 700 years, for the Crab is a long-enduring tree.

Except for making a very sharp vinegar, it does not appear that crab apples have been much utilized in modern days, but a friend familiar with rustic appliances, tells me that in winter time before turnips and mangel-wartzel were much cultivated, cattle and sheep were frequently in hard frosts reduced to the miserable fare of gorse and ivy-leaves, when they got into a poor, lean and emaciated state. As a remedy for this, old Tusser in his quaint work on Husbandry prescribed the following recipe, in which " Verjuice " is the chief ingredient :—

> " From Christmas till May be well entered in,
> Some cattle are faint and look poorly and thin ;
> And chiefly when prime grass at first doth appear
> Then most is the danger of all the whole year.
> Take *Verjuice* and heat it, a pint for a cow,
> Buy salt, a handful, to rub tongue ye wot how,
> That done with the salt, let her drink off the rest.
> This many times raiseth the feeble up best."

The same quaint writer under " October's Husbandry," says :—

> " Besure of Virgis, (a gallon at least),
> So good for the kitchen, so needful for beast :
> It helpeth the cattell so feeble and faint
> If timely such cattle with it thou acquaint."

But Verjuice is not so much needed now, and except perhaps as an application to sprains and bruises is but little used even in the country, though for pickling it is better than any other vinegar.

Pippins in a rotten state, and no doubt Crabs would have done equally well, are mentioned in one of Beaumont and Fletcher's Plays as a cure for bruised, or black eyes :

> " Bring in rotten Pippins
> To cure blue eyes, and swear they came from China."
> *(The Honest Man's Fortune.)*

With regard to the old approved custom of " wassailing " the orchard trees, under the idea that they would bear better for the operation, that ancient almost worn-out ceremony appertains more to the Folk-Lore of the Apple ; but Johns in his " Forest Trees of Britain," mentions one curious Wassail observance in which Crabs are introduced, and it therefore deserves insertion, as showing that Crabs are still roasted, though for a less enjoyable purpose than hissing invitingly in a bowl of spiced ale. Johns says—after mentioning that in passing through Devonshire on the night preceding Twelfth Day, he had been alarmed by the report of fire-arms, and was told that it pro- ceeded from farm men who were firing in an orchard at the Apple-trees in order that they might bear a good crop the next season—he goes on to say, " In certain parts of this country superstitious observances yet linger, such as drinking health to the trees on Christmas and Epiphany eves, saluting them by throwing roasted crabs or toast from the Wassail bowl to their roots, dancing and singing round them, lighting fires, &c. All these ceremonies are supposed to render the trees productive for the coming season." (*Forest Trees of Britain, Vol. I., p. 303).*

It is generally admitted that all the excellent cultivated varieties of the Apple are derived from the Crab, a belief that is strongly confirmed by the great tendency shewn by the seedlings from Apples to degenerate back to their origin. It is curious, however, to notice that with the Pear, this connection with its wild representative is by no means so clear. Decaisne believes that they too are all produced by cultivation from the small wild forms of *Pyrus communis*, of which one variety, the *Pyrus cordata* of Desvaux has been found growing near Plymouth. An excellent paper on these "Small Fruited Pears," English and Foreign, has been published by Dr. Maxwell T. Masters in the Journal of Botany (1876). It affords no proof however of the derivation of the cultivated varieties of Pears from them ; and the half-wild Pears, figured by Mr. Wilson Saunders in the Journal of the Horticultural Society (1872), do not show any resemblance to them. The wild Pear-tree in England is much more sparingly distributed in woods and coppices than is the Crab tree, and it is very rare to find it in bearing. Herefordshire and the adjoining Counties have been closely examined by Botanists and only one instance has been recorded of its occurrence. Mr. J. Tom Burgess has also stated that the only wild Pear tree to be found in Warwickshire grows on the Fosse-way (a Roman road) at Chesterton Camp. When fruit is found on these trees, like that represented by Mr. Wilson Saunders, it rather resembles a renegade from cultivation, than an improved form of the Small Fruited Pear.

Crab-trees apparently of great age, may often be seen on old forest ground, but in hedges they are now seldom observable.

EDWIN LEES, F.L.S., F.G.S.,
Vice-president of the Malvern and Worcestershire Naturalists' Clubs.

"STRATA JACENT PASSIM SUA QUÆQUE SUB ARBORE POMA."
Virgil, Ecl. VII., 54.

With falling fruits and berries paint the ground.
Dryden.

"THE FRAGRANT STORES, THE WIDE PROJECTED HEAPS
OF APPLES, WHICH THE LUSTY HANDED YEAR,
INNUMEROUS, O'ER THE BLUSHING ORCHARD SHAKES;
A VARIOUS SPIRIT, FRESH, DELICIOUS, KEEN,
DWELLS IN THEIR GELID PORES; AND, ACTIVE, POINTS
THE PIERCING CIDER FOR THE THIRSTY TONGUE."
Thomson's "Seasons."

"WE HAVE ALSO LARGE AND VARIOUS ORCHARDS AND
GARDENS, WHEREIN WE DO NOT SO MUCH RESPECT BEAUTY, AS
VARIETY OF GROUND AND SOIL, PROPER FOR DIVERS TREES AND
HERBS: AND SOME VERY SPACIOUS, WHERE TREES AND BERRIES ARE
SET, WHEREOF WE MAKE DIVERS KINDS OF DRINKS, BESIDES THE
VINEYARDS. IN THESE WE PRACTISE LIKEWISE ALL CONCLUSIONS
OF GRAFTING AND INOCULATING, AS WELL OF WILD TREES AS OF
FRUIT TREES. AND WE MAKE, BY ART, IN THE SAME ORCHARDS AND
GARDENS, TREES AND FLOWERS TO COME EARLIER OR LATER THAN
THEIR SEASONS; AND TO COME UP AND BEAR MORE SPEEDILY THAN
BY THEIR NATURAL COURSE THEY DO."
Bacon, "New Atlantis."

THE ORCHARD AND ITS PRODUCTS.
CIDER AND PERRY.

"Nec vero terræ ferre omnia possunt." .
(Virgil. Geor. II. 109.)
Not every plant in every soil will grow.
(Dryden.)

"Would'st thou thy Vats with gen'rous juice should froth?
Respect thy Orchats ; think not that the Trees
Spontaneous will produce a wholesome Draught
Let Art correct thy Breed." *(Philips' Cyder.)*

"We had also a drink, wholesome and good wine of the grape, a
kind of Cider made of a fruit of that country, a wonderful pleasing
and refreshing drink." *(Bacon.)*

The variable and temperate climates of Northern Europe are better suited to the growth of
the Apple and the Pear-tree than to that of the heat-loving Vine : and thus in olden times, when
communication was difficult or almost impossible, and each locality was very much dependent upon
its own productions, Cider and Perry became the natural drink of the inhabitants. It is not however
in every soil and situation that the juice of the Apple and Pear are sufficiently rich to produce
fermented liquor of high flavour and quality ; and it is curious to observe how limited are the
districts to which the experience of centuries has restricted the growth of Cider and Perry Orchards.
In England it is only the Western Counties which are noted for their Orchards. The West
Midland district comprising Herefordshire, Worcestershire, Gloucestershire and some parts of
Monmouthshire ; and the South-western district comprising the Counties of Devonshire, Somersetshire and part of Dorsetshire. Cornwall, also possesses many Orchards ; and the fame of Kent is
widely spread for its extensive production of dessert and table fruit. In Ireland some fair Cider is
made in the Counties of Waterford and Cork, but not to any considerable extent.

In Normandy Cider Orchards may be traced back to the 11th Century. They were much
more extensively planted between the 13th and 16th Centuries, and at the present time considerable

quantities of Cider are still produced there. Pear Orchards seem never to have been much planted in Normandy, and Perry to have been lightly esteemed. In Germany on the contrary Perry is more highly valued than Cider, and it is made largely for distillation. Cider has been known in Spain from a very early period. A graphic description is given of the Cider of Biscay by Nasagerus in the Journal of his Embassy from the Republic of Venice to the Emperor Charles V., in the early part of the 16th Century. It now forms the ordinary drink of the inhabitants of the northern provinces of Spain and Portugal. In Jersey much Cider is made which has a high repute for its strength. In many parts of the United States of America Cider is the common drink of the country; but the manufacture of Perry is chiefly confined to the Eastern States, where it is produced in considerable abundance.

It was not until the end of the 17th century that the English Orchards began to be much planted. The Civil War with all its troubles had passed by : Continental wars prevailed for the most part : and as Foreign Wines ceased to be imported, it became an object of national importance —a patriotic duty—to encourage the home production of Cider and Perry in every possible way. Poets and Writers extolled their praise : Esquires and Yeomen vied with each other in their efforts to meet the national want : and the great care and attention resulting from all this enthusiasm culminated in a success so remarkable as to outstrip all former efforts, and, as we read the accounts, to make us lament the more the neglect of later years.

Cider and Perry were then made in large quantities of an uniform superior quality, and met with a ready and highly remunerative sale. They formed the household family drink—varied on festive occasions with home made wines, in the excellence of which all good housewives prided themselves. The farm labourers or hinds, who were at that time usually boarded in the house, had to be content with "ciderkin" or "purr," a weaker cider made by the addition of water to the must as it was passed again through the mill. This was allowed to the men in almost unlimited quantities during hay time and harvest and formed a wholesome and harmless drink.

This was the golden age for Orchard culture and for Orchard produce. Cider was never so highly esteemed. Philips, the cider poet calls it :

> " Nectar ! on which always waits
> Laughter and Sport, and care beguiling Wit,
> And Friendship, chief Delight of Human Life.
> What should we wish for more ? or why in quest
> Of Foreign Vintage, insincere and mixt
> Traverse th' extreamest World ? why tempt the Rage
> Of the rough Ocean ! when our native Globe
> Imparts from bounteous Womb, annual Recruits
> Of wine delectable, that far surmounts
> *Gallic* or *Latin* Grapes, or those that see
> The setting Sun near Calpe's tow'ring Height.
> Nor let the *Rhodian*, nor the *Lesbian* Vines
> Vaunt their rich Must, not let *Tokay* contend
> For Sov'ranty ; *Phanæus* self must bow
> To th' *Ariconian* Vales."—*(Philips' " Cyder ").*

This great prosperity of the Orchards was not destined to continue for any lengthened period.

Agriculture was soon called upon with greater urgency for the more essential articles of food, and it was found more profitable to produce corn and cattle, and thus the chief attention of the farmer was drawn from his fruit trees and was given to these objects. Orchards are uncertain in their yield, the fruit requires much care and attention, and with all this, a good season is as necessary for superior Cider and Perry as it is for fine Wines : whereas the grain crops are much more to be depended upon, and the area of their production is practically without limit.

The farmers grew rich, their farms kept increasing in size, and less and less attention was given to the Orchards, until at last they began to be looked upon sometimes as a nuisance. This neglect, extended through a series of years, became disastrous : failing trees had their places supplied by worthless varieties : little care was given to the management of the fruit, or the making of the liquor, beyond the two or three hogsheads required for the household use. Thus year by year, enormous quantities of Cider and Perry, of a very indifferent quality were produced, and in consequence of this deterioration they could not be sold at a price worth consideration. They were therefore given the more freely to the labourers on the farm, thus inducing habits of indolence and intemperance, and of course lessening their wages.

The quantity produced was however far too great to be consumed locally, and hence arose the need of the Cider-Merchants, "Cidermen," or "buyers of sale liquors" as they were called at the end of last century, who bought up everything by wholesale and almost at their own prices. There can be no question but that, with some honourable exceptions, these middlemen have done more to damage the reputation of Cider and Perry than all other causes put together. In ordinary seasons many thousands of hogsheads passed through their hands and were submitted to various processes calculated to destroy rather than to regulate fermentation. They were next fined, flavoured and fortified to suit, in their estimation, the public taste. They were then sent to London and Bristol, in those days the two great centres of trade : the best in bottles to *mis*represent pure whole-some cider in the home market ; but the greater part to find its way, it is said, to the Continent and return again to this country in the shape of cheap Hamburgh Ports and Sherries ; or more probably to be manipulated at home for these purposes. Not a little of this nefarious traffic it is to be feared goes on at the present day.

There were other causes also which tended from an early period to limit the production of Cider and Perry. Taxation was very soon imposed, sometimes on the Orchards, but generally on its produce. It was often most oppressive and caused many Orchards, not protected by the landlord's agreement or lease, to be uprooted. The obnoxious visits of the Supervisor continued to the commencement of the present century but have now happily ceased.

Foreign Wines soon again began to be introduced in the intervals of war, and their importation has continued to this time in ever increasing quantities with the improved facilities of transport and the diminution of duty. These, with malt liquors have at all times been formidable rivals for public appreciation, and it is a standing proof of the excellence of Cider and Perry, that they should have been able to hold their own, as well as they have done, in spite of so much general deterioration, and in the face of such general competition.

The same falling off in the quality of Cider of late years has been observed in other countries.

In France it has been strongly commented on in the Report of the Congress appointed by the French Government to consider this subject.   1. It says, "The Cider of which the old authors wrote in such glowing terms are scarcely to be met with now.   Such for example as the *Ecarlatin* prepared from the *Ecarlate* (Scarlet) apple, which yields an excellent Cider, red as wine, sweet, piquant and as aromatic, as if sugar and cinnamon had been used ; or such as the *Muscadet*, which recalls the colour, scent, and taste of the *Muscadelle* wine, of which the old French soldier song says :

> "Il vaut mieux, près beau feu, boire la *Muscadelle*
> Qu' aller sur un rampart faire la sentinelle."

Or, lastly the Cider furnished by Apples, called *d'Espice*, which is as superior to ordinary cider, as the Vin d'Orleans is to Vin Ordinaire.   The late King Francis the Great, in 1532, passing through the district, gave orders that some barrels of it should be carried in his train, and drank of it himself as long as it lasted."   (2) A similar compliment was afterwards paid by King Charles I to the excellence of Herefordshire Cider in 1645.   *(See page 78)*.

The same neglect was observed in America some half century ago, when Thacker called attention to their Orchards. (3). His warning would seem to have been effective, since of late years a marked improvement has shewn itself in all kinds of American Apples and Pears, whether for dessert, for culinary purposes or, for the production of Cider and Perry.   " American farmers are now beginning," says Mr. Downing, " to recognise the fact that no farm is complete without a well selected and well cultivated Orchard." (4).

The wonders effected in commerce by the great discoveries of the present century have completely thrust aside the results of all former experience.   The power of the steam engine by land and by sea, enables space now to be overcome by rapidity of movement, and lessens expenditure by cheapness of conveyance, and thus wider markets are offered for all articles of trade.   Nor have these advantages by any means reached their limit.   Every year sees some new economy effected, some fresh article of commerce introduced into new districts to compete with those already in the field.   Competition thus becomes world-wide and according to the inevitable laws of trade, the best and the cheapest must prevail in the end.   The benefit to humanity at large is unquestionable, but to individuals and localities the result is often ruinous.   Agriculture is now tried to its uttermost to contend with these great changes, and the struggle still goes on with increasing severity in all the articles of its production.   The result cannot be otherwise than to compel every locality and every district to produce the articles, for which it is specially adapted, in the best possible form, or in other words by the highest cultivation.   If free-trade in Corn, and the introduction of live and

(1). "*Le Cidre*" by M. M. L. de Boutteville and A. Hauchecome.   Published at Rouen, 1875, giving the results of the work of the Congress appointed by the French Government to study the Cider Fruits during the years 1864 to 1872— a scientific and comprehensive work of the highest value.

(2). " *Traité du Vin et du Cidre*" par Julien de Paulmier.   Caen, 1589.

(3). *The American Orchardist.*   By James Thacker, M.D., Boston, 1822.

(4). " *American Fruits for Farm and Garden.*"   1871.

dead meat, restrict the profits of the farmers, happy are they, who as in the fruit districts of England, have their orchards to help them.

Two hundred years ago, it was the necessities of isolation that caused the Orchards to be looked to as a chief source of profit : in these times it is a world-wide competition that makes the same demand : and thus it has come to pass by a curious revolution in the cycle of commerce, that the careful cultivation of English Orchards has again become a necessity, and every effort must be made to improve their condition, and to make them, as they can be made, one of the main sources of the profits of the farm.

The fruit districts of England in all ordinary seasons should afford the chief supply to the English markets ; but they do not do so.   American and Continental Apples and Pears are brought year by year in larger quantities to supply our great centres of population.   They are always noted for those first two marketable qualities " size " and " beauty of colour," and are often also excellent in flavour and quality.   In bad seasons, as in 1879, American apples are brought moreover into our own apple districts, and this competition will for the future have always to be considered, and it must, and it may be met successfully by care and attention.   Of late years table and kitchen fruit, "pot fruit" as the local name has it, have been much more extensively grown in our Orchards, and they must still be grown in increasing quantities and in improved quality.   This change however will not prove the universal panacea for agricultural prosperity that has so recently been thought.

The English Orchards however afford a better resource.   The products in which they are unrivalled,and for which they need not fear competition, are Cider and Perry of superior quality.   Here is the speciality that requires the immediate attention of our fruit growers, and it will well repay all the care they can bestow upon it.   For many years past the Cider and Perry of first quality has been chiefly made by the small holders of land.   They have looked to their Orchards for their rent and livelihood : and by unremitting attention to their trees, have received a liberal and just reward. The holders of the larger farms, and larger Orchards, must follow their example.   It does not answer to produce a drink of inferior quality, when it is possible to produce better : and it may assuredly be said now, as truly as it ever could have been said, that so long as the quality is superior, however large the quantity may be, a ready market will always be found for it at highly remunerative prices.

The writers of the 17th and 18th centuries produced many excellent practical works on Orchard culture and the manufacture of Cider and Perry.   They are for the most part the result of personal experience, and vary greatly in their views : indeed they also show signs of isolated culture.   The Orchardist whose land is variable, and but little of it good, thinks "soil" is the one thing essential : he whose land has been undrained and whose trees grow unkindly, with rugged mosscovered branches, lays great stress on "drainage" : he whose Orchards are on low ground exposed to night fogs, and whose hopes have been again and again cast down by spring frosts destroying the fertility of the bloom, dwells fondly on the importance of a "sunny, airy, upland situation " : he whose land is everywhere good and well adapted for orcharding, throws all his energy into the absolute need of selecting "the best varieties of fruit " for cultivation : whilst, lastly, he who happily possesses all the foregoing advantages, considers that "the management of the fruit and its proper fermentation," are the requisites supremely essential for the production of Cider and Perry of the highest quality and excellence. All are right from their experience, though all are wrong in the restriction of their views. The careful personal attention of the cultivator must be given to each and every one of these points with patience and perseverance when it will only remain for favourable seasons to insure success.

The present condition of the English Orchards is far from satisfactory. They shew sadly the result of long continued neglect. It is the object of the present paper to give a brief practical review of the requirements for their proper cultivation and management.

I. THE ORCHARD.

SOIL.—The Apple and the Pear tree are very hardy. They will grow and flourish in almost every variety of soil, producing in abundance their most useful fruits. The Apple tree certainly prefers a Sandstone wherever it is found, as the Pear tree rejoices in calcareous soil. It has been universally observed however that the same trees will produce fruit varying much in size and quality on different soils. "Every variety of the Apple" says Thomas Andrew Knight "is more or less affected by the nature of the soil it grows upon. On some soils the fruit attains a large size and is full of juice, on others it is dry and highly flavoured."

When fruit is required for Cider making, the proper quality of the soil on which it is grown is all important. As the poet has well said :

> " Next let the Planter, with discretion meet
> The Force and Genius of each Soil explore ;
> To what adapted, what it shows averse :
> Without this necessary care, in vain
> He hopes an Apple Vintage, and invokes
> *Pomona's* aid in vain." (*Philips* " *Cyder.*")

Happily however the rough handed experience of every day life has been able to get on in advance of Science. The practical farmer has not to wait for the chemist to tell him which of his fields are most productive. The dairyman, for example, soon finds out from which of his meadows he gets the best milk, the richest cream, and the most valuable cheese ; and his next object is to get the best breed of cattle to graze them, or in other words to find the cows that will best perform their part in dairy produce. So it is with the Orchardist, the liquor in his vats will soon point out to him the particular Orchards which afford him Nature's best laboratory for the production of the finest and strongest Cider ; and his efforts must then be directed to get them provided with the best varieties of fruit. It is with Orchards moreover, as it is so remarkably with Vineyards, that some portions of the ground will produce much finer liquor than the rest, although the soil apparently is the same throughout. The fact is undoubted, but the reason seems inscrutable and beyond the powers of chemistry to define.

The Cider and Perry from the English Orchards are admitted to be far superior to those liquors from other countries and thus our Orchards should shew the soils best suited to their production. The evidence from history on this point is not quite satisfactory, for all the authorities of the 17th century agree in recommending light sandy soils, such as are usually termed 'Rye lands' :

> " Look where the full-eared Sheaves of Rye
> Grow wavy on the Tilth, that Soil select
> For Apples." (*Philips* " *Cyder.*")

Knight says, " The excellence of the Cider formerly made from the *Redstreak, Golden Pippin,* and *Stire* apples in light soils seems to evince that some fruits receive benefit from those qualities in the soil by which others are injured." Marshall gives the instance of the once celebrated *Stire* which in the limestone lands of the Forest of Dean yielded an incomparably rich and highly flavoured Cider, but when grown in the deep rich soil of the vale of Gloucester afforded a liquor only useful for its strength and roughness. The *Hagloe Crab* again, another celebrated apple in its day, required the calcareous rock called " Dunstone," to give full flavour and richness to its liquor. The *Foxwhelp* on the other hand yields the Cider so remarkable for its strength and *gusto,* or that peculiar flavour for which it is so highly esteemed, from deep clay Sandstone loam, and if the trees are grown on light, or too sandy a soil, its Cider is then thin and very inferior in flavour. The same might be repeated again and again of many other varieties.

It is a curious fact, and certainly more than a coincidence, that the practical experience of so many generations of men should show that the two English counties which have chiefly given its high character to English Cider, viz. Herefordshire and Devonshire, are both remarkable for the same character of soil, that is, for the deep clay loam of the Old Red Sandstone. This experience of centuries is fully borne out in our own times, and even in these favoured counties the districts specially noted for this character of soil are equally remarkable for Cider of the highest flavour and quality. The light soils will not now give superior Cider, and he who would plant a successful Orchard must choose a deep stiff Sandstone loam if he has the opportunity of doing so.

The following analysis of Herefordshire soil was made by Mr. G. H. With, F.R.A.S., in 1877 :

ANALYSIS OF THE CREDENHILL MARL, OR CORNSTONE.

Organic matter and Combined water	...	...	2·261
Silica and insoluble Silicates	...	...	... 56·068
Tricalcic Phosphate	...	...	... ·391
Lime Carbonate	...		... 26·098
Magnesia Carbonate	...		... 2·211
Peroxide of iron	...		... 5·170
Alumina	...	...	... 3·600
Chloride of Potassium		...	... 1·070
Chloride of Sodium	...	...	... ·427
Peroxide of Manganese Sulphuric, acid, and Loss			2·704
			100·000

Credenhill is noted for its Orchards, and their fertility is due in great measure to the supply of Lime, from the Marl, or Cornstone, which surrounds this hill, as it does so many others in Herefordshire.

The Pear tree is still more hardy than the Apple tree ; it will grow on the dry clay itself. The celebrated *Taynton Squash* draws its finest liquor from the heaviest soil : and that popular Pear the *Bare-land Pear*, takes its name from the coldness and poverty of the soil it grows on : thus Perry might be produced to great profit and advantage on many a soil that will scarcely give back the labour spent on it in other ways. The old proverb tells unfortunately against it :

> " He who plants Pears
> Plants for his heirs."

and the patriotism which should plant Perry Orchards is not always to be found.

SURFACE.—The question of turf or tillage as best adapted for orcharding has been much discussed ; and pasturage has been commonly favoured under the idea that the soil beneath the trees was thus kept more cool and moist during the heat of summer. This is not the case ; for the crop of pasture, or hay, or green crops of any kind, not only require much moisture for their own growth which they take from the soil, but they also exhale much more moisture during the heat of the day than is compensated for by the dew that falls on them by night ; and thus, in both ways, the trees are robbed in dry weather of the moisture necessary for their healthy and fruitful growth.

The old orchard writers are therefore right in giving preference to tillage, rather than to pasture land, for an Orchard. Thomas Andrew Knight, and most other Herefordshire authorities, think there is no more suitable place for a young Orchard than a Hopyard ; and the most approved method in Kent at the present day is to cultivate the Orchard as a Hop garden until such time as the fruit trees are large enough to yield a paying crop. The trees profit by the cultivation and the protection given to the hops ; they grow more freely ; bear finer fruit ; and yield, it is said, a longer-keeping Cider. As the trees grow large, the hops must be uprooted or other green crops given up, and the field laid down to permanent pasture.

In America roots are almost always grown in the first five years in new Orchards, and the soil deeply ploughed every year at a proper distance from the trees. They consider grain crops as too exhausting and injurious to the soil. The home Orchard attached to most Herefordshire and Devonshire farms must be pasturage of necessity, for the great convenience it affords for the ewes and lambs in spring, or the ordinary farm cattle at all seasons.

DRAINAGE.—A due amount of moisture in the soil is absolutely necessary to the proper growth of the higher forms of vegetation, but it should not be in excess, and above everything, it must not be stagnant. A want of good drainage is fatal to an Orchard. The temperature of water-logged soil is always low. The warm rains of spring run off the surface without mixing with the cold water left there by winter ; and it is very late in the year before the sun can lessen its quantity by evaporation, and impart warmth to the soil. If water moreover remains long stagnant in contact with any vegetable matter it soon becomes impure by the formation of noxious gases, and thus is rendered positively injurious to the trees growing there. An Orchard in this condition is a miserable sight ; the trees are rugged and stunted in growth, their boughs are weak, covered with lichen or moss, and can seldom produce much fruit ; and yet it is a sight that is by no means uncommon.

A good Orchard must therefore be well drained by art, if not by nature. The excess of water should flow off gradually, so as to leave the soil porous and ready to receive from the atmosphere quickly its own warmth, that the roots may be stimulated early in the season, to take up from the soil all the principles necessary for the healthy life and vigorous growth of the trees.

ASPECT, CLIMATE, AND SITE.—The Aspect and Site of the Orchard involve its Climate ; and on no subject do the writers of the 17th and 18th centuries differ more, for though all agree in preferring the South, they embrace nearly every point of the compass. The "*Compleat Planter and Cyderist*" (1690,) recommends a South, South-East, or South-West, aspect protected from the North, North-East, and North-West winds by buildings, woods, or higher grounds. Dr. Beale in his "*Tract on Herefordshire Orchards*" (1656,) preferred a South aspect inclining rather to the rising than to the setting sun. Mortimer in his "*Husbandry*" recommends any site from East to West, Thomas Andrew Knight also thought any site from the East by South to the West, favourable for Orcharding.

The general belief is that a Southern aspect with an inclination to the East, is best adapted for the Orchard ; thus following the popular idea of the health giving properties of the morning sun. In other words this aspect gives a better supply of light and heat and therefore affords a better promise of healthy vegetation and fruitful crops. This belief holds good for Herefordshire where the West winds are apt to prevail with very great violence ; but apart from such special circumstances, any aspect, tending Westward, is the proper one for an Orchard. It is well known that if plants are exposed to the direct influence of the rising sun when they are frozen they will suffer, and in some cases altogether perish. But if the same plants are shaded till they are gradually thawed by the increasing temperature of the atmosphere, they recover from the effects of the frost, and are rarely injured. Hence it is that an Orchard, if exposed to the direct influence of the morning sun, is almost sure to suffer after an attack of spring frost when the trees are in blossom, or when the fruit is setting ; whereas one with a Western aspect, which does not receive the direct rays of the sun till he has risen, and the temperature of the atmosphere has risen also, and dispelled the frost, it escapes, and the fruit crop is saved. We frequently find one side of an Orchard, or one side of a tree, bearing fruit abundantly, when the other side is quite bare, and this very generally arises from the same cause. If frozen blossoms could be shaded till the sun had diffused its warming influence and dispelled the frost, before its rays reach them, the blossoms would be saved.

It is sometimes found advantageous to have plantations in different aspects so as to secure crops in variable seasons. Marshall had an Orchard in a North-West aspect fully fruited in 1783 when the Cider fruit was cut off in every other aspect that year. The same fact was happily experienced in 1879, by Mr. Hill of Eggleton, and some other growers.

Orchards are often planted too low in the valleys, for though they may yet have more rich alluvial soil and better protection from wind there, they have to encounter the cold damp fogs of night which are often so destructive to the blossom in spring and are apt to check the free growth of the fruit. The best situation, where the soil is good, is one that is raised well above the level of the night fogs, on the low ground.

Worlidge has these quaint and consolatory remarks on the best position for the Orchard :

"for the distinguishing thereof there are many rules, but he that is seated and fixed in any place and cannot conveniently change his habitation, must be content with his own, and if any defect or disadvantage be in it, it may be that it hath some advantages that others want."

Wherever the Orchard may find itself, it is desirable to give it the protection of buildings, high quick hedges, woods, or higher grounds to keep off the dangerous spring frosts and blight, and afford as much shelter as may be from strong winds, for thus the blossom is often saved from destruction, and the crop of fruit when full-grown kept secure.

MANURING.—Apple and Pear Trees, whether in arable land or pasture, are very insufficiently manured. The trees often become weak and exhausted from the heavy loads of fruit they bear, and yet their ungrateful owners forget to feed them. This neglect no doubt often gives the explanation why so many trees only bear fruit on alternate years. On arable land they take their share of the manure supplied for the green crops grown there ; but on pasture land they have only to share with the grass the manure from the animals that graze beneath them and enjoy their shade. A careful farmer in the neighbourhood of a town may sometimes scatter a few ashes over the Orchard to help the grass, but it very seldom occurs to him to think that the trees would be equally grateful for some better nourishment.

The kind of manure best suited for the Orchard may be learnt from the consideration of the solid constituents of the tree itself and its fruit, since this analysis must shew the inorganic ingredients they demand from the soil. Professor Emil Wolff, of the Royal Academy of Agriculture, Hohenheim, Wirtemberg, has made a most careful investigation into the ingredients of the ashes of plants, and has published the following results :[1]

ANALYSIS OF ASH OF APPLE TREE WOOD.		ANALYSIS OF ASH OF THE APPLE ITSELF (whole fruit.)	
One Hundred Parts by Weight, gave :		One Hundred Parts by Weight, gave :	
Potash ...	... 12.0	Potash ...	... 35.7
Soda ...	... 1.6	Soda ...	... 26.1
Magnesia ...	... 5.7	Magnesia ...	... 8.8
Lime ...	... 71.0	Lime ...	... 4.1
Phosphoric Acid	... 4.6	Phosphoric Acid	... 13.6
Sulphuric Acid	... 2.9	Sulphuric Acid	... 6.1
Silica ...	... 1.8	Silica ...	... 4.3
Chlorine ...	... 0.2		
	99.8		98.7
Loss	2	Undetermined Matter, and Loss	1.3
	100		100

---

(1) From Professor Wolff's "*Mittlere Zusammensetzung der Asche aller land und forstwirthschaftlich wichtigen Stoffe*."
                                                                                    Stuttgardt, 1865.
Adapted for English use by Professor A. H. Church and W. T. Thiselton Dyer.   Macmillan & Co., 1869.

Professor Wolff has also given the following result of his examination of the fruit of the Pear :—

ANALYSIS OF THE ASH OF THE PEAR (whole fruit).

Potash ...	..	...	...	...	...	54·7
Soda ...	...	...	...	...	...	8·5
Magnesia	...	...	...	...	...	5·2
Lime ...	...	...	...	...	...	8·0
Phosphoric Acid		...	...	...	...	15·3
Sulphuric Acid ...	...	...	...	...	5·7
Silica ...	...	...	...	...	...	1·5
						98·9
Undetermined Matter and Loss					...	1.1
						100

The amount of Phosphoric Acid contained in Apples and Pears, is shewn by these analyses to be so considerable that they have been considered as specially adapted to sedentary men, who work with their brains, rather than with their muscles ; for Phosphorus is thought to be the best brain food. However this may be, it has been thus demonstrated that the essential inorganic ingredients for the healthy growth of the trees and their fruits are : Potash, Lime, Soda, Phosphoric and Sulphuric Acids, and these must all be contained in good Orchard soil : but the mode in which they act and re-act on each other, so as to present themselves in a soluble form that can be selected and taken up by the rootlets—to be again modified by the action of the atmosphere in the leaf structure, is not clearly known. Science tells us these principles must be furnished to the plants by the soil, and experience proves the necessity of supplying the loss to the soil, and the great advantage of doing so in the increased health and fruitfulness of the trees.

The best means for replenishing the soil with these materials is not difficult to point out, but they are not readily to be obtained on the spot. The ordinary farm-yard manure is deficient in Potash and Phosphates. It is too stimulating, and therefore more likely to cause the production of weak succulent wood, than of hard fruit-bearing spurs ; and it is all wanted moreover, for the green crops on the farm, and for these it is eminently suitable.

There should be a special place at every farm assigned to Orchard Manure. Its foundation might well be road parings and scrapings, with ditch and pond cleanings, mixed freely with lime, and to this should be added the "must" from the cider mill. This material, useless for any other purpose and now only burnt, or wasted, should always be returned to the Orchard. It is not great in quantity, but it would always serve to indicate the Orchard Manure heap.

The following materials will be found admirably adapted for orchard fertilization, whether to encourage the vigorous growth of young trees, or to restore the weak and exhausted state of those which have borne large crops of fruit :—

Bone Dust ...	...	...	...	...	1 part
Pure Dissolved bone	...	...	...	...	1 part
Kainit	..	...	...	...	2 parts
Charcoal dust, or fine Coal Ashes		...	...	20 parts

If these materials, carefully mixed, were lightly forked into the surface of the soil around the trees, the amount required per acre would be something under a ton, and the cost be about two pounds, at present prices ; a moderate sum, when the value of the apple crop is considered.

PLANTING.—The young trees selected to furnish the Orchard should be stout and well grown, not less than 8 or 10 years old.   They should be planted at equal distance from each other at spaces varying from 15 to 40 feet apart, according to the habit of growth of the variety, or to the further use it is proposed to make of the ground.   Mr. Knight was in favour of close planting whether in arable or pasture land.   Those planters who wish to have the largest return at the earliest period, should plant the trees at 15 feet apart in the rows, cutting away every other tree, as soon as they approach each other, taking care to keep the rows 30 feet apart from the first.   Dr. Beale advises that the crab stocks "be settled in the ground 30 feet apart, and after three years time to let the artist be sent for to graft them with the best fruit."   Mortimer would have "all trees and rows at 40 feet apart and pruned to grow like a fan."   The trees certainly should stand so clearly apart from each other as to allow of their full growth, since a large tree turns off not only more, but better fruit than a small one.   They should be planted carefully in lines for the convenience of cultivation, and their roots should be kept as near the surface as may be ; that is, they should not be planted too deeply in the ground.   The soil beneath should be double dug, and if some roughly broken bones could be dug in at the same time, say a peck to a tree, they would form an enduring support to the young trees.

Trees of a similar variety, or a similar habit of growth, and which ripen their fruit at the same period should be planted together ; for thus there will be a better certainty of uniform space for light and air ; the general appearance of the Orchard will be greatly improved ; and much time and labour will be saved in gathering the fruit in Autumn.   It is thought desirable also to have a mixture of early and late blooming varieties in the same Orchard, that if a part of the crop is cut off by any adverse circumstances, such as frosts, storms, or blight, there may be a better chance of saving some portion of it.

When the trees are planted they should be well staked, and if in pasture land, they should be safely protected from cattle or sheep ; and lastly the Orchard itself should be well fenced in, for it is but too often an inclosure only in name, and its fences badly kept and much trespassed on.

II.   ORCHARD TREES.

"Let sage Experience teach thee all the Arts
Of Grafting and In-eying ; when to top
The flowing Branches ; what Trees answer best,
From Root or Kernel."   *(Philips "Cyder.")*

It is the common result of experience in all countries, and on every soil, that the quality of the Cider and Perry manufactured depends very greatly upon the varieties of the Apples and Pears cultivated.   It was Thomas Andrew Knight's opinion that "Herefordshire is not so much indebted for celebrity as a Cider county to her soil, as to her valuable varieties of fruit."   So too does the French Commission in its admirable Report, "*Le Cidre de France,*" lament, again and again, the absence in these days of that intelligent industry in the selection of the best varieties of fruit for cultivation, which so distinguished the planters of the last century. There is much force in these observations, though they do but present a one sided view of the true cause of the decadence in the quality of Cider and Perry. The present state of our Orchards is most unsatisfactory in this respect, since they contain so large a proportion of varieties, without name, or character, or merit.

SEEDLINGS.—Every Orchard farm, properly cared for, has a nursery for young trees in some out of the way corner of the garden, or field. Here young Crab stocks are procured by placing the " must" or squeezed pulp from the Crab Apples, after verjuice has been made, in rows beneath the soil, when the pips, uncrushed by the mill, spring up, and in four or six years, after a few careful transplantings, become strong enough to graft with varieties of fruit, whose merits are established.

The most approved method of raising young stocks is to separate the pips from the " must " by washing, so as to obtain clean seed. Mix this with moist sand, or light mould, and set aside until February or March. Then sow it in drills an inch deep, on a firm, well manured soil, made as for an onion bed. The seed should be sown thinly, so as to get the young plants an inch or two apart. A few will vegetate immediately, but it will generally remain a year in the ground before the full crop appears. The seedlings are apt to grow unequally, but by the end of the second year they will generally be ready to transplant into rows a foot apart, and three or four inches from each other. Here they must remain for two years, when they will be strong enough to plant out in the nursery in "quarters," as it is termed, that is on ground well trenched, two spades deep, and ‚heavily manured. They should be planted in rows two feet six inches apart, and one foot from each other, when they will be ready for budding the following August. Seedlings should always be transplanted early in autumn as soon as the leaf falls, and never later than the beginning of November.

It is however still more common to grow the young seedlings from the pips in the "must" from the cider mill. There can be no question, that these young Apple seedlings often escape grafting altogether. They have often been found to bear a good looking, "eyeable," fruit, and were then planted out to supply the vacancies that are so constantly occurring in the Orchard, and it is by this careless practice, that worthless varieties are now found to prevail so extensively.

BUDDING AND GRAFTING.—Budding is much more practised in these days than formerly. It presents greater economy in material, in labour, and, above all, in time. The young seedlings may be budded about the third or fourth year, and if in the following Spring the buds should fail they can then be grafted, and the chance of blanks on the bed be very greatly diminished. Whichever process may be adopted it should be done in the nursery where the growth of the scions may be well protected and regularly superintended. The young trees should not take their place in the Orchard until they have gained strength, and have got a good outline of head, and this will rarely be before the tenth or twelfth year of the age of the stock.

A custom has arisen in the Orchards of late years which is often practised with good effect ; it is to regraft trees which show a diminution of fruitfulness, or are altogether unproductive, although they may be of considerable age. The scions should be of some strong variety which succeeds well in the locality, and they should be carefully grafted as near the ends of the branches as possible, when they will come the more quickly into bearing. They will want careful protection from the wind.

ORCHARD TREES : VARIETIES OF THE 17TH CENTURY.—The names of the celebrated varieties of fruit of the 17th century, have been carefully handed down to us, in prose and verse, by the writers of the period—and it is very interesting to keep them in remembrance. First and foremost in those days stood the *Redstreak*, with its varieties, the *Summer, Winter, Yellow, More-green, and Red Redstreaks*. The *Bromsberrow Crab* and the *Westbury Crab* (a Hampshire Apple) stood next in order of merit, for making a long lasting cider. The *Foxwhelp* then just rising into repute; the *Coleing* (about Ludlow); the *Underleaf;* the *Arier Apple;* the *Olive* (another Ludlow fruit); *Gennet Moyle*, as renowned for its cooking properties, as for its Cider; the *White*, and *Red Must Apples ;* the *Oaken Pin;* the *Summer* and *Winter Fillets* or *Violets ;* the *John Apple* or *Deux-ans*. Then follow the *Pearmains* and *Pippins* in great variety ; of which the most celebrated, even in those days, was the *Golden Pippin* as well for the long life of the tree, as for the long keeping of its Cider; the *Stocking Apple ; Elliot; Harvey;* Devonshire *Bitterscale* and *Deaus Apple;* the Salopian *Otley; Nonsuch; Mangold* or *Onion Apple; Summer and Winter Queening ;* the *Woodcock Apple ; Richards* or *Grainge Apple; Claret Wine Apple ;* Gloucestershire *Heming;* with "all (both *Russettings* and *Greenings*) which have a relish of agreeable Piquancy and Tartness."

> "There are, that a compounded Fluid drain
> From different Mixture, *Woodcock, Pippin, Moyle,*
> Rough *Elliot,* sweet *Pearmain,* the blended streams
> (Each mutually correcting each).create
> A pleasurable Medley, of what Taste
> Hardly distinguish'd." *(Philips' " Cyder.")*

Since this time some other apples have obtained an established repute in Herefordshire Orchards, such as ; the *Friar;* the *Cockagee; Royal Wilding ;* the Devonshire *Dufflin; Bennet Apple; Forest Styre; Best Bache; Dymock Red; Cowarne Red;* the *Pawsan; Garter Apple;* the *Bromley; Hagloe Crab; Stead's Kernel; Skyrme's Kernel;* and the many so-called Normans. There is however no history of their origin for the most part, or of their introduction into the Orchards.

The Pears named for Perry making by the old 17th century writers, are the *Barland Pear ;* the *Horse Pears, Red* and *White; Taynton Squash;* divers *Choak-pears,* whereof the red coloured yielded the strongest liquors ; the *Red* and *Green Squash Pears; John Pear ; Money Pear ; Lullam Pear ;* and others with local names and merits.

VARIETIES OF THE 19TH CENTURY.—The names of the varieties, considered at the present day most worthy of cultivation in Herefordshire, may be given as a comparison with the lists of the old varieties, although it is not here that their several merits can be discussed. There are still happily in our Orchards many of the varieties of established merit, whose names have already been given. The *Foxwhelp*, which has been the favourite apple for some hundred and fifty years past, still lives and is still propagated ; so too, do the *Royal Wilding ;* the *Styre Wilding ;* the *Yellow Styre ;* the *Cowarne Red ; Dymock Red ; Skyrme's Kernel* ; *Garter Apple ; Summer* and *Winter Queenings* and several others named before, whose varieties are still grafted in the localities where their merits are most appreciated.

Many valuable additions have also been made of late years although the mode of their appearance and their general prevalence in the Orchards may be rather difficult to explain. The *Cherry Norman ;* the *Strawberry Norman ; Handsome Norman ; Red Norman ; Yellow Norman ; Black Norman ; Cwm Norman ;* are all very favourite varieties and their introduction is comparatively of recent date. The *Ladies Finger ; Black Kingston ; Pym Square ; Upright* and *Spreading Redstreaks ; Eggleton Styre ; Wilding,* and other *Bittersweets.* There are others as *White Buckland ;* the Devonshire *Staverton ;* with *Grittleton ; Red-budd ; Black Eyed Pippin* and many others with local names, which bear well, and fill the barrel, but when enquiry comes to be made into their value, have no other merit to save them from condemnation. The day for common rough Cider has happily passed away.

The Perry Pears now most in favour in the Orchards are ; of the early varieties : *Taynton Squash* (a very favorite pear supposed to be nearly worn out, but now being again grafted successfully); *Barland ; Yellow* and *Black Huff-Cap ; Pint Pear* and others with local names and of doubtful merit. Amongst the late Perry Pears at this time are : *Oldfield ; Moorcroft ; Blakeney Red* (a Gloucestershire variety); *Red Pear* and its varieties; *Longland ; White Longland ; Holmer ; Staunton Squash ; Butt Pear ; Gregg Pear ; New Meadow Pear,* &c., &c.

There can be no question but that there is a very large percentage of fruit trees in Herefordshire Orchards at the present time, which, if they are past the age for re-grafting the ends of their branches with valuable varieties, should be "grubbed up" as the country phrase hath it. They are useless for making superior Cider by themselves, and they serve now, but to spoil that which is made from other and better Apples in the Orchard. It would be economy in every sense to turn them into faggot wood.

PRUNING.—Orchard pruning is very apt either to be neglected altogether, or to be carried out in excess. In the one case the boughs grow matted together and bear fruit, small in size and deficient in quality, from a want of light and air : or in the other, whole boughs are mercilessly lopped off close to the trunk, leaving those great round scars, commonly called "owls' faces," to offend the eye of every good Orchardist ; since he knows how deeply they injure the trees and shorten their lives. It would sometimes seem, as if the trees were left until the almost finished state of the faggot stack suggests the expediency of "a turn at pruning," when the ordinary pruner is sent for, who slashes away at the cost of the strength of the poor victims, who possibly have not recovered from the last raid.

Apple and Pear Trees when full grown require very little pruning. *"The compleat Planter and Cyderist* " says, "while your tree is young bring it into a handsome shape and order, and when it comes to bear fruit forbear pruning, unless in case of broken, or such boughs as grow cross, or gall and fret others." Mortimer gives similar advice, and adds, " thin most at the outmost branches, or where they are thickest." Thomas Andrew Knight also lays great stress on judicious pruning, for he did not fail to observe the injury done in the Orchard from the wholesale lopping off of large branches. The scar does not grow over, it decays, and the tree becomes hollow and is broken off with the wind, or split down the middle, and the term of its natural life and vigour is materially shortened ; and yet it is not difficult to remove even large branches without injury if it is carefully done.

The late Mr. Chandos Wren Hoskyns, in a paper on " Pruning" in the volume of the Woolhope Club's Transactions for 1867, has so well explained the true principles on which Pruning should be done, that a short abstract of his paper is here presented.

The trunk of a tree is fed by its branches, just as a river is fed by its tributaries. It is not nourished by the sap taken up by the roots from the soil, until it has been acted upon by the atmosphere in the leaves ; and thus its growth is downwards from the foliage, and not upwards from the roots. Every branch of a tree has smaller branches of its own, and is in fact to them a tree. Now, supposing a branch to be condemned, instead of proceeding by capital punishment, (which admits of no repentance *except to the inflictor*), the humane process is this. Select a branchelet which happens to grow in the most favourable direction, and at the point where it springs, cut off the main branch obliquely in the direction of the growing branchlet, undercutting at first to prevent spaltering, and prune the wound as much as possible into symmetry with the direction of the new leader. In another year or two serve the new leader in the same way, and the process may be repeated if requisite. The result is this. The growth of the original condemned branch is entirely stopped without its being itself *killed*, and, as the trunk of the tree increases, its size gets less in proportion, and may generally in a few years be removed entirely without injury, or eye sore, close to the stem, that is to say when the proportionate size of the scar to the stem is such that it will heal perfectly in two or three summers.

Trees grow in very different forms, some varieties are upright, some spreading, some straggling in growth and others altogether irregular. The careful pruner will take the peculiarities of each variety into consideration and leave in each as much bearing wood as possible, always remembering the great physiological truth, that in a healthy tree the extent of root surface must be balanced by the extent of foliage, to produce a well grown fruitful tree. Mr. Knight deplored the system of pruning in his day, which consisted in eliminating every branch in the middle of the tree until at length " small tufts of branches were left at the extremities of long and large boughs." This is not altogether the fault of the pruner, for in the growth of spreading mop-headed trees the middle of the tree is thrown completely in shade, and the smaller boughs, if not removed, could never bear healthy fruit. It is more commonly the result of leaving the trees too crowded in the Orchards.

Cutting off main branches should only be required in young trees, and when this is properly done, no leading branch should afterwards be touched, and the trees should be left to live out the natural term of their lives and fruitfulness.

TREE ENEMIES.—A Volume might be written on the many enemies that attack Apple and Pear trees in health and disease, but without much avail, since few of them admit of the ready application of a remedy.　A brief notice must yet be given of those which most commonly and most persistently affect them : such as Mistletoe, Canker, Insects, Fungus, and other vegetable parasites.

MISTLETOE *(Viscum album).*—The health and vigour of the trees in an Orchard will generally denote the attention given to them by the owner.; but neither care nor attention can altogether keep off the parasite, Mistletoe, from a Herefordshire Orchard. The thrushes and some other birds eat the Mistletoe berries. The seeds they contain pass through their bodies and are thus sown on the branches of neighbouring trees. The young seedlings send their roots into the tissues of the tree, and live at its expense for the future. There is a common impression among Orchardists that the Mistletoe renders the supporting tree more fruitful, and thus does but little injury. This idea is a very mistaken one ; the parasite may and often does throw the tree into bearing. The tree makes the effort with the knowledge as it were, that it is attacked by a vital enemy, which will never leave, until it has destroyed it branch by branch. The tree begins to shrivel and decay and the fruit becomes smaller year by year, albeit the tree may keep up the struggle for a human life-time. The "baleful Mistletoe" as Shakespeare truly terms it.

Something may be done to help the trees. The Mistletoe should be attacked boldly, and all established plants be broken off or cut closely, year by year. If this is done before Christmas the berried branches will readily sell at any Railway Station at £4 the ton. The only effectual remedy however is to destroy the young seedlings. The silvery seeds are deposited by the birds on the branches, and the first rain washes them to the underside, where the glutinous matter causes them to adhere. Here the Tits and Finches happily eat many of them, but the careful quick eye of the Orchardist should see many others which his spud would remove at once. If the young Mistletoe seedling escapes these dangers it will send its root down the inner bark and throw out its first leaves the second or third year; nothing now can be done, but to remove the branch close to the trunk, or if the young Mistletoe itself is near the trunk, it is hopeless to attempt to destroy it, and the place of the Apple tree should be supplied by one of the supernumeraries from the nursery.

CANKER.—The terror of all Orchardists, and the bane of most Orchards, is always due to direct injury, but from whatever cause this injury may arise, weakness is at the bottom of the mischief. The tree is old ; or the variety very old, or very delicate ; the soil is not sufficiently drained, or it is too poor ; or for some cause does not suit the variety ; in all these cases there is a want of vitality ; the young wood may be weak and not well ripened ; when a sudden frost, especially after rain, ruptures the vessels of the bark and thus forms the chief cause of canker. Any direct injury, however, to the bark of the tree, as from the friction of one branch upon another, the pressure of a clothes-line tied from tree to tree, or injury from the ladder in fruit gathering, may all cause it, even in healthy trees. Canker commences with enlargement of the vessels of the bark—more apparent by the way in Apple than in Pear trees—and continues to increase until in the course of a year or two, the *alburnum* dies, the bark cracks, rises in large scales, and falls off leaving the trunk dead and ready to break off with the first wind, if not before removed.   The Canker shews itself quickly, and if the cause is sought for it will often admit of a remedy.   The one most usually effective is a good supply of nourishment to the trees affected, together with removal of the parts injured.

AMERICAN BLIGHT *(Aphis lanigera).*—The common Apple-tree *aphis* in Spring is often very destructive to the apple blossom when the weather is unfavourable, but its ranges are too widely spread to admit of any effective artificial remedy.   This *aphis* attacks the young foliage and clusters of blossom ; but the *American Blight* attacks the woody parts of the tree, and is still more fatal.   It is the most important of all Insect Blights, and is known only too well.   It attaches itself to any part of a tree on which the cuticle is broken.   The insect is viviparous like most other *aphides.*   It lives on the sap of the tree and by its irritating presence it causes excrescences of growth and eventually the death of the branches beyond.   It is the habit of this *aphis* to retire into the ground during the winter, and cluster in the crevices of any roots it may find suitable. The pest is difficult to get at, but the remedy consists in applying a weak mixture of petroleum with soft soap, say an ounce of petroleum, and half a pound of soft soap boiled gradually in a gallon of water, apply with a brush or syringe wherever the woolly insect shews itself.   This remedy has the additional advantage of attacking its winter quarters at the foot of the tree, as it is washed there by the rain.   The petroleum emulsion is very troublesome to keep well mixed, and when the blight is not very extensive a strong solution of soft soap, or of agricultural salt, is much more easy of application, and often very effective in destroying the insects.

RED SPIDER *(Gamasus telarius)* is occasionally very destructive to the leaves of Apple and Pear trees.   It is believed to be due to the condition of the soil in which the tree grows.   It may be too light or too poor for it ; and this belief points out the direction in which the remedy must be sought.   Many other Insects attack Apple and Pear trees, such as *Episema cæruliocephala ; Cheimatobia Brumata ; Porthesia auriflua ; Lozotænia rosana ; Tortrix heparana ; Tortrix ribeana ; Tinea corticella ; Curculio vastator ; Semasia Wœberana ;* with several other species of *Aphis, Acarus,* and *Coccus.*   The visits of these enemies however, are for the most part local, and their presence can only be met by the partial remedy of smoking to windward, when plenty of damp straw or mouldy hay at hand, gives the opportunity of doing so.

FUNGUS GROWTHS, are always unwelcome guests in an Orchard.   A botanist may admire a fine *Polyporus hispida* or other *Polyporus,* or rejoice in a magnificent cluster of *Agaricus Pholiota squarrosus,* with its leopard-like spots and colour, growing from the bole or at the foot of an apple tree, as it so often does ; but these with all their tribe do but indicate decay within.   They must of course be cut away at once, but the disease on which they have fed will exist there still.   There are yet some microscopical funguses, which are so frequent and injurious as to require special notice.

MILDEW.—Blight, or Mildew (a microscopic *Oïdium*) generally growing on the young leaves and shoots of the tree.   It may appear at any time from Spring to Autumn.   It causes first a white mealy appearance of the young shoots and leaves, which then curl up—grow black and

PART IV.

PRICE 21s.

THE

HEREFORDSHIRE POMONA,

CONTAINING

COLOURED FIGURES AND DESCRIPTIONS OF THE MOST ESTEEMED KINDS OF

APPLES AND PEARS,

CULTIVATED IN GREAT BRITAIN,

EDITED BY

ROBERT HOGG, L.L.D., F.L.S.,

Honorary Member of the Woolhope Naturalists' Field Club; Secretary of the Royal Horticultural Society;
Author of 'The Fruit Manual'; 'British Pomology'; 'The Vegetable Kingdom and its Products', &c., &c.

" *Hope on.   Hope ever.*"

" Ζεφυρίη πνείουσα τὰ μὲν φύει ἄλλα δέ πέσσει,
ὄγχνη ἐπ' ὄγχνη γηράσκει, μῆλον δ' ἐπὶ μήλῳ,
αὐτὰρ ἐπὶ σταφυλῇ σταφυλή, σῦκον δ'ἐπὶ σύκῳ."

Homer Odyssey vii. 119–22.

" THE BALMY SPIRIT OF THE WESTERN GALE,
ETERNAL BREATHES ON FRUITS UNTAUGHT TO FAIL.;
EACH DROPPING PEAR, A FOLLOWING PEAR SUPPLIES,
ON APPLES APPLES, FIGS ON FIGS ARISE."

Homer, Odyssey vii.—Pope.

LONDON: DAVID BOGUE, 3, ST. MARTIN'S PLACE, TRAFALGAR SQUARE.
HEREFORD: JAKEMAN AND CARVER, HIGH TOWN.

1881.

drop off to the great detriment of the trees, if the Mildew is at all extensive upon them. This fungus appears under certain atmospheric conditions, such as moisture, with the sudden prevalence of cold winds checking growth. Its remedy is known to be sulphur, when it admits of proper application, which can seldom be the case in an Orchard. The common practice of white-washing the trunks of the trees, if they would but add to every gallon of whitewash a handful of soot to sober down the colour, and a handful of sulphur to be exhaled by the sun during the heat of summer, might possibly also render good service in checking such fungus blights.

RUST, *(Helminthosporium pyrorum)*, is another microscopic fungus which in cold wet summers, as in that of 1879, is most destructive in the Perry Orchard. It appears in patches on the leaves of the Pear trees and on its fruit, and seldom ceases as long as a leaf or a pear is left on the tree. *Ræstelia cancellata* and some other microscopic plants could also be named, but their presence and power of destruction depend more on the season, than on any other cause ; and they admit of no remedy that care can supply, over the extent of an Orchard.

LICHENS.—These plants are of several kinds, and form the grey mosses which often completely cover the great and small branches of the trees. They seem to be the attendants of a damp atmosphere (that is to the want of more air and sunshine) and derive their sustenance chiefly from it ; indeed the Orchard itself is seldom well drained where these plants abound. The only injury they occasion the trees is by preventing the access of air and warmth to the branches, and harbouring the numerous leaf-eating and other Insects always ready to prey upon them. Drainage and Pruning afford the only known means of prevention ; and when once the lichens and mosses exist on the trees, scraping them off and washing the boughs with a strong solution of soft soap, or with lime water is the simplest plan of checking them, whenever it may be thought worth the trouble.

OTHER TREE ENEMIES.—The old writers dwell at considerable length on many other Orchard enemies, such as : Cattle, Hares, Coneys, Moles, Water-rats, Birds, Snails, Caterpillars, Pismires and Ants. These must be met, as they occur, by the ingenuity of the Orchardist. The most real, are the Hares and Rabbits, which in severe weather, when the ground is covered by snow, and other food is scarce, will soon destroy an Orchard by barking the young trees. The best immediate remedy is the lime and sulphur wash. Furze if at hand may be tied round the tree stems, but wire netting is the only effectual remedy, where these animals abound. The use of grease, tar, petroleum, so often resorted to, are better avoided for they are apt to be themselves injurious to the young trees.

III. FRUIT MANAGEMENT.

The customs which prevailed in the Orchard two hundred years since are very different from those prevailing at the present time. The early ciderists divided their fruit into three classes : the first consisted of such apples as would make a summer Cider for immediate drinking; as the *Codlings*, *Jenettings*, *Spice Apple*, *Summer Queening*, and all the early summer fruits. The second class consisted of those that made the best and richest, and longest keeping Cider, and embraced all the established varieties of cider fruit, as the *Redstreak, Broomsberrow Crab, Golden Pippin, Gennet Moyle, Westbury Apple, John Apple, Underleaf,* with the *Musts, Fillets, Elliots, Stocken Apple, Oaken Pin, Nonsuch, &c., &c.* Lastly, the third class contained all such fruits as were useful for the tables "making a pleasant, sweet, acceptable Cyder, though not long lasting"; such for example as *The Pippins, Pearmains, Gilliflower, Marigold Apple, Golden Rennetting, Harvey Apple, Winter Queening, &c.* The early ciderists thus recognized the fact, that in the cider districts, Cider could be made from all varieties of Apples; but, at the same time, they shewed the keenest appreciation of the varying qualities of the Cider made from the different varieties of fruit. In these days of cheap and easy transit, the first and third of these classes find a more lucrative sale in the markets for domestic consumption, and they are only used for making Cider in some exceptional year, or for some peculiar reason. The Apples used now for making the best qualities of Cider, and the same may be said of Pears for Perry, are especial varieties grown for the purpose, and are not worthy of consideration for use in any other way. They vary as a matter of course as to their season of maturation and are therefore practically divided into early and late varieties; and thus in well regulated orchards the mill is supplied in convenient succession. In the Channel Islands, in Germany and sometimes in America, however, it is still the custom to use the best varieties of dessert fruit, both of Apples and Pears, for the manufacture of Cider and Perry, but it can scarcely be said that the result justifies the practice.

> "Fruit gathered too timelie will taste of the wood,
> Will shrink and be bitter, and seldome proue good :
> So fruit that is shaken, or beat off a tree,
> With bruising in falling, soon faultie will be."
> Tusser.—*Points of Good Husbandry.*

FRUIT GATHERING.—The first care of the orchardist is to gather the fruit when sufficiently ripened, and this period will vary considerably, not only according to the season, but also according to the varying aspects of each individual tree. The ripeness of the fruit is generally indicated by the change of colour, by the perfume and flavour of the fruit itself, by the blackness of the pips, and by the fact of its beginning to fall from the tree; but the experience of the fruit grower enables him easily to recognise the proper time for gathering it, even in the varieties in which these signs may not be very marked. The earlier kinds of Pears, and also of Apples, will generally be ready about the end of September, and with this early fruit it is generally customary to mix such of the windfalls as may be in good condition, and thus clear the ground and prepare the way for the better qualities of fruit. The gatherings from which the best cider is made usually occur about the second or third week in October, and by the end of the month the trees should be cleared of even the latest varieties.

> "The moon in the wane gather fruit for to last,
> But winter fruit gather when Michael is past."
> *(Tusser).*

The mode of gathering the fruit also demands attention. The better kinds of fruit, such as are required for the market, or for domestic use, must of course be carefully hand picked, since every bruise will injure it ; but this extreme care is not necessary for the varieties required for Cider and Perry now under consideration. These may be gently shaken from the trees. When the weather is fine and dry the fruit may be collected. A layer of straw should first be laid under the tree (unless the grass is abundant there), and then coarse cloths or pieces of sacking should be placed upon it, as well to save the fruit from being too much bruised, as also for the ready convenience it gives of removing the fruit from time to time. The simple plan recommended by Marshall cannot be surpassed. As soon as the spontaneous fall of fruit begins to take place he recommends the first gathering to begin. The boughs should be gently shaken by means of a pole with a hook attached to it, but the fruit that sticks firmly to the tree must be left to become more mature, and be shaken off at a subsequent period. This practice is still followed in the best orchards, where the trees are thus gone over three or sometimes four times at intervals of ten or twelve days, until the whole crop has been matured and collected. The fruit which falls the second time is considered the most favourable for the best and strongest liquor required for bottling.

> " The farmer, with foreseeing view,
> Prepares himself for the forth coming spring ;
> Nudged by the ripen'd fruit that silent falls
> On the long grass beneath ; at early morn
> He clears the orchard boughs, and piles the fruit,
> And the press gushes with the pleasant juice."
>   Partridge's " *English Monthly*"

APPLE HEAPS.—As the fruit is gathered from the trees it is placed in heaps, until it becomes sufficiently ripe and mellow to be crushed. There has been much discussion as to the position and formation of the Apple heaps. The common practice is to place them on the plain ground in the orchard itself, or in some convenient place by the homesteads. They are usually made from about eighteen inches to two feet six inches in thickness, and are left without any protection either from the sun, from the rain, or from frost, not to mention the fowls and wild birds. Thus they remain for some two or three weeks, to as many months with the later varieties, to suit the convenience of the cider maker. Marshall recommends that the fruit after being collected perfectly dry should be laid up under cover, in an open shed, or where a thorough current of air can be had, in heaps not more than 10 inches thick. The best writers of the 17th century gave the same advice :—" The fruit should be laid out of the sun, and the rain, not abroad but in a heap on a sweet dry floor, on straw to sweat for about a fortnight ; and harder Apples, like the *Redstreak*, a month or more. The longer they lie the better, so that too many of them begin not to rot." (" *The Compleat Planter and Cyderest.*") —Marshall admits that this practice was not followed in his day any more than it is in our own, nor is it ever likely to be followed in large extensive orchards, although the advice is both good and sound.

The object of placing the fruit in heaps is to allow it to become uniformly ripe and mellow for the mill, and in order to insure its equal maturity with greater certainty, the different varieties of Apples should always be placed in separate heaps. It is better to do this even when the quantities are small, so as to insure their not being sent to the mill until each variety is sufficiently mellow. With the exception of a very few varieties of noted strength and flavour, Cider is made from different sorts of Apples mixed together, and this custom has ever been most popular :—

> " There are, that a compounded Fluid drain,
> From different Mixture. *Woodcock, Pippin, Moyle,*
> Rough *Elliot,* sweet *Pearmain,* the blendid streams
> (Each mutually correcting each) create
> A pleasurable Medly, of what Taste
> Hardly distinguish'd."--Philips' " *Cyder.*"

But all the fruits should be well mellowed in the heaps. The good judgment of the Ciderist here comes into play, in mixing the varieties which will improve each other. The *Foxwhelps, Redstreaks, Styres, Cowarne* and *Dymock Reds,* &c., will give flavour and strength in return for the sugar and mucilage they receive from the *Wildings* and the best of the *Norman* Apples.

When placed on the ground in the open air the Apple heaps may be allowed to be from one to two feet in thickness without fear of the fruit becoming heated, but on a dry floor the thickness should not exceed one foot. The heaps should most certainly be protected from all changes of weather, which cannot fail to be injurious to it. When placed in the orchard, therefore, the heaps should be made in rows that can be protected by thatched hurdles resting on a pole, running the whole length of the heap, which are at all times readily moved or replaced, and covered with cloths or tarpauling, if frost should set in.

The sun causes the fruit it falls upon to ferment unequally, though it seldom shines sufficiently, at least in England, to do much mischief in the autumn. Rain, which is so frequent at this time, injures the quality of the fruit very seriously. If any one doubts this, let him put a whole and sound Apple in a glass of clear water, and let it remain there for seven or eight hours. By this time the water will have taken a rosy hue with the sweet taste of the Apple, whilst the Apple itself will have lost much of its flavour. The explanation is, that by the natural laws, always in operation between fluids of different density, the water has kept passing into the Apple, and the juice has passed out into the water, greatly to the injury of the fruit. Frost is also very injurious to fruit, for after it has been frozen it will never ferment properly. A French chemist found the loss to be about one and a half per cent. of alcohol with fruit that had been frozen.

It is most desirable therefore that the fruit heaps should be well protected even if it may not be thought advisable to place them in some open shed, or wash house. Protection of the fruit from frost, is as little thought of in Herefordshire, as it is from rain. During the Winter 1878-9, and 1879-80, though fruit was scarce and both winters exceptionally severe, it was a rare circumstance to see the Apple heaps about the orchard in any way protected.

Cider makers in all ages have agreed that to make prime cider the best fruit must be used in its best condition. When the fruit has become in good order, it must be carefully looked over as it is put into the baskets to be carried to the mill, and all that is unripe, inferior, or rotten, must be scrupulously rejected. Unripe fruit contains neither sugar nor flavour :—heated or frost bitten fruit will not ferment properly : bruised and rotten fruit introduce the elements of injurious fermentation ; indeed all watery and inferior fruit should be ground by itself for the inferior liquor it must of necessity produce. The very common practice of mixing all sorts of fruit in the heaps, and of carrying them altogether to the mill, whatever may be their sort and condition, is fatal to the production of good Cider. The quaint remark of Worlidge on this careless custom is as true at this time, as it was in his own days. "This error or neglect hath not onely been the occasion of much thin, raw, phegmatical, soure and unwholesome Cider, but hath cast a reflection on the good report that Cider well made, most richly deserves;" and he adds very sensibly "better lose part of the cider than spoil the whole."

Pears are not considered to require so much care and good judgment as Apples do until they are carried to the mill. The early varieties may be taken at once from the trees to the mill, and the usual custom is when the fruit begins to fall, freely to shake off the remainder of the crop, and grind the whole without delay. The long keeping varieties require to be placed in heaps as Apples are, and are much improved by being allowed to become uniformly ripe.

> "Lo! for Thee my Mill
> Now grinds choice Apples, and the *British* Vats
> O'erflow with generous Cyder."
> (Philips, "*to his friend Harcourt in Italy.*"

THE MILL.—The modes of extracting the juice from Apples and Pears, to make fermented liquors seem to have been of the rudest kind until a comparatively recent period. The fruit was grated, or crushed in any rough and simple way, and since the quantity required was but trifling and labour cheap, it answered sufficiently well. Worlidge writing near the end of the 17th century, says "The operators did beat their fruit in a trough of wood or stone with beaters like unto wooden pestles with long handles, whereby three or four labourers might beat twenty or thirty bushels in a day." When a larger quantity of fruit was grown, and Cider and Perry became important articles of commerce it was necessary to find out some process more economical and expeditious. The happy idea occurred to some one—whose name is lost to a grateful country—to make the trough of a circular shape and roll round a heavy cylinder in it. This mill originally was of a very rude construction, and both the wheel, or cylinder, and trough were made of wood studded with hobnails. The wooden cylinder soon gave place to stone for the advantage of its weight, and this entailed the necessity of making the trough of the same material. A mill thus constructed worked with one horse, crushed the fruit so rapidly as to make from two to three hogsheads daily.

> "Blind Bayard worn with work and years
> Shall roll the unweildy stone from morn to eve."
> Philips "*Cyder.*"

Dr. Beal in Evelyn's Pomona speaks of some mills so large as to grind half a hogshead at a time. The construction of such a mill required the heaviest and most durable stone. In Herefordshire the Millstone Grit from the Forest of Dean soon came to be noted as best suited for this purpose. Such a mill was necessarily very expensive in construction, and so efficient that one mill often served for the district ; the grist, in the shape of Apples and Pears, being brought to it from all the surrounding Orchards. In course of time every large farm had its own mill, and these mills are very numerous at the present time, and many of them regularly used. The great fault of the Stone

Mill is that it is apt to roll the pulp too quickly before it and the fruit thus escapes been evenly crushed, and this is not altogether obviated by having the stones grooved diagonally in the usual way. The difficulty of removing the pulp from the trough is another disadvantage belonging to it.

About the year 1689 Worlidge invented a moveable iron mill, which he called the " Ingenio," a name borrowed from the Cubans who, curiously enough, grind their sugar at the present day with a machine thus called. With this mill he tells us that " two labourers, one feeding and the other grinding, can manage eight bushels an hour by interchanging all the day, with ease and delight. The " Ingenio " was introduced into Somersetshire many years before it was introduced into Devonshire, Gloucestershire, or Herefordshire, but it is only of late years that its use became general in Hereford-shire, for the Stone Mills on the spot were as durable as they were effective and kept in high favour. It is curious that at the end of the 18th century Marshall speaks of the old Stone Mill as "an unfinished machine," whilst Thomas Andrew Knight some twenty years later, at the begining of the 19th century, considered that much of the celebrity of Herefordshire Cider was due to the perfection to which the Stone Mill had been brought ; this feeling in its favour still exists more or less through the county.

The French have paid great attention to their fruit machines. They have one, the *Ecraseur*, " Salmon and Bergot " which grinds seventy-five bushels per hour with ease ; and this has now been surpassed by the " *Ecraseur Universal*," which with only one pair of granite cylinders will grind two hundred bushels of fruit per hour, besides being ready at all times to do the whole work of pulping roots, &c., which may be required on the farm.

Various mills have been invented of late years. Mr. Davis, of Linton, near Ross, has introduced an admirable machine, in which the crushing power, by a clever application of the French principle, is very considerably increased by causing the two stone cylinders to rotate at different degrees of speed. Indeed, in these days, machinery has reached so great a degree of perfection that a traction steam engine draws the mill and an attendant press into the orchard ; grinds up the fruit heaps at a rapid rate, and presses the pulp forthwith. The math or cake is rejected on the spot, and the casks all filled at once with the must. The whole process is completed with an economy of time and labour that can scarcely be exceeded. This economy, however, is false ; it exists only in the rapidity of the work. It would soon vanish if the mill had to be taken from time to time to the orchard, as the different varieties of fruit ripened; and thus it comes to pass that all the Apples are ground up at once.—early and late varieties,—ripe and unripe,—and are forthwith submitted to the press. No time is allowed for the pulp, or " pommage," as the old writers called it, to absorb the oxygen from the air, or for the juices set free to extract the full flavour of the fruit from the rind, the pips, and its more solid parts,—thus the Cider must inevitably be of inferior quality, and thus too the so called economy defeats itself.

GRINDING.—The degree of fineness to which the fruit should be reduced into pulp has been much discussed. The writers of the 17th century considered that the fruit need not be ground very small, though they stated that it was the common practice to do so, in order to obtain more Cider. Marshall says that in the South, and everywhere except in the Cider Counties, it was thought that the cellular juice of the fruit alone formed the necessary ingredient of good Cider. In Herefordshire it was the common belief in those days, as indeed it may still be said to be, that the flavour of the Cider was chiefly derived from the kernels or pips, and the colour from the skin of the fruit ; and it was therefore considered important that the pips should be crushed in the mill. M. Berjot, a distinguished French chemist,—who has studied the subject of Cider making, and who invented a very valuable *Écraseur*, or crushing mill, and who, moreover, took a prize for an essay on the "*Chemical Analysis of the Seeds of Apples*,"—proves, by numerous experiments, that for Cider of the best quality it was better not to crush the pips, because the diffusible odour of the essential oil they contain, spoilt the delicate flavour of the Cider : but with fruit of an inferior quality, deficient in good flavour, that it is rather an advantage to do so, since the pips then gave their own flavour to the Cider, and took away the earthy taste it is otherwise so apt to have. M. Berjot's mill, therefore, was specially designed to tear up and crush the fruit without bruising the seeds.

M. Hauchecorne also distilled the spirit from Cider made with the pips, and from that made without pips, and obtained excellent brandy from both, though the flavour was very different. The judges pronounced them equally good.—"*Le Cidre*," p. 341.

The common belief, therefore, that it is necessary to crush the pips to obtain the best quality of Cider is not correct ; and the impression also that its colour is derived from the skin is equally wrong, for, as was pointed out by Marshall, the palest coloured Apples often produced the ruddiest Cider. He instanced the *Hagloe Crab*, and it is equally observable in the Cider from the *White Must*, the *Royal Wilding*, and several other Apples, that have but little or no colour themselves.

In grinding the first portions of fruit, especially in a dry season, it is necessary to sprinkle water over the Apples "to wet the mill" as it is termed. The first grinding should be immediately pressed and the expressed juice or must, used to give moisture to the succeeding grindings. The facility with which water may be added, however, is much to be lamented, for in this way the character of the cider is much deteriorated. It has become the custom, especially in bad seasons when fruit is scarce, to add a considerable quantity of water to the must, and the result becomes a water-cider, cyder kin, or purre, as our ancestors called it. The value of the so-called "cider," is of course lessened in proportion to the quantity of water added, and though it may thus be sold at a cheaper price, and the adulteration doubtless is made to suit the pockets of customers, it is a sad mistake all the same ; increase of bulk increases the trouble and therefore the expense, and the adulteration prevents the possibility of obtaining the price a better cider would command.

The solid portions of the pommage, that which remains in the pressing bags, now called the "math," "cake," or cheese;" and by old writers "powz" or "mure," is often re-ground at the mill along with the inferior fruit and with the addition of a considerable quantity of water. There seems every reason to believe that this practice is a good one when the second grinding quickly follows the first. In this way an inferior cider, or cyder kin for home use is legitimately made.

> "Some when the Press by utmost Vigour screw'd,
> Has drained the pulpous Mass, regale their swine
> With the dry Refuse . thou more wise shalt steep
> Thy Husks in Water and again employ
> The pondrous Engine."'		(Philips "*Cyder.*")

When not required to be re-ground, the math, or cake, is sometimes mixed with chopped straw and given to the cattle. The animals enjoy it and it is thought wholesome and good for them. In small holdings it is frequently used when dry as fuel ; or lastly, the math may be thrown to the manure heap for the purpose of being returned in this form to the Orchard.

By common consent, Pears for Perry require comparatively but little grinding.

IV. FERMENTATION.

CLEANLINESS.—In all the varied processes in the manufacture of Cider and Perry, from the beginning to the end, the most scrupulous cleanliness is required. The mill should be well cleaned before the fruit is brought to it ; if of stone, it must be scrubbed throughout ; the iron clamps, and especially the leads which fixes them, must be carefully cleaned : if it is an iron mill, not only the stone rollers should be cleaned, but any rust which may have collected upon the framework must also be rubbed off, and the surface, if possible, kept bright. The juice of fruit will not dissolve the metal itself unless it is left long in contact with it ; but it will readily dissolve at once the dull grey powder which forms on lead, and the brown rust of iron, which are the oxides of the metals ; while the salt formed, being quite soluble, is carried through the process of fermentation, and remains in the Cider. The salt of iron, if it were sufficiently strong to be injurious to health, would spoil the Cider, so no more need be said of it : but the salt of lead is much more dangerous, and, since it sweetens the Cider, it is not to be detected so easily. In almost all the Cider districts the most painful cases of colic frequently occur from want of care to prevent any contact with lead. Sometimes the pulp is left in the mill for many days, when the lead is dissolved from lead soldering of the clamps, which connect the stones together ; or again, when it is in a still more dangerous form, the lead is dissolved by the Cider from the casks when white lead has been most improperly used by the cooper as caulking, to prevent leakage.

> " Evil is wrought by want of thought,
> As well as want of heart."
> Hood's " *The Lady's Dream.*"

Or lastly, and this is perhaps the most frequent cause of all, the Cider takes up the lead from the cider engine at the bar tap ; and the "Boots," who drinks the first jug drawn in the morning, instead of throwing it away, as is the general rule, gets a most serious illness. Too much care cannot be taken to prevent the contact of Cider with lead, either in its manufacture or its preservation.

The barrels or casks must also be examined, and if not perfectly clean and sweet they must be made so. Scalding with boiling water is the common practice, and some first clean the barrels by passing in through the bung hole a yard or more of stout iron chain, with a cord attached to one end for its removal, and rolling the barrel well about. A powerful jet of steam thrown into the barrel is far more effective, for obvious reasons, than scalding water, where circumstances admit of its application. Should these efforts not make it perfectly sweet, the barrel should be fumigated with the sulphur match, and scalded well after it. It is far better, however, to take out the head of every foul or even doubtful cask, that the cleansing may be thorough and effectual. This excellent practice is followed by some of the best Cider makers as regularly as the season comes round.

It would be well, however, whenever the cider making house admits of being sufficiently closed, to place all the vessels and implements used there, and then to fumigate it altogether by burning sulphur within it, and thus destroy any germs of injurious fermentation that may exist from previous operations.

COMPOSITION OF THE FRESH JUICE.—An analysis of the exact composition of the fruit of the Pear has already been given (see page 128). The Chemists appointed by the French Congress for the study of Cider fruits have given the following analysis of Fresh Apple Juice, as the mean of many examinations of juice from the best varieties of fruit ; their density varying from 1067 to 1080.

One Thousand Parts of Juice contained :

Water	800
Sugar capable of being converted to Alcohol	173
Tannic Acid, or Tannin	5
Mucilage, or pectosine (soluble pectine, gum)	12
Free Acids (Malic, Tartaric, (&c.) ... '	1.07
Albumen and Fermentable Matter	5
Saline Matters (Lime, Mallates of Potash and of Lime, Phosphate of Lime) ..	1.75
Pectic Acid, Coloring Matter, Fixed and Volatile Oils, and Insoluble Substances in suspension	2.18

They found the juice from inferior fruits contained the same materials, but in very different proportions, with the exception of the Albumen, Fermentable Matter, and the Salts of Potash and Lime, which were in much the same proportions in all qualities of juice. These inferior juices, having a low density, had one-third less of sugar ; the Tannin was only 1 part instead of from 4 to 6 ; and the amount of Mucilage was only 4 instead of 12 parts in a thousand.—" *Le Cidre,*" p. 111.

From an examination and comparison of the best Cider fruits of France, England, Germany, and America, the Congress states that the general characters they should possess are, good Perfume, slight Bitterness, and very little Acidity, with a notable quantity of Tannin and of Mucilage, and a very large amount of Sugar. Sugar, and the Alcohol formed from it, are the most important elements of Cider, and the best varieties of fruit are essentially necessary for their production in sufficient quantity to enable it to travel without injury. The best and soundest Cider should contain from 8 to 10 per cent. of Alcohol, (the French say 12 per cent) ; with from 2 to 3 per cent. of sugar still unreduced, to give it the highest commercial value.

M. Pasteur gives as the result of many analyses that 100 parts of the Sugar of fruits capable of being convertible into Alcohol yield :

Carbonic Acid	46.67
Alcohol	48-46
Glycerine	3.23
Succinic Acid	0.61
Matter yielded to the ferment ...	1.03

Tannin, or *Tannic Acid,* is the next most important element in the fresh juice. It makes the liquor "fine" more readily by causing the Albumen, the Pectine, and the Yeast Plants to be deposited ; and thus acts indirectly as an antiseptic, regulates the action of fermentation, and prevents the after tendency to ropiness, so apt to appear in the liquor from fruits of great richness.

It is believed also to possess the great virtue, by its astringent qualities, of moderating the action of the Alcohol on the system in the wine and other liquors containing it, and thus rendering them at once less exciting and more strengthening.   The French chemists state that it requires from 2 to 3 parts in 1000 to produce its full effect in the manufacture of Cider and Perry, and that from 2 to 3 more thousandths should be present for its wholesome qualities.

The *Mucilage*, or *Pectosine*, when abundant, forms another element of distinction between good and bad juice.   It renders the fresh juice more thick and viscous, and eventually gives softness and body to the liquor.   It helps also to preserve the alcohol by opposing the acetous fermentation,' and is thus always present in long keeping Cider and Perry.

The *Malic* and *Tartaric Acids* give the refreshing character to Cider and Perry, which is so desirable in a summer beverage.   The Malic acid abounds most in Apples, and Tartaric acid in Pears, and their too great abundance is to be feared rather than their failure.   The *Acidity* these acids give, together with the *Perfume* and *Bitter Principle* in the juices, which also tend to render the Cider and Perry more pleasant and agreeable, are difficult to determine chemically ; but their proper quantity can be estimated, with sufficient practical accuracy, by smell and taste.   An excess of acidity is the chief characteristic in fruits of poor character.

THEORY OF FERMENTATION.—The natural saccharine juices of all fleshy fruits, if left to themselves at a temperature from 50 to 80 degrees, will immediately begin to take on vinous fermentation, and especially if they are acid, as is usually the case.   This fermentation, moreover, takes place without the addition of any substance to bring it on.   Thus, if the pulp, or pommage, from ripe Apples, or Pears, as taken from the mill, be left to itself, at the ordinary autumnal temperature, minute bubbles are soon observed to rise to the surface and form a white froth ; its bulk will be increased ; and, if the thermometer is plunged into it, its temperature will be found to have risen.   These changes are due to the commencement of alcoholic fermentation, the bubbles contain Carbonic Acid gas, and, if the juice is tasted, it will be found to have acquired a spirituous fragrance due to the formation of Alcohol.   Why should the simple crushing of ripe fruit lead up to a series of changes so curious, and yet so certain ?   The distinguished Frenchman, M. Pasteur, has occupied much of his life in attempting to answer this question.   He has conducted a long series of experiments, requiring the utmost patience, with the closest attention to minute details throughout, and he happily possesses the genius which has enabled him to arrive at many very interesting and important results.

M. Pasteur has succeeded in proving that on the external surfaces of all fleshy fruits, when they become ripe, there exist certain minute particles or germs, which when brought into contact with the ripe juices of the fruits develop into minute plants, which grow with great rapidity.   These plants are the yeast plants, which belong to a great family of microscopic funguses.   They are called *Saccharomyces*, or " Sugar-eating Funguses," from the peculiar power they possess of decomposing, and living upon the saccharine principle of plants, the grape sugar, or glucose, as it is termed by the chemists, and thus causing their elements to be re-arranged into Alcohol, Carbonic Acid Gas, Glycerine, Succinic Acid, Volatile Acids, and other products.

M. Pasteur obtained these corpuscules, or germs, by washing ripe fruit—(grapes he first used)—with pure distilled water. The water was rendered slightly turbid by the presence of an infinite variety of minute particles ; many of them were shapeless atoms of dust, scales of epidermis, or spicules of crystalline matter, but many others appeared to be organized corpuscules resembling the spores of funguses. These organized corpuscules differed considerably from each other, and when M. Pasteur cultivated them, with all due care in saccharine fluids, he found them to swell and germinate at different times and in different ways. In an hour, and often in less time, he observed a copious formation of new cells, whilst small bubbles of Carbonic Acid gas were given off, shewing that the formation of Alcohol had already began. They were thus proved to be true yeast plants, or *Saccharomyces*. M. Pasteur traced the growth of several species of yeast plants under the microscope, all differing in their size of cells, shape and mode of budding, and general growth. The most common of these plants, to whose growth in the natural saccharine juices of fruits the formation of Alcohol is chiefly due, he described minutely in 1862, in the *Bulletin de la Société Chimique*, p. 67. These observations were fully confirmed by Dr. Rees, a German Physician and Naturalist, in 1870, and he first attached to them the following specific names :—

Saccharomyces apiculatus, which is the first of them to grow, and the most minute in size :

Saccharomyces Pastorianus, which is by far the most active and abundant, and which Dr. Rees named after M. Pasteur :

And *Saccharomyces ellipsoideus*, which is the slowest in growth, but most persistent, and which is the ordinary ferment of wine.

M. Pasteur describes minutely, from his observations, the life history of these *Saccharomyces*, their several modes of rapid development and reproduction, together with the chemical changes they effect by the decomposition of Glucose, such as the production of Alcohol, Glycerine, &c. These plants frequently take different forms, according to the varying circumstances under which they grow : for example, one form of *Saccharomyces Pastorianus* is so small that it was thought to be a different ferment, and was called by Dr. Rees, *Saccharomyces exiguus ;* whilst another form was named by M. de Bary *Dematium pullulans.*

Certain it is, that the mode of life of these plants is essentially different from that of all other living organisms, and the resulting chemical action is equally exceptional. Most organised beings live and grow by absorbing oxygen from the air and setting free carbonic acid : so do the *Saccharomyces* in the first stage of their existence ; but the air of the fluid in which they live is quickly exhausted, and when this happens, they obtain the oxygen essential to their growth from the glucose ; and in decomposing the glucose they set free more oxygen than they require ; and this, uniting with the hydrogen and carbon present, forms the various products of the fermentation they occasion.

There are numerous other microscopic funguses, whose minute germs are always present in the air, ready to take their life growth in the decomposition of saccharine fluids; such as various species from the families *Mucedines, Mucorina, Torulæ,* &c. These fermentations, commonly called " after fermentations," M. Pasteur calls " diseased," because their propagation and development is always attended with the loss of the Sugar or Alcohol, and with the production of some unpalatable result. The Cider or Perry may thus become acid, viscous (ropy), or be altogether spoilt, according as the germs of the several funguses which produce these results have been able to develop themselves within it.

The fungus which causes the dreaded ACETIC FERMENTATION is the *Mycoderma Aceti.* Its germs are so minute as to be only perceptible with a powerful microscope, when they are diffused in the liquor; but when aided by exposure to air and a high temperature, they are quickly developed into chains and chaplets, that soon appear as a film of grey mould floating on the surface, and this is commonly called "*Flowers of Vinegar.*" When this film has grown thicker, and becomes submerged, it takes on a gelatinous form of surprising toughness and lubricity, and is then called "*Mother of Vinegar,*" or the "*Vinegar Plant.*" The *Mycoderma Aceti* requires a warm temperature, and a much more abundant supply of air than do the *Saccharomyces* which effect the Alcoholic Fermentation, and the more freely it is supplied the more rapidly the plant grows, and the more quickly is the vinegar produced. The *Mycoderma Aceti* has the power of decomposing sugar and alcohol singly or in combination, producing Acetic Acid and water without the evolution of Carbonic Acid gas. When the access of air is prevented, as should always be the first care of the Wine or Cider maker, its action is extremely slow. It is sure, nevertheless; for the germs that find their way into closed vessels and well corked bottles will prevail in the long run. You may have an excellent bottle of wine or cider, but it will end in a bottle of vinegar, though it may take half a century to produce it.

VISCOUS FERMENTATION, or "ROPINESS" is also caused by the rapid growth of the minute spherical germs of a fungus, not as yet specifically named. It quickly develops itself into chains of vesicles, and in this process changes the glucose into Gum and Mannite, with the evolution of Carbonic Acid gas. In some seasons Ropiness is very troublesome, and remedies in abundance have been recommended to check it, in accordance with the prevailing belief as to its cause.

PUTRID FERMENTATION, it need scarcely be said, is not due to the growth of fungus plants, but to the presence of Bacteria, Vibriones, and Infusoria in general, whose germs are also always present in the air, and when deposited under circumstances favourable to their growth develop themselves with great rapidity, to the destruction of the liquor.

M. Pasteur, having thus proved that the vinous fermentation of saccharine fluids was caused by the plants growing from germ cells found on the surface of ripe fruits, next endeavoured to account for their presence. Infinitesimal as they are, and only perceptible by the aid of the microscope, he concluded that they formed part of the dust wafted about in the air. The germs themselves, and their mode of growth, he found to resemble the spores and habit of growth of certain funguses in the family group of *Dæmatiei,* which are common on dead wood during the autumn months. Some species of this family, there is reason to believe, produce two forms of germs cells, the one set adapted to aërial growth, and the other capable of living when submerged in fluid by decomposing the substances with which they come in contact. Thus Alcoholic Fermentation may be briefly defined as "A CHEMICAL ACTION RESULTING FROM THE DECOMPOSITION OF GLUCOSE BY THE GROWTH OF CERTAIN CELLULAR FUNGUSES."

These results of M. Pasteur's labours have met with general acceptance, and, so far as they have gone, they have completely changed the theories of fermentation formerly believed in. They require confirmation, however, and to be carried much farther, before the minute and complicated changes which are ever going on in the decomposition of organic substances—acting and re-acting on each other as they do—can be fully understood. It is happy for mankind that, guided by practica. experience alone, it has been able to enjoy the results of fermentation without the necessity of waiting for Science to give the explanation of the various stages of the process.

> "Come let us hye, and quaff a cheery bowl,
> Let Cyder new wash sorrow from the soul."
> Gay's " *Fifth Pastoral.*"

PRACTICE OF FERMENTATION.—It is agreed on all sides that the pommage, or pulp of the fruit, should be removed from the mill as soon as the grinding is finished—that is, as soon as the Apples or Pears have been reduced to one uniform pulp : but there has been much discussion as to how long it should be allowed to remain before being submitted to the press. The old writers state that it was the general practice in their times to press the pommage at once from the mill, and fill their barrels from the press; but they are unanimous in advising that the pommage should be placed in open vessels, from twenty-four to forty-eight hours, before the must is expressed from it. Thomas Andrew Knight held the same opinion. In America the pommage is allowed to remain in an open vat from twenty-four to forty-eight hours, or longer, according to the prevailing temperature, and an instance is given in Kenrick's "*New American Orchard*" (1844), where a Mr. Price won the first prize in Concord, Massachusets, and great distinction, for Cider made from Apples, whose pommage had been exposed for eight days before being brought to the press. In Germany, and in some parts of Normandy, Professor Schlipf states the pommage is left in open vats from five to twelve days, until fermentation is well established and the lees begin to settle, when the thin liquor is drawn off, and the remainder submitted to the press. The French do not follow this custom, and, except when they very occasionally use table fruit, they press the pommage at once from the mill.

The advantages to be derived from leaving the pommage undisturbed for a certain period are very great. The juices of the fruit, set free by the crushing, are enabled to re-act on the peel, the kernels, and the more solid tissues, aided doubtless, by the alcoholic fermentation which so quickly sets in ; and they thus have time to extract the full flavour, perfume, and colour of the fruit, which are all so essentially required to give character to the Cider. The common practice (becoming unfortunately still more common) of pressing the pommage direct from the mill, is therefore very disadvantageous.

The pommage from the mill should be placed in large wooden vessels, and filled to within a foot or eighteen inches of the top. The vessels should be covered with a cloth or board, and allowed to remain untouched for a couple of days, and the pommage may safely be left for three or four days if the weather is cool. A gentle fermentation quickly begins, and within a few hours minute bubbles rise to the top, and soon form a white froth there. As the Carbonic Acid gas escapes, it spreads over the surface at the top of the vat, and thus keeps off the action of the outer air from the pommage, although it may be left for some days.

> " Yet even this season pleausance blithe affords,
> Now the squeezed press foams with our Apple hoards."
> Gay's " *Fifth Pastoral.*"

PRESSING THE POMMAGE.—When a sufficient time has elapsed successive portions of the pommage are taken from the vats, and placed upon close textured rough horse-hair cloths, with the ends folded over. Several of these cloths are placed over each other, a dozen or more at a time, and are all pressed together. In Devonshire, successive layers of reeds, or of fresh drawn clean straw, are often used in the press instead of the horse-hair cloths. The press is precisely similar in principle to that used for making cheese, but its machinery of late years has been considerably improved; and, indeed, the whole process of what is technically called "making the cheese" simplified and accelerated. Pressure should be made very gradually at first; for the juice which first exudes is turbid, and the latter portion runs clear. It is put altogether into large hogsheads, generally holding 100 to 115 gallons each in Herefordshire, but in Devonshire the hogshead invariably contains 50 gallons. The barrels are not quite filled up, a slight "ullage," as the unfilled space at the top of the barrel is termed, being left. They are placed in a cool cellar or draughty outside building, to undergo the more active stage of fermentation. If the temperature is favourable, ranging from 60° to 70°, very evident signs of increased action will soon appear. The bubbles of Carbonic Acid gas begin to rise quickly, and a constant hissing noise will be heard. These bubbles of gas carry up with them to the surface many of the lighter particles of the cellular tissues of the fruit that have passed through the press, and thus form a thick scum on the surface, to which cells of the yeast plants are gradually added in considerable quantity. This scum thus becomes a thick spongy crust, sometimes called the "upper lees," which is supported on the surface by the Carbonic Acid gas arising beneath it, so long as this gas is generated in sufficient quantity. At the same time that this action is going on the more solid particles of tissue sink through the fluid, accompanied with a considerable portion of the mucilage, and an abundance of yeast cells. This deposit forms the "lees," or "lower lees," at the bottom of the barrel.

As the fermentation declines the hissing moderates, since less Carbonic Acid Gas is generated ; the floating crust gets dry on the surface, cracks, and losing its buoyancy falls in fragments, to increase the amount of the lees below. By this time the liquor will have become moderately clear, or "dropped bright," as the phrase goes. It should then be racked off immediately and the temperature kept low. This is the crucial point of the whole process, and requires close observation and care ; for any delay at this stage incurs the risk of injurious secondary fermentation. The clear liquor should be racked, or run off from the lees into a fresh cask, perfectly clean and sweet, by means of a syphon, so as to prevent any unnecessary exposure to the air. A considerable "ullage" should be left in the barrel, and the bung is usually left open for some days, or even weeks. It is better, however, to close the cask with a bung through which a curved tube passes, one end open below the bung, and therefore in the "ullage" space of the cask ; and the other end bent down and up again so as to hold a tablespoonful or two of water, placed in the outside bend ; or the outer end of the bent tube may itself be put in a shallow cup of water without being turned up again ; so that if any excess of gas should be formed in the barrel, its pressure would force it to escape through the tube and water, whilst the outer atmospheric air would be prevented from passing into the barrel by the water in the tube itself, or in the vessel it terminates in. This tube may be made of zinc, or other metal, but it is better and equally cheap to have it of glass, because the amount of pressure from the gas within the barrel can then be seen at a glance.

If at the end of a week the liquor remains quiet, and becomes more clear, an ounce of dissolved isinglass should be added to each hogshead, and the vessels permanently closed. The isinglass should be allowed to dissolve gradually in some of the liquor without the application of heat ; and this will require a period of two or three days to take effect properly. In January or February the bungs may be tightly driven into the barrels, and they may remain until the liquor is required to be racked in the spring months into purchasers' casks.

The process of fermentation should have been conducted thus far at a temperature as uniform as possible. It should never exceed 70 degrees, and it should never be below 50 degrees. After racking it becomes very advantageous to keep it below 40 degrees. The barrels into which the liquor has been racked should therefore be placed in a cold cellar, and kept at a temperature as low and as even as possible.

Although active fermentation may be said to cease when the hissing noise is no longer perceptible, yet it still continues to go on quietly, for the quantity of Alcohol slowly increases, and the Sugar decreases in proportion ; whilst at the same time the liquor becomes more clear, and requires a higher aroma as well as additional strength.

When the fruit has been well ripened on the trees and well mellowed in the heaps, there is generally but little difficulty in managing the fermentation, and still less fear of the liquor not fining properly.

The American method of fermentation, as described by Downing, consists in placing the newly filled casks with the bungs out either in a cool cellar or in the open air and as the scum works out the barrel is kept filled every day with some of the same must, kept for this purpose. In two or three days the rising will cease and then the first fermentation is over, the bung is now closed and in two or three days driven in firmly, leaving a small vent hole open which is also to be stopped in a few days. The clear liquor is now racked off by syphon into a clean cask, and if in a few days it is found to remain quiet, a gill of finely powdered charcoal is added to each barrel, when it is closed and left until spring. In March they rack again and if the cider is not quite bright three quarters of an ounce of dissolved isinglass is then added to each barrel. In a few days it will be fit for bottling, and this may be done at any time up to the end of May.

The French method of fermentation is as follows. They remove the must at once from the press into large oak casks well cleaned and prepared for it. They are filled to within 3 or 4 inches of the brim, and placed in rows in a cellar with a minimum temperature of 12° centigrade, or 53° Fahrenheit ; and if fermentation is slow they increase it to 25° centigrade, or 77° of Fahrenheit, by moveable stoves. When the active effervesence begins to subside, and the cider, to use a technical phrase, is between the two lees, the density of the fluid will be found to have decreased from 1067 to 1035. This is the proper time to rack it, which they do into casks which have been well cleaned and are quite free from any bad smell or taste. The oxygen of the air is exhausted by burning a little spirit in the cask, or if its condition is in the least doubtful, it is sulphured. Sometimes a small portion of Alcohol is now added to each cask, and almost invariably, they also add eight ounces of Catechu previously dissolved in cold water to every hundred gallons of cider, and then fill up and slightly bung the casks. When the density of the liquor is reduced to 1022, the bungs are to be tightly closed. An ullage of one or two inches being allowed to each cask.

In Jersey and the Channel Islands the active fermentation is permitted to take place in open vessels, covered only by cloths. the scum or upper lees being skimmed off as it forms. As soon as the liquor becomes clear and the fermentation subsides it is racked into sulphured casks, and this process is repeated some three or four times.

The time over which sensible fermentation should extend is necessarily variable, since it depends on the density or richness of the fruit, and the temperature of the air. It is most favourable when it is active and regular, but if it is too violent the liquor will overflow and waste, and if it is too slow, it will be imperfect and develop the desastrous "after-fermentations."

> "*Perry* is the next liquor in esteem after Cyder, in the ordering of which, let not your pears be over ripe before you grind them ; and with some sort of pears the mixing of a few crabs in the grinding is of great advantage, making *Perry* equal to the Redstreak Cyder."   Mortimer.

The manufacture of Perry in its earlier stages differs somewhat from that of Cider. The fruit contains more Sugar and Mucilage. The must after pressure is allowed to remain in open vessels, lightly covered, to undergo active fermentation. As soon as this has subsided, the liquor between the upper and lower lees should be sufficiently bright to be drawn off and treated as in the case of Cider : but as a matter of fact Perry can seldom be made so easily. The amount of Mucilage renders it necessary, almost invariably, to follow the tedious process of dropping it through bags carefully made of a rather course flaxen material, called "forfar." The liquor must be stirred up each time the bags are filled, for the more turbid it is, the brighter it will run through the bags when the process is carefully managed. The filtered liquor is put forthwith into hogsheads. From one to two ounces of isinglass, previously dissolved in some of the cold liquor, is added to each hogshead, the amount varying according to the condition of the liquor.

The casks are generally placed on their sides, but some think it is more safe to place them on end, but in either case a considerable ullage, to the extent of at least a couple of gallons is left. Then close up tightly to exclude the air, cement the bung, but leave a vent tube through it, the one end open in the ullage space, and the other end bent down and dipped into a cup of water as before explained. Should the Perry remain quiet for a week the bent tube may be removed, and the hole it passed through, quickly and effectually closed, or as is sometimes done, the tube may be allowed to remain until Spring. If it should not remain quiet, and syphon racking become necessary, it would be a great misfortune for the Perry.

ORCHARD BRANDY.—A spirit may readily be obtained from the refuse of Apples and Pears when it may be thought desirable to do so, just as it is from that of grapes after wine making. The cakes from the press are added to the lees on the first racking with a sufficiency of water and refermented. As soon as the active fermentation is over and the lees settled to the bottom, the spirit may at once be distilled from the fluid portion, or it may of course be distilled from the first fermentation of the fruit. The distillation should be effected by means of the water bath, or the brandy will have a burnt rancid taste. The brandy will vary in flavour and in strength according to the goodness of the fruit, and the care which has been taken in its manufacture.

In the early part of last century an extraordinary Cider was made which received the name of "Royal Cider," and during the wars with France was extolled to the skies as eclipsing the finest

French wines. The whole secret consisted in distilling the Alcohol from one hogshead of cider and adding it to another; thus "fortifying" it, as brandy is used to fortify grape wines for exportation.

In years of great abundance of fruit, when the barrels are all filled with cider, and tons upon tons of fruit are still left to rot away in the Orchards, a great economy would be effected if the fruit could be crushed, fermented, and the spirit distilled from the liquor; for with good fruit a brandy of superior character would be obtained. The great obstacle consists in the uncertainty of the crop. Marshall mentions that "in 1788 there were men who would make 100 hogsheads that in 1783 did not wet the press;" and many of our readers will remember that in the years 1856-7-8-9, a succession of bad seasons, there was not half a hogshead of cider made in several famous fruit districts in Herefordshire, whereas in 1867-8, hundreds of bushels of fine fruit were lying in heaps in March which could not be sold, and were left there to be absorbed by mother earth.

Well fermented Cider of good quality should contain from 5 to 10 gallons of Alcohol in 100 gallons of liquor, and the French chemists say as much as 12 per cent. Good Perry is stated to yield 7 per cent. of spirit. The practical rule for estimating the strength of the must of Cider or Perry, and for all Saccharine unfermented liquors, is to allow 1 per cent. of Alcohol for every five degrees of density. For the sake of comparison it may be added of the grape vintage, that Claret of first quality should contain from 13 to 17, Sherry from 18 to 20, and Port from 24 to 26 per cent. of Alcohol.

DIFFICULTIES OF FERMENTATION.—The combination of favourable circumstances necessary to perfect fermentation cannot always be commanded by the most careful managers; but with the carelessness so general in the orchards it is often positively prevented. The ordinary sources of difficulty are many. The season may have been bad and the fruit not well ripened; the varieties of fruit may be poor with weak watery juices; the fruit may not have been well mellowed in the heaps; it may have been overheated and frostbitten; or it may have been crushed indiscriminately from the heaps; the prevailing temperature at the time, may delay injuriously, or hurry on too quickly the fermentation; or lastly, there may be a want of cleanliness in the Cider house, the vats or the implements used. Such circumstances for the most part inevitably result in the production of inferior liquor, but by good and proper management it can be kept from getting so bad as it otherwise would do.

The knowledge that all Fermentation is connected with the growth of Fungus Yeast Plants in the fermenting fluid, at once affords the explanation of many of the difficulties that arise in the process, and points out the means best adapted to meet them successfully. Circumstances which encourage the rapid growth of these plants, such as a warm temperature and juices rich in Saccharine principle, produce an active fermentation: Whereas, a low temperature and thin watery juices, deficient in Glucose, cause them to grow so weakly, that a low fretting fermentation sets in and creates great difficulty, at first to increase its activity and afterwards to arrest it. This would often be impossible without the addition of certain substances which have the power of stopping fermentation at once. These are commonly called "anti-ferments," but we now know that they stop fermentation simply by destroying these Yeast Plants. Bearing these facts always in mind the difficulties most commonly met with may now be briefly considered.

Active Fermentation.—When the weather is hot, and the juices rich, the fermentation soon becomes very active, and may cause both waste and trouble by out-pour from the barrel. In its earlier stages however, fermentation can scarcely be too active if it is not too long continued ; and all that need be said about it here is, that the windows of the cider house should be thrown open and cold water sprinkled about for evaporation, and every thing be done to cool the temperature.

Dilatory Fermentation.—This is a much more frequent and troublesome difficulty when cold weather sets in suddenly as it so often does in late Autumn ; and especially when the juice is of inferior quality. If it is simply a matter of temperature, and the tight closing up of the cider house is not sufficient, the introduction of a small stove or two will be the best remedy. This will be aided too by drawing out two or three gallons of the juice from the cask, warming it up to 70° and returning it again, when the must should be well stirred up. The French recommend this stirring up to be done frequently with a long rod of birch twigs through the bunghole. There is a fancy often, followed, of adding a little old Cider or Perry to the cask, and some go so far as to add a little ordinary yeast from malt liquor, but this is a proceeding of very doubtful benefit.

Persistent Fermentation.—The first fermentation will sometimes continue in a more subdued form, when its active stage is over. This is called " fretting fermentation," and if allowed to go on, it will quickly exhaust the Saccharine principle, while the liquor will lose its sweetness and strength and become acid. The practical cider maker judges by the smell and taste of the liquor when this period has arrived, but the Saccharometer is more to be relied upon, and when this shows that the density is below 1040, the fermentation must be at once arrested or the quality of the Cider will be injured. It then becomes necessary to use one or other of the anti-ferments or plant destroyers, such as Sulphur, Sulphurous Acid water, Bisulphites of Lime or Soda, Salicylic Acid, &c., &c. The two first named are the most safe, and the most effectual, and indeed they form the base of most of the others. They are easy of application, economical, and ought not eventually to produce any perceptible effect on the liquor, either to smell or taste. They are the only ones that need be alluded to here.

Sulphur has been used to arrest fermentation from time immemorial in all the great Wine districts of the Continent ; and in all the Cider and Perry districts of England ; and it may be said, that the custom of late years has prevailed universally. When everything is ready for the racking, the fresh clean cask is filled with the fumes of burning Sulphur, "stummed " or "stunned " as it is termed, (a contraction doubtless from brimstoned). It is done in the following manner :—A strip of clean canvas cloth, or linen, some 10 or 12 inches long by 2 or 3 wide, having been dipped into melted Sulphur and allowed to harden, is attached to a long piece of wire. This cloth-match is lighted, and immediately passed into the barrel, the wire being fixed by the bung. It is thus suspended until it has exhausted the atmospheric air and filled the barrel with fumes of Sulphurous Acid Gas. The match is removed when it goes out, the syphon introduced, and the fermenting liquor racked into the barrel without allowing the fumes to escape.   The liquor absorbs the Sulphurous Acid Gas, and thus the yeast plants are destroyed. It is at first made thick and muddy by the process, but in a short time it becomes clear and remains so, without any taste or smell of the Sulphur, if it has been carefully done.   Should the hissing begin again in a few days the process is repeated.   The fumes of Sulphurous Acid are readily absorbed by water and a saturated solution is sometimes used instead of the ordinary gaseous fumes from burnt Sulphur.

Salicylic Acid has many advantages as a yeast plant destroyer, and has of late become in more frequent use to arrest continued fermentation.   It is a powerful remedy, and requires much care.   In proper proportions, it is quite harmless, tasteless, free from smell and does not change the colour of any liquid to which it is applied ; so long as it is not brought into contact with any metallic substance, particularly with iron which would at once give it a black colour.   It is used in solution, and is thus more easily applied than Sulphur.   An ounce or an ounce and a half to one hundred gallons is all that is required, and it is simply poured into the fermenting liquor immediately after it has been racked.   It is very effectual and leaves no sensible effects on the liquor.

Persistent fermentation is the great difficulty to be encountered with juices of inferior quality, whether this may have arisen from bad varieties of Apples or Pears ; imperfect management of the fruit ; or from the indifferent nature of the soil on which the trees have been grown.   The French chemists have had much experience in the endeavour to make good liquor from such poor materials, and have attained an amount of success that demands special notice.   They have established the fact that juices of inferior quality are deficient not only in Glucose, but also in Tannin and Mucilage. When the first fermentation is over, they rack into a cask filled with the Sulphurous Acid fumes to check further fermentation—they supply the deficiency of Alcoholic fermentation by the addition of Alcohol in the shape of brandy to "fortify" or preserve it—and add half a pound of the extract of Catechu (previously dissolved in some of the cider,) to every 100 gallons of liquor, which they believe not only aids in fining and preserving it, but also in making it more wholesome.   This may be so, but it requires some education in an ordinary palate to be able to enjoy the peculiar astringent flavour it produces.

Much more might be said on this subject as, for example, on the addition of Bitartrate of Potash, or Cream of Tartar, &c., &c., but the attempt to make good liquor by chemical means from bad juices can never be successful, and should never be encouraged.   The axiom might be laid down "*that Cider or Perry is the more pure and wholesome in inverse proportion to the amount of chemicals employed in its manufacture.*"   The best cider-makers in good cider districts do not happily require their use.

WANT OF CLEARNESS is the last difficulty to be considered, and it is one so very frequent in every quality of fermented liquor that careful cellarmen seldom trust to nature alone, however favourable the process of fermentation may have been,   The richer the juice and the more abundant the Mucilage, the greater is the difficulty of obtaining a clear bright liquor.   When the active fermentation is over, and the liquor is racked from the lees into a fresh cask, various substances are added to it for the purpose of "fining" or clarifying it.   To the best qualities of Cider or Perry,

an ounce or an ounce and a half of Isinglass, which has been previously allowed to dissolve gradually
in some of the cold liquor, or in milk, is added to each hogshead. Fish glue, in about the same
proportion, will answer equally well. Various other materials are often used : some add the
whites of a dozen eggs to each barrel ; roasted apples beaten up ; a pound of powdered charcoal ;
chips of fir ; oak, or beech wood are thrown in ; a lump of clay is sometimes ground in the mill
with the fruit, to make it " fine " the more readily ; a quart of wheat or barley is thrown in it ;
quick lime is added ; and many other heterogenous substances such as blood in large quantities,
are often used for the Albumen they may contain by trade cider makers. So long as they
afford Albumen in a cheap form, it matters not to them how disgusting the material may be which
contains it.

> " As *Cider* is for some time a Sluggard, so by like
> case it may be retained to keep the *Memorials* of many
> *Consuls ;* and these smoaky bottles are the *nappy Wine.*"
> " Dr. Beale, in Evelyn's *Pomona.*"

PRESERVATION OF CIDER AND PERRY.—When the liquor is made, and firmly and closely bunged
down in the casks, it will improve and keep good for a period which will vary according to its
strength. In former times it was drunk much sooner than it is now. It was never expected to
keep long, and would not do so, since very little bottling was practised. The cooling and Summer
fruit cider was ready to drink in a month ; that made from the *Gennet Moyle, Pippins, and
Pearmains* after the first frost ; whilst the *Red Streak* and Winter fruit cider barrels were not tapped
until the winter was well advanced, and were then drank through the following Spring and Summer.

The strongest and best cider will keep good in casks for four or five years. It was the custom
in the last century not to bottle it until two years old, and up to within the last twenty years, it was
not usually bottled until the late Autumn of the following year, when about a year old. It has now
become the custom to bottle all Cider and Perry in the early Spring of the next year, and by this
means a much greater amount of richness is obtained, although the risk of loss from the bursting of
bottles is greatly increased.

When a cask of Cider or Perry is to be bottled, the bung should be removed the evening
before, that the free gas it contains may escape, and the bottles should be all filled if possible before
any of them are corked, that the liquor may become still more flat, and thus the risk of loss from
bursting be lessened. For the same reasons, the bottles and especially the corks should be of
the best quality and be carefully wired. It is better also, when the amount of cellar space admits of
it, to let the bottles remain on end for a few weeks or even until the following Autumn before laying
them down in the bins.

A certain amount of insensible fermentation or molecular change continues to go on in Cider
or Perry, long after all signs of active fermentation have gone by. Thus they improve up to a certain
time in the cask, and they improve still more, and of course last for a much longer time, in bottles.
The Alcohol slowly increases and the Sugar decreases in the same proportion, whilst at the same

time, the liquor becomes more clear and acquires additional aroma with its strength.
Our ancestors well understood this, as the poet Philips shows :—

> "Cyders in Metal frail improve ; the *Moyle*
> And tasteful *Pippin*, in a moons short Year
> Acquire compleat Perfection : Now they smoke
> Transparent, sparkling in each drop, Delight
> Of curious Palate, by fair Virgins crav'd.
> But harsher Fluids different length of time
> Expect : thy Flask will slowly mitigate
> The *Elliots* roughness, *Stirom*, firmest Fruit,
> Embottled, (long as *Priameian Troy*
> Withstood the *Greeks*) endures e'er justly mild.
> Softened by age it youthful Vigour gains,
> Fallacious drink ! Ye honest men beware
> Nor trust its Smoothiness ; the third circling Glass
> Suffices Virtue."   Philips " *Cyder.*"

Nor does the poet in any way exaggerate either the durability, or the strength of Cider. A
supply of good Foxwhelp Cider, made in a good year, would have refreshed the warriors for twice
or thrice the duration of the siege of Troy. It will retain its full flavour for 20 or 30 years, and
a strength moreover that would require the three glasses of the poet, to be small ones.

Cider or Perry in cask, of ordinary quality, does not travel well. It is apt to undergo renewed
fermentation, and lose all its chief virtues. The Cider made in Normandy is chiefly used for sea-
faring purposes, and the French chemists have had great difficulty in enabling it to bear the rolling
of the ships at sea. It is with this view, as we have seen, that they add Tannin, and a small portion
of Alcohol to the liquor after its first racking. Economy prevents them adding sufficient Alcohol
to preserve it, and after a number of elaborate experiments, they found that the next best plan was
to bring the Cider up to a high temperature by artificial heat, and they have established furnaces for
this purpose in all their great manufacturies. The process however does not improve the quality of
the liquor, though it does not render it less effective in checking any tendency to scurvy.

Good, well made Cider should however travel in cask any where in reason ; and it will safely
do so, if its quality is what it always might be in Herefordshire. In bottle it travels well in cool
weather.

V.—THE ORCHARD IN ITS COMMERCIAL ASPECT.

The quantity and value of Apples and Pears grown in this country are very insufficiently
appreciated, for the good reason, that there are no statistics from which such information can be
accurately drawn. The " *Agricultural Returns*" last published by Parliament, for 1880, show that
the amount of orcharding in England, that is " *The Acreage of Arable or Grass Land, but used for
Fruit Trees of any kind,*" is 175,200 acres. Herefordshire stands highest in the list with 26,683
acres ; Devon is next with 25,758 acres ; then comes Somerset with 22,993 ; Worcester, 15,854 ;
Kent, 14.685 ; Gloucester, 14,178 ; and then, with a wide difference, Cornwall, 4,678 ; Dorset,
3,716 ; Monmouth, 3,618 ; Middlesex, 3,249 ; Salop, 3,248 ; and the remainder is divided between

twenty-nine other counties. This "*Return*" affords a basis for calculation, from which a rough estimate may be derived of the value of the Fruit Crop, but since it embraces all the hardy Orchard Fruits, it will be better to limit the enquiry to Herefordshire where the fruit acreage is the highest, and where the only Orchard Fruits grown are Apples and Pears.

Herefordshire contains,.as we have seen, 26,683 acres of Orcharding. Of this amount, in these days of cheap and rapid transit, when all Apples with size and colour meet with a ready sale as "Pot fruit," as it is called, that is fruit for edible or culinary purposes; not less than one sixth must in this way be first accounted for. Thus the product of 4,470 acres of "Pot fruit," at the low estimate of 60 bushels to the acre, and at the equally low price of 3/- a bushel, would produce £39,930. The remaining five sixths, or 22,213 acres, for the production of Cider or Perry would yield, on a low annual average, 2 hogsheads of 100 gallons each per acre; and this at the low price of 6d. a gallon would produce £111,065 :—and thus at this computation, purposely made so low, the yield for this County, would be at the rate of £5 13s. per acre of orcharding, per annum; and if the best fruit was grown, and the best Cider and Perry made, the profit would be very much greater as a matter of course.

It must·also be remembered that "pot fruit" is grown in almost every garden through the County, and this does not appear in the "*Return*." Its amount could scarcely be estimated at less than the Orchard "pot fruit," or £39,930 in addition to the sum already named.

The total annual value of the Herefordshire Apple and Pear crop, thus reaches, according to these estimates, the very large sum of £190,925. As a matter of fact however, it is not easy to determine the actual produce of English Orchards ; for there are no published records of the exact crops they yield year by year. As a general rule, the trees of "pot fruit," or "table fruit" as it is better called, bear a full crop every alternate year ; but this is not the case with the varieties grown for making Cider and Perry. These trees will bear profusely for some two or three years in succession, but after these great "hits," they seem to become exhausted, and, with the exception of a few individual trees. are apt to yield only a sprinkling of fruit for the next three or four years ; which leads to the direct inference, that with proper care, and a more liberal supply of manure, they would bear much more regularly.

The French have published a few systematic observations on this point. In the report of the French Congress, "*Le Cidre*" so often quoted, it is stated (p. 339—40,) that M. Varin-Simon, the proprietor of the celebrated Orchard for Cider fruit at Yvetot, kept an exact register of the annual yield from 105 fruit trees, for thirty-eight years in succession. His books show that each tree, from 5 to 20 years old, gave an annual average, over this series of years, of 216 litres (or 40 gallons) : and each tree from 20 to 80 years old yielded 307 litres (or 57 gallons) : or taking all the trees together during the 30 years preceding 1869. each one gives, the annual average of 2 hectolitres. 6 litres (or 45 gallons). This return of course denotes the highest cultivation and an excellent climate, but it is still so extremely favourable on the annual average, that we may well believe the popular saying in Normandy "*Le dessus vaut mieux que le dessous*," the trees yield more than the ground beneath them. The· actual return would, at this rate, amount to about 10 hogshead's per acre, even if the trees were 60 feet apart, which is double the distance of a thickly planted orchard.

Little information is handed down from early times as to the commercial value of Cider and

Perry. Evelyn speaks of *Redstreak* Cider which sold for sixpence the winequart, "not for the scarcety but for the excellency of it," and he mentions also that it was sometimes exchanged on equal terms for the best French wines.

In the Household Accounts at Holme Lacy in 1662, the price of the Hogshead of Cider is set down at £1 14s., whilst Beer cost only £1 4s. the hogshead.

In a letter dated "Bristoll, 20th November, 1691"—addressed by one Thomas Wattmore, a vintner, to Sir Barnabas Scudamore "at his seate neare Citty of Herriford," the writer states that he bought "six hogshatts of *Red Strike* Sider and never tasted them at all but gave you a noate under my hand to pay 25£ 15s. 00 for them." The Cider turned out badly and he demands a repayment. At the end of the letter he adds "I bought 50 hogshatts last yeare at Dimmock and they are as rich as new Canary. I cannot sell bad Sidor, &c., &c." This letter is quoted to show the high price of the best Cider at that time.

It appears from the Household Accounts of the Right. Hon. James, 3rd Lord Viscount Scudamore, also at Holme Lacy, that in the years 1703 and 1704 apples were bought at 2s. 3d. the bushel, and in a bill, without date, but of about the same period, a hogshead of *Red Streak* Cider was sold for £2 10s. : hogsheads of Cider were brought from Amberley and Marden for £1 2s. 6d. each ; a hogshead of *Golden Pippin* Cider cost £1 7s. 6d. ; a hogshead of *Quince* apple cider £1 6s., and a hogshead of Cider from Rotherwas cost £1 5s. It may be mentioned also that the price of labour for cooperage, cider making, grafting, &c., was 1s. per day at that time.

Batty Langley who wrote at the beginning of the 18th Century (1713) mentions that the *Devonshire Royal Wilding*, (a variety at the present time unknown in that county) "would fetch five guineas per hogshead, while common Cider goeth for 20s."

In Herefordshire, celebrated varieties seem always to have created a market, when inferior ones failed to do so. Marshall mentions the *Hagloe Crab* and the *Stire* Cider as worth at the press, from £5 to £15 per hogshead, but he adds that the ordinary price of Cider "on a par of years" is 25/- per hogshead. In 1720 bottled cider fetched 6d. a bottle, a sum equivalent to about 3s. at this time.

In Smith's "Dictionary of Commerce" it is stated that in 1833-4-5 the best cider ranged from 1s. to 1s. 6d. the gallon :—family cider for the farmer's own use, or for public houses 4d. to 10d. a gallon ; whilst the Cider-kin, or water Cider of the labourer when sold, ranged from 2½d to 6d. a gallon ; and these prices seem to have amply remunerated the producer.

The market prices of Cider at the present time are as follows : For the best quality of Cider sold by the Manufacturers in cask, from 1s. to 2s. the gallon; and the same quality meets with a ready sale, when fresh bottled, at from 8s. and 10s. to 12s. the dozen :—for Cider of the second quality, to which usually more or less water has been added, for family use on draught from the cask, at from 8d. to 10d. the gallon, according to its quality : whilst the common Cider for farm house use, varies from 4d. to 8d. The price of Perry ranges from 6d. to 1/6 the gallon.

These prices are those which generally prevail immediately after production ; but for special varieties of fruit, and for Cider a few years in bottle, the prices are much greater. At a publick auction a short time since, *Foxwhelp* Cider was sold freely at 30s. a dozen—*Taynton Squash* Perry fetched 28s. a dozen,—and either of these varieties, and some others too of good age and quality, will always command high prices. Twenty dozen of *Oldfield* Perry, in a good season, have been sold for as many guineas, from the Glebe land of Credenhill parish.

As a general rule perhaps, the small orchardists make better Cider than do the large farmers, and for the very reason that they give their chief thought to it. It is their main harvest, and it is not too much to say, that many of them get from their trees, not only the rent they pay, but a considerable help towards their livelihood too. A rough calculation may easily be made. At 30 feet apart there will be 50 trees to the acre, and with a fair " hit " of fruit, 40 of them should yield, at the very least, 6 hogsheads of liquor. The Cider or Perry, at the low rate of sixpence a gallon, will fetch £2 10s. the hogshead, or £15 altogether ; but some of it should be worth much more than this. Then there is the "pot fruit" from the 10 remaining trees, still left to be sold in the market ; and in addition, the profit to be derived from the produce of the ground. An acre or two at this rate, would give a handsome return ; and if the occupier be a Smith, a Tailor, a Shoemaker, or a Mason, as not uncommonly happens ; and if,—for there is one other important " if "—he does not drink too freely from his own vats, this addition to his trade earnings will put him in easy competence, and enable him to educate and place out his family to great advantage.

The fruit trees on farms of higher pretentions should also contribute much more towards the rent than they usually do.

VI.—RENOVATION OF THE ORCHARDS.

The condition of the Orchards at the present time is most unsatisfactory and the closest attention will be required to restore their value. A century of neglect has caused the loss of many of the best varieties ; and whenever substitutes have been wanted, they seem to have been procured at haphazard, that is at the least possible expense ; so that a large number of unproved, or chance seedlings, and other worthless varieties abound in most Orchards.

The first step towards their improvement will be to subject the Orchards to a thorough revision. Stock should be taken of every individual Apple and Pear tree on the farm, and its condition and character carefully considered. Such trees as are mere cumberers of the ground should be cleared off at once, root and branch : and such varieties as are proved to be unmistakably inferior must have their places supplied by those which are known to be good. If these last trees are vigorous and healthy, they should be beheaded as far from the main trunk as possible, and each spur of not more than two inches in diameter, should be grafted with strong growing scions, so as to bring them into bearing again with the loss of only two, or at the most three seasons : whilst if the trees of bad varieties are not vigorous they should of course be uprooted. In every instance the vacancy must be supplied whether by grafting or by replanting, with well-proved varieties; for it must never be forgotten that *when once planted, the best fruit trees do not require any more care or expense than the worthless ones.* This complete revision of the Orchards will require perhaps some years to effect, but it is a work of great interest and will well repay, by its success, all the time given to it.

TEST OF QUALITY.—The commercial value of any fruit for the manufacture of Cider or Perry can be definitely ascertained by any one who will take the trouble to do so. It is only necessary in a good season to crush out the juice from five or six well ripened Apples, of the variety he wishes to examine ; filter it through white blotting paper; and having procured a small glass instrument, called a " Saccharometer," (to be had for a few shillings from any druggist), let him float it in the fresh juice. The scale marked on the instrument will give the density of the juice with exactitude, as compared with the standard of water, which is placed at 1,000. This density is chiefly caused by the saccharine matter the juice contains, and chemists have proved that in general terms, every five degrees of density shown on the scale of the instrument denotes a spirit producing power equal to one per cent., or one gallon of Alcohol in one hundred of the liquor. Now, moderately good Cider should not contain less than six per cent. of Alcohol ; and about one fourth or one sixth of its density should still be left of unreduced sugar, to give it sweetness and body. The density of the juice therefore for moderately good Cider should not be less than 1040. The richer the juice, the higher the density, and the greater its value. Juice which has a density below 1040, though it may make Cider, or Perry, if the fruit has been grown on good land, can never give the superior quality it is so desirable to produce. *The Saccharometer will thus point out all the varieties of fruit trees which should be got rid of.*

The following Table shows at a glance the exact amount of spirit producing power contained in juice of any given density, according to the experiments of the French Chemists :—

TABLE SHEWING THE AMOUNT OF SUGAR CONTAINED IN THE FRENCH " LITRE," EQUAL TO ONE AND THREE QUARTERS PINT (OR 35 OZS.) OF FRESH APPLE JUICE, AND THE PERCENTAGE OF ABSOLUTE ALCOHOL IT WILL PRODUCE ON FERMENTATION.— " *Le Cidre*" (p. 130).

Density.	Sugar in 1¾ pint, or 35 ounces.	Absolute Alcohol per cent.	Density.	Sugar in 1¾ pint, or 35 ounces.	Absolute Alcohol per cent.	Density.	Sugar in 1¾ pint, or 35 ounces.	Absolute Alcohol per cent.
	Drs. Oz. Avr.			Drs. Oz. Avr.			Drs. Oz. Avr.	
1010	1.098	1.138	1038	2.273		1066	5.124	
—12	1.141		1040	2.350	4.85	—68	5.181	
—14	1.183		—42	3.006		1070	5.274	9.50
1015	1.203	1. 93	—44	3.084		—72	5.372	
—16	1.223		1045	3.114	5.64	—74	6.013	
—18	1.274		—46	3.161		1075	6.051	10.51
1020	1.322	2. 48	—48	3.252		—76	6.090	
—22	1.353		1050	3.332	6.43	—78	6.183	
—24	1.400		—52	3.409		1080	6.261	11.33
1025	1.415	3. 3	—54	4.065		—82	6.338	
—26	1.431		1055	4.102	7.26	—84	6.431	
—28	2.040		—56	4.143		1085	7.030	12.14
1030	2.071	3. 57	—58	4.220		—86	7.072	
—32	2.118		1060	4.308	8.11	—88	7.149	
—34	2.114		—62	4.386		1090	7.242	12.98
1035	2.169	4. 12	—64	5.031		92	7.324	
—36	2.195		1065	5.077	8.76	—95	8.210	13.86

It will be observed that the Sugar increases in relative amount the higher the density becomes ; which is explained by the fact that the Mucilage, to which the density in the lower ranges is partly due, does not increase as the Sugar does, in the higher ranges of density.   The Alcohol is little more than one per cent. for every 10 degrees of density, instead of one per cent. for every 5 degrees, which is the scale of the English Excise Office.

It is proposed before the conclusion of this work to give a list of the best varieties of fruit for the press, both of Apples and Pears, showing the density of their several juices, when a favourable season shall have enabled it to be taken to advantage.

TABLE FRUIT.—The varieties of fruit suitable for cooking or dessert purposes, "pot fruit," or "table fruit," should be grown much more freely than is usually the case, for the supply of our home consumption.   Size and colour are essential for ready sale in the market, and the longer the apples will keep the better and more valuable they will be.   This fruit will have to compete, and should be able successfully to compete, with American and other foreign fruit, and it will therefore require to be grown in orchards near the homestead where it can be protected.   As much as ten pounds an acre for fruit of this character upon the trees, has not unfrequently been given, but its market value will depend very much upon the season.   As a general rule "table fruit" is not well adapted for making Cider.   The French say "*petites pommes gros Cidre*," small apples rich Cider, and so too the finest wines are produced by the smallest grapes.   Large apples have too much Mucilage by themselves, though they may sometimes be added to others with advantage.

PEAR TREES.—On the poorer soil of the farm, pear trees should occupy the orchards.   They are much more slow of growth.   "He who plants pears, plants for his heirs" as the old saying goes, but they are much longer lived than apple trees.   They are very hardy, the flowers resist well the Spring frosts, and the trees bear abundantly.   The fruit ripens early and is more sweet and juicy than that of the Apple ; and Perry is always saleable to the merchants for a variety of purposes. The trees too are more productive than Apple Trees, and as they grow more upright, they interfere less with the cultivation of the ground beneath them.   A good hit in an orchard of Pears of a valuable variety, is sometimes worth the fee simple of the land the trees grow upon.

SEEDLING TREES.—The advantages of seedling fruit trees are very great, if the varieties are good.   They are more robust and hardy, and consequently bear more freely.   They should never be planted in the orchard until the fruit has been carefully examined by the Saccharometer and its juice found to possess the requisite density.   Seedling trees for the most part are quite worthless, and they must therefore be tested with much care.   A special exhibition of seedling fruit trees was held at Yvetot in Normandy, when 172 varieties were sent for examination, nine only of these, (which had themselves been selected from many others), furnished a rich juice of high density.   And again M. Legrand out of 65 carefully grown seedlings had only one single variety worth cultivating.   Mr. Thomas Andrew Knight met with the same result, for amongst the many thousands of seedlings he grew, but very few indeed proved to be of any value.

Seedlings must always be grown nevertheless, for difficult as it may be to get valuable varieties, it is the right way to seek for them.   The attempt is always interesting, and a philosopher has said, that " he who provides a new fruit renders a greater service to mankind than he who wins a battle." It requires great patience, perseverance, and fortitude too, for it is not every one that can with equanimity bear to be told that the seedlings he has grown himself and watched and petted for years, are worthless, and only good as stocks for grafting.

THE CIDER HOUSE.—The want of suitable buildings is a very serious drawback to the proper storage of fruit, and to the manufacture of Cider and Perry in perfection. Marshall and other writers have pointed out the saving in time and labour that would be effected if every orchard farm had a well arranged Fruit and Cider House, furnished with simple machinery, and with the usual mechanical fittings. Such buildings should be so constructed as to command a low, or still better, different degrees of temperature at will. They need not necessarily be expensive—thick walls of stone, or hollow bricks, and a good thick straw thatch, with due arrangement for free ventilation, is all that is essentially required. By these means it would be possible to regulate and prevent those sudden changes of temperature, which so frequently prevail in Autumn, and which are often so injurious to the liquor ; at one time suddenly checking fermentation, and at another exciting it again, when it is most important to avoid it.

In America, fruit growers find the greatest advantage in their refrigerating houses, in which by simple and ingenious mechanical and chemical appliances, they preserve their Apples and Pears, at a temperature a little above freezing point, in the finest condition for exportation throughout the entire year. In the manufacture of Cider and Perry these houses afford the utmost advantage. The details of their management are given in full in Downing's " *American Orchardist*," and when it is considered that these appliances are only required during the end of Autumn and in the early Winter months, it should be a matter of serious consideration for the landlord and the tenant in a fruit district, whether suitable buildings of the same character should not be provided.

Mr. Thomas Andrew Knight at the commencement of the present century felt so much the necessity of commanding a low temperature for his Cider that he built a cellar on the hill side at Wormsley Grange in the bed of a small stream, so that he could keep it filled with running water, and thus prevent refermentation. The theory was good, but the practical inconveniences connected with this means of carrying it out, proved to be greater than the advantages derived from it.

DISTRICT FACTORIES.—The establishment of large Cider and Perry Factories in the immediate vicinity of the Orchards has been often advised. Marshall and other old writers recommended the plan, and it is very probable that they would have been established very much more generally, if the manufacture had not gone into a state of lamentable neglect. There are private cider makers now, who will buy up such of the superior varieties of Apples as they require, but they will not purchase the enormous amount of poor fruit which pervades the Orchards. The farmers therefore have to make the Cider and Perry themselves, as best they can, and sell it in bulk, at a very low price, to the so called " Cider Merchants." From their hands it passes on to other manipulators, and eventually it is believed to reappear, as Hock, Champagne, Sherry, or Port, according to the prevailing demand of the market at the time.

The establishment of Cider and Perry Factories would prove of the greatest advantage in the Orchard districts. A ready home market for the best fruit would soon cause the inferior kinds to be got rid of; and the manufacture of Cider and Perry of superior quality, would restore their character to the outer world, and render them properly appreciated. Under present circumstances when a great "hit" of fruit occurs, the Apples and Pears are scarcely saleable at any price, the home barrels are all filled, and the waste is enormous. It sometimes happens at these times, that a barrel of Cider is placed in the yard, ready tapped, and with a mug standing by it, that all comers to the house may help themselves. Such prodigal hospitality is by no means desirable, and if the demand for Cider were as great as it might be, its value would soon put a stop to such wasteful use.

It is precisely in these good seasons, when fruit is so abundant and well ripened, that the best liquor can be made. It would be the golden opportunity for a Factory, supported by capital. Very great quantities of Cider and Perry could be made, and laid by in cask, and in bottle, to meet the failures of future years. With good management, a Company formed for the manufacture of Cider and Perry in this way, could not fail to give a very handsome return to its proprietors, and at the same time to increase greatly the value of the Orchards.

VII.—ORCHARD PROSPECTS.

English Agriculturists have now to meet the competition of the world, and it is desirable, on every account, that they should enlarge their sphere of action. Instead of confining themselves so much to corn and cattle, as they have hitherto done, they should pay closer attention to the growth of other products which will command a constant and lucrative market in our own populous and wealthy towns; such as hops, where the soil is suitable; poultry and eggs; milk, butter and cheese; fruit of all kinds ; and such vegetables as local circumstances may require, or good judgment determine. Happy, in these times, are they, who living in districts specially adapted for the growth of hardy fruits, can turn their efforts in this direction. Our Orchards ought to supply economically, and profitably, the markets of our towns and cities with an abundance of Apples and Pears ; and be able to meet there, moreover, an active competition from America, from the Continent of Europe, and even from Australia. It is true that the rent of land is dearer and the fruit seasons much more uncertain in England ; but these disadvantages are almost balanced by the greater expense of labour, (at least in America, our greatest rival) ; by the increased expense of packing ; the cost of carriage ; the liability to injury; and by the still more serious item of profit to the middlemen, or importers. The importation of fruit must always be more difficult than that of grain, and the cost greater ; this cost, moreover, must increase so soon as the commercial depression of the last 4 or 5 years passes off, and ship freightage returns to its ordinary rates. There is every reason therefore to believe that steady perseverance in Orchard culture will meet with a successful reward.

The occurrence of favourable seasons affords the great opportunity for remunerative Orchard management. At these times, in addition to increased cellar storage for vintage fruit; and the sale of fruit in the market, just treated of ; there is great scope for individual energy in the preservation of table fruit. This may be done in a variety of ways. Apples and Pears may simply be dried whole ; as, for example, the *Herefordshire Beefing*, the *Norfolk Beefing*, &c. : they may be peeled, cored, and the flesh dried in the form of " apple chips," " apple cuttings," or " apple rings," as the Americans call them : they may be preserved with syrup in tins ; or they may be converted into jelly. All these preparations may be easily made without the need for any great capital ; and when made, they will give a good profit, and keep well to supply the deficiencies of failing years.

Herefordshire, Devonshire, and Somersetshire, and other districts, capable of producing Cider and Perry of good quality, have a peculiar advantage in the possession of a branch of agricultural industry that may be made very remunerative. It is one the least likely to be interfered with by the fluctuations of ordinary trade, and has therefore with proper care, only the seasons to contend with. The present state of our legislature is most favourable to its extension, since there are no longer any restrictions on its produce by taxation ; nor yet on its sale direct from the Orchards : whilst as regards foreign competition, there is no probability that the supply for our home consumption can be seriously interfered with, for this, if for no other reason, that beverages which only contain so slight a proportion of Alcohol are readily susceptible of refermentation, from the constant shaking incident on conveyance from a distance.

It would greatly conduce to the improvement of Orchard culture if the Agricultural Societies in the special fruit districts would take up the subject, hold annual exhibitions of fruit, and offer prizes for the best collections. Agricultural meetings are always held conveniently in the Autumn, and such exhibitions of fruit could scarcely fail to prove attractive, and they would certainly spread a knowledge which would lead to the growth of the superior varieties of fruit.

The theory and practice of horticulture and fruit growing might also be introduced with great advantage as a Science subject into our country Elementary Schools, as was most successfully done many years since by the late Professor Henslow, in his village school in Cambridgeshire. In these respects English Schools are far behind those on the Continent. There, elementary instruction in Horticulture is aided, as it should be, by manual work in the garden ; and to the growth of vegetables and herbs and flowers is added the practice of budding, grafting and pruning of fruit trees. This excellent practice could not fail to produce a much more extended interest in the production of the best varieties of fruit ; as the knowledge of how to bud a rose briar has introduced many of the most beautiful roses into the Herefordshire cottage gardens.

Landlord and Tenant are alike interested in the utmost development of our Home Industries. The greatest attention must be paid to the special products of every district. Great competition must be met by high cultivation, by economy, and by intelligent, persevering industry. The land must be managed, if not in the letter, yet in the economic spirit of John Stuart Mill, who pointed as an illustration to the cabbage of the French proprietor, so carefully dug round, watered, and manured ; so individualized in short, as though the whole profit of the farm centred in that one single vegetable. By thus paying greater attention to minute details, the farm may become, what it ought to be, in these days of competitive agriculture in both hemispheres, a duplicate of the garden on a large scale.

A flask of prime Cider is the crowning enjoyment in Tennyson's charming description in " *The Pic-nic :*"

> "There on a slope of orchard, Francis laid
> A damask napkin, wrought with horse and hound ;
> Brought out a dusky loaf that smelt of home,
> And, half cut down, a pasty costly made,
> Where quail and pigeon, lark and leveret lay,
> Like fossils of the rock, with golden yolks
> Imbedded and injellied ; last with these
> A flask of cider from his father's vats,
> Prime, which I knew ; and so we sat and ate."

CHARLES HENRY BULMER, M.A.

APPENDIX.—ORCHARD MANURE.

The following formula for artificial manure has been published by the " Hereford Society for Aiding the Industrious," under the name of " *With's Universal Manure.*" In these exact proportions, it has been found to possess excellent fertilizing qualities, and it has the additional recommendations of being cheap and easy to make.

TAKE:—		Cwt.	qrs.	lbs.	Cost (about). £	s.	d.
Finely sifted Dry Earth	... ...	15	0	0	0	3	0
Finely sifted Coal Ashes	... ...	10	0	0	0	3	0
Kainit, finely pounded	... ...	0	3	0	0	3	6
Nitrate of Soda, finely pounded	... ...	0	3	0	0	12	0
Best Peruvian Guano ...	... ...	2	0	0	1	8	0
Best Bone Meal	...	1	2	0	0	12	0
Pure Dissolved Bone ...	...	3	1	0	1	12	6
Superphosphate of Lime	...	1	0	0	0	6	0
Coprolite, or Phosphorite powder		1	0	0	0	4	6
		35	1	0	£5	4	6

The ingredients must be of the best quality and mixed thoroughly together. Pass the compound through a quarter inch screen. The cost per ton, at present prices, including labour, will be about £3 5s. ; and something less than half a ton per acre every third or fourth year will suffice, since its effects will be found very durable.

When the composition of this manure is compared with the proceeds resulting from the analyses of the ash of the tree and fruit of the Apple and Pear (see pp. 122—3), it will be seen that it is admirably adapted to supply their requirements, and can scarcely fail therefore to prove very beneficial in the orchard. So far as experience has gone, this has been found to be the case. It may be added that for single trees as pyramids or espaliers, the quantity required would be from 5lb. to 6lb., forked in to the depth of a few inches, in early spring around the trees.

Foxwhelp

[PLATE I.

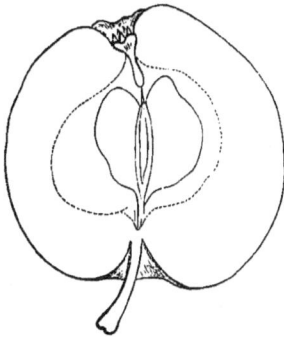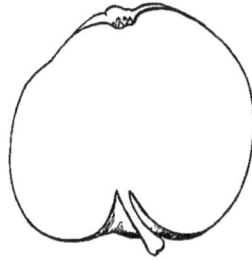

Fig. 1.   Fig. 2.

THE FOX-WHELP APPLE.

"Cider for strength and a long-lasting drink is best made of the Fox-whelp of the Forest of Deane, but which comes not to be drunk till two or three years old."—*(Appendix to Evelyn's " Pomona." Edit. 1706).*

The Fox-whelp Apple is the favourite cider Apple of Herefordshire. Its origin and its singular name are alike obscure.

The earliest record we have of the Fox-whelp is by Evelyn in his " *Pomona*," which is an Appendix to the *Sylva* " concerning fruit trees in relation to cider." This was first published in 1664, and at that time and long after, the great Apple of Herefordshire was the Red-streak. The Fox-whelp is disposed of in a few words—" Some commend the Fox-whelp." Ralph Austen, who wrote in 1653, makes no mention of it when he says, " Let the greatest number of fruit trees not onely in the orchards but also in the feilds be Pear-maines, Pippins, Gennet-Moyles, Red-streaks, and such kinds as are knowne by much experience to be especiall good for cider." Neither is any notice taken of it by Dr. Beale in his " *Herefordshire Orchards*, written in an epistolary address to Samuel Hartlib, Esq.," in 1656.

The first notice of it after Evelyn is by Worledge in 1676, who merely says, "The Fox-whelp is esteemed among the choice cider fruits." In Evelyn's time it seems to have been regarded as a native of Gloucestershire, for Dr. Smith in the " *Pomona*" when writing of " the best fruit (with us in Gloucestershire)" says, " the elder of the Bromsbury Crab and Fox-whelp is not fit for drinking till the second year, but then very good ;" and in the quotation at the head of this paper "A person of great experience " calls it " the Fox-whelp of the Forest of Deane."

[PLATE I.

Its great merit as a cider apple seems to have been quickly recognised, but its cultivation up to this period could not have been on an extensive scale or it would have been more generally known. Even Philips in his celebrated poem entitled "*Cyder*" seems as ignorant of its existence as most of the writers on orchards were at that period. A highly appreciative notice of it is found in a letter to a friend written by Hugh Stafford of Pynes in Devonshire, Esq., bearing date 1727. He says, "This is an Apple long known, and of late years has acquired a much greater reputation than it had formerly. The fruit is rather small than middle-sized, in shape long, and all over of a dark red colour. I have been told by a person of credit that a hogshead of cider from this fruit has been sold in London for £8 or eight guineas, and that often a hogshead of French wine has been given in exchange for the same quantity of Fox-whelp. It is said to contain a richer and more cordial juice than even the Red-streak itself, though something rougher if not softened by racking. The tree seems to want the same helps as the Red-streak to make it grow large. It is of Herefordshire extraction." Mr. Knight in the "*Pomona Herefordiensis*," published in 1811, also thought it "certainly a true Herefordshire Apple," and this of late, has been the prevalent belief, derived probably from the opinion of the two last named writers.

The merit of its production thus rests with the Forest of Dean, on the authorities we have given, but there is no record of the origin of its singular name. It may readily be supposed however that the stray seedling sprung up near a fox's earth, and thus when it had shown its character it obtained its name. Some devoted admirers think they see in the eye of this Apple a distinctive resemblance to the physiognomy of a young fox, but here surely the name has guided the imagination. Wherever the young seedling may have grown, the brilliant colour of its fruit would render it conspicuous, and its rough peculiar flavour with a judge of Apples would proclaim its merit. It is probable that a fox-hunter found and named it, and certainly none appreciate more highly than fox-hunters the merits of its cider.

Description.—The fruit is roundish, inclining to conical or ovate, with an uneven outline, caused by several obtuse ribs on the sides, and which terminate in ridges round the eye; in good specimens one side is generally convex, while the other is flattened. Skin beautifully striped with deep bright crimson and yellow; on the side next the sun it is more crimson than it is on the shaded side, where the yellow stripes are more apparent. Eye very small, set in a narrow, shallow, and plaited basin; segments short, somewhat erect, and slightly divergent. Calyx-tube funnel-shaped. Stamens marginal. Stalk three-quarters of an inch long, obliquely inserted by the side of a fleshy swelling, which pushes it on one side and gives it a curving direction. Flesh yellow tinged with red, tender, and with a rough and acid flavour. Cells of the core wide open. It belongs to group 10 of Dr. Hogg's New Classification of Apples.

The surface of the Fox-whelp Apple is usually marked by small dark coloured, circular scabs or patches, which are thought by some growers to be characteristic of the Fox-whelp, but this is not so. The round patches are formed by the microscopic fungus, *Spilocæa pomi*, and are commonly to be found on the apples of very aged trees of all kinds of fruit. Like all fungus growths this is much more abundant in some seasons than in others.

The coloured plate of the Foxwhelp Apple was drawn from fruit grown on the estate of W. H. Apperley, Esq., of Withington. The trees are believed to have been planted by one of his ancestors about the year 1690, and are still in fruitful vigour.

PLATE 1.

The form of the fruit varies according to the age of the tree, and this is the case with most varieties. The sections are taken from fruit grown by John Bosley, Esq., of Lyde, and exhibit the result of successive graftings from one graft taken originally from one of the old trees of the Fox-whelp. Fig. 1 represents a fruit from a tree which is the result of four successive graftings, the scions being taken in each instance from the tree grafted the previous year; and Fig. 2 is that from a tree which has had the grafting repeated five times. The outline, Fig. 3, at the end of this paper is from an apple grown at Credenhill in the orchard of F. W. Herbert, Esq.

A Fox-whelp Apple of good size and colour, grown in the year 1876, yielded 7½ drachms of a strongly acidulated juice with its own flavour, and of the specific gravity of 1068; and others of smaller size gave 5½ drachms of juice with a specific gravity of 1074. Mr. Knight gives the higher specific gravity of 1076 to 1080, which perhaps might be due to a more favourable year.

Analysis.—The following analysis of the juice of the Fox-whelp, grown in the year 1877, has been specially made for these pages, by Mr. G. H. With, F.R.A.S., in the Laboratory of the Experimental Garden at Hereford :—

FOX-WHELP APPLE.

Density of fresh juice	1.068
Density after 24 hours	1.070

In 100 parts by weight of juice.

Sugar	. 14.400
Tannin, Mucilage, Salts, &c.	. 8.500
Water	. 77.100
	100. —

It must be stated however that the absence of sun and the great rainfall of the summer of 1877, made it a most unfavourable season for the growth of any fruit in perfection.

The home of the Fox-whelp Apple, be its origin what it may, is in the deep clay loam of the Old Red Sandstone in the central districts of Herefordshire, and especially in the valleys of the rivers Lugg and Froome. The chief orchards are to be found in the villages of Lugwardine, West-hide, Withington, Lyde, Moreton, Sutton, Wistaston, Marden, Bodenham, Burrup, Wellington-on-the-Lugg ; and those of Weston Beggard, Yarkhill, Tarrington, Stoke Edith, Stretton Grandison, Eggleton. the Froomes, the Cowarnes, and the other villages on the Froome, are seldom without a few old trees, of the Fox-whelp Apple.

The broad valley of the Wye does not generally present so good and rich a soil. The river has been so erratic in days gone by, that large beds of gravel and marl are to be met with in all directions, and the orchards of repute therefore are only to be found on the rising slopes of the valley out of the river's reach. There are many excellent orchards from King's Caple and Holme Lacy by Credenhill to Kinnersley, Sarnesfield, Dilwyn, and the Weobley district ; the Fox-whelp may be found in any of them, and wherever it is found, it is treasured greatly for its valuable fruit.

The Fox-whelp Apple tree is upright and handsome in growth where age has not rendered it rugged and gnarled. It is a slow growing tree and a shy capricious bearer, and this may perhaps

PLATE I.

partly explain why fruit growers should prefer to propagate those sorts which grow more freely and are more certain croppers. The tree is hardy, and its fruit is in great demand. There is yet a want of young trees generally, for, be the reason what it may, grafts of late years have not succeeded well. The orchardists however have only to apply themselves to the cultivation of the Fox-whelp, and resolutely determine to perpetuate this precious variety, when the same success will crown their efforts in the future, which followed those of their predecessors in the past.

The Fox-whelp cider when pure is of great strength, and always has a peculiar aroma, so marked that it can be detected directly the cork is drawn from the bottle. In taste it is generally rough and strong, with a peculiar vinous musky flavour, which gives its aroma. In ordinary seasons, unless made with great care, it is not sweet enough to be acceptable to strangers, and the taste which enjoys its peculiar flavour fully, must in such circumstances, perhaps, be acquired ; but in a favourable year—a year of sunshine and genial showers, when the fruit has been ripened to perfection—happy is he who has a good hit of it. If he carries it well through the process of fermentation and keeps the flavour of the fruit, and its sweetness too, he has cider in perfection—a cider that will sell readily in its own district at a guinea a dozen ; and a cider moreover, that will unquestionably improve in quality for some three or four decades of years. It will not all be sold, however, for it is the pleasure and pride of the cider-growers of Herefordshire to have always ready for a friend a bottle of good Fox-whelp cider of a good year.

The juice of the Fox-whelp Apple is, however, most used to give strength and flavour to the cider of mixed fruit, and when this is well made it is perhaps more generally popular than the very strong and pure Fox-whelp. A cider of this kind, excellent in quality, can be got at 1s. a bottle from the growers. The Fox-whelp cider has the character of changing colour very quickly on exposure to the air, and even at the table if not drunk quickly the dusky greenish tint will show itself. Some other strong ciders have also this peculiarity, which is certainly not a virtue.

The Fox-whelp, beyond all question, is the most valuable cider apple, and by intelligent perseverance in propagating it, it will long continue to be so.

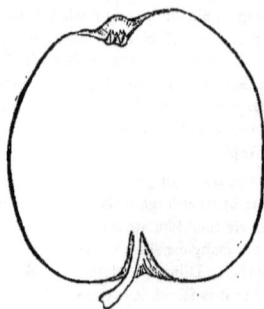

Fig. 3.

[*For all the varieties of the Fox-whelp : the " New," the " Bastard," the " Red," and a description of their several merits, see Plate VIII.*]

1 Pomeroy.

2 Winter Pomeroy

PLATE II.

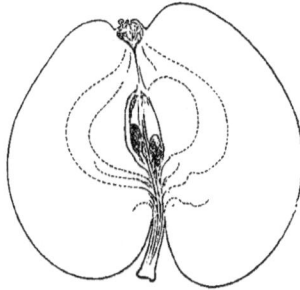

1. POMEROY.

SYNONYME.—*Herefordshire Pomeroy.*

There are several varieties of Apples distributed throughout the country that are known by the name of Pomeroy and which differ in many respects from one another. In the third edition of the "*Catalogue of Fruit cultivated in the Garden of the Horticultural Society of London*," published in 1842, three varieties are mentioned. The Early Pomeroy, the New Pomeroy and the Old Pomeroy, the first of three appears from the descriptions to be the Pomeroy of Herefordshire. The first is described to have the fruit conical, striped, and ripe in October; the second ovate, russet, and in use from November to December; and the third, which is also called the Taunton Pomeroy, is said to be conical, brownish yellow, and in use from November to February. The one which comes nearest to our Pomeroy in respect of season is the first or Early Pomeroy, and it also agrees in shape and in colour, ours being both conical and striped. The second is evidently our Winter Pomeroy; and the third may be the variety recorded in a MS. description of Forsyth's, which he had from Kirke of Brompton, and which he says is "a middle-sized apple, pretty pleasant flavour, rather sharp. Ripe in January." But in his "*Treatise on the Culture and Management of Fruit Trees*," Forsyth describes the Pomeroy as "ripe nearly as soon as the Juneting; and though not so beautifully coloured, is larger and much better tasted. There is a variety which is a Winter Apple." He cites Langley's figure as identical with this description, but upon what other ground than the mere name it is difficult to imagine, for Langley gives neither description nor the season of ripening. This description of Forsyth's is identical with that of Switzer's which was published 80 years previously, and excites a suspicion that the one is a mere copy of the other, and not made from original observations.

PLATE II.

Although the name appears to indicate that this is originally a French Apple, it does not follow that it is so, and there is nothing to warrant such a conclusion. In the most searching investigations that have been made, no mention of the name can be found even in records of the oldest French varieties, reaching back to the 13th and 14th centuries. It does not appear in the list of Olivier de Serres in 1600, or of Lectier in 1628, nor can any trace be found of it from that period downwards in any French work, ancient or modern. The probability is that it is a true English Apple of great antiquity, and that its name was given at a time when Norman French was prevalent in this country, for we find the name applied to other than apples, as for instance, that of a place in Devonshire called Stockleigh Pomeroy.

The earliest record we have of the Pomeroy is in Leonard Meager's list published in 1670, but it must have been in existence long before. Langford mentions it in 1681, and so does Ray in 1686. The author of "*The Compleat Planter and Cyderist*" (1690) says, "The Pome Roy hath a good taste, a pulpy substance, and not yielding much juice, yet that which is, is very good," but these all agree in calling it a Winter Apple, and it may be our Winter Pomeroy. It is also named by Philips in his poem on "*Cyder.*" Coming to more modern times we find Rogers in 1837 speaking of the Pomeroy or "Pomme Roi" as a very good culinary apple, usable during October, November and December, and is much esteemed in the county of Sussex, to which locality it is probable it may have been received from the opposite coast of France." But from the great lack of proper descriptive characters by all our early writers on fruits, it is impossible to identify with certainty any of these varieties with the Pomeroy of which we now treat.

Description.—Fruit; rather below medium size, short conical, or ovate, with ribs on the sides which become more prominent as they approach the eye, and around which they form ridges. Skin; greenish yellow, marked with some traces of russet and streaks of red where shaded, but on the side exposed to the sun, it is covered with a large patch of dense cinnamon-coloured russet, and the whole surface is strewed with large russet dots. Eye; set in a pretty deep and angular basin; the segments are connivent and reflexed at the tips; the tube is short conical, and the stamens are inserted near the margin of the tube. Stalk; short, sometimes about half-an-inch in length, and deeply inserted nearly the whole of its length. Flesh; yellowish, tender, juicy, sweet and richly flavoured. The cells of the core are open and symmetrical.

This is a dessert apple of great excellence, and with flesh remarkable for its delicacy of texture. It ripens in September and lasts only for a short time, as it begins to decay before the middle of October.

PLATE II.

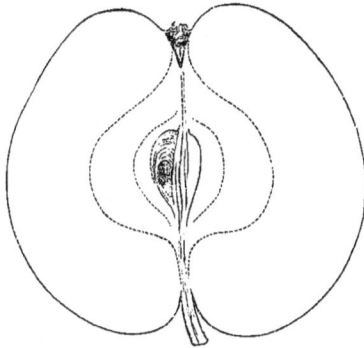

2. WINTER POMEROY.

Description.—Fruit; medium sized, roundish, inclining to ovate, flattened at the base, and distinctly five-sided, especially towards the apex, where the angles form distinct ridges. Skin; smooth, deep yellow on the shaded side, when ripe, and strewed with large russet dots; on the side next the sun, it is coloured with a bright red cheek. Eye; open, set in a moderately deep basin with somewhat connivent segments; the tube is short conical, and the stamens are inserted in the middle of the tube. Stalk upwards of half an inch long, very slender, and inserted in a deep narrow cavity, which is lined with a patch of pale brown russet. Flesh yellowish, firm and crisp in texture, with a pleasant subacid flavour. The cells of the core are open and regular.

This is an excellent culinary apple, in season from October to December.

The tree is hardy and bears well. Both these Pomeroys are to be found in most of the leading garden-orchards of Herefordshire, though for the most part their names are forgotten. The Winter Pomeroy is the most frequent, which may perhaps be accounted for by the fact that a tree exists in the old Nursery grounds of the late Mr. Godsall at Hereford, with whom this variety was a great favourite, and who propagated it extensively. It is always brought to the Christmas market, and sold under the absurd name of "Green Blenheim," which is of course merely a tribute to the popularity of the Blenheim Orange.

PLATE III.

MISCELLANEOUS DESSERT APPLES.

"Studded with apples a beautiful show."—*Wordsworth.*

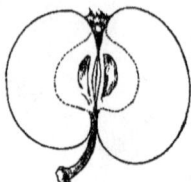

1. JOANNETING.

[Syn.: *Ginetting; Juneting; Early Jeanneting; White Juneating; Juneating; Owen's Golden Beauty; Primiting.*]

The St. John Apple is the earliest apple of the year. It is one of our oldest apples, but though generally known and popular, it escaped the notice both of Miller and Parkinson. Rea first mentions it in 1665, and describes it as "a small, yellow, red-sided apple upon a wall, ripe the end of June."

The derivation of its name has given rise to much discussion. Abercrombie was the first to write it "June-eating," as if in allusion to the period of its maturity. Dr. Johnson, in his dictionary, writes it "Gineting," and says it is a corruption of "Janeton," signifying in French Jane or Janet, having been so called from a person of that name. Ray says, "*Pomum Ginettinum, quod unde dictum sit me latet.*" (*Hist. Plant. II., 1447*).

There can be no doubt, however, that the Joanneting owes its name to its ripening about St. John Baptist's day (June 24th), which it might very well do against a wall in some seasons, particularly when we remember that at the time the name was given the old style of reckoning time was in use, and that the 24th day of June, O.S. would be the 5th of July in the new calendar. But supposing it to have been a variety imported from abroad, as its name would lead us to suppose it was, then there is nothing remarkable in its being ripe even in the open ground on the 5th of July. Curtius distinctly says the "Joannina" are so called "*Quod circa divi Joannis Baptistæ nativitatem esui sint*" (*Hortorum. p. 522*). J. B. Porta also says of it, "*Est genus alterum quod quia circa festum Divi Joannis maturescit, vulgus 'Malo de San Giovanni' dicitur.*" And according to Tragus, "*Quæ apud nos prima maturantur 'Sanct Johans Opffell,' Latine, Precocia mala dicuntur.* (*Hist., p. 1043.*)

In the Middle Ages it was customary to connect the festivals of the Church with events which took place at the same periods, and the practice has continued to our own days with reference to the sowing of crops and ripening of fruits, &c. Other apples have derived their names in the same way; thus we have the Margaret Apple, so called from being ripe about St. Margaret's Day (July 20th); the Maudlin, or Magdalene, from St. Magdalene's Day (July 22nd). There is also an old French

1 Joanneting

2 Summer Golden Pippin

3 Court of Wick.

4 Devonshire Quarrenden

5 Borsdorffer

6 Worcester Pearmain

7. Kerry Pippin

8 Early Spice

PLATE III.

pear, "Amiré Joannet" (Wonderful Little John), which Merlet informs us was so called because it ripened about St. John's Day. This is precisely analogous, for we have only to add " Joannet " the termination " ing," so common amongst apples, and we have our " Joannneting."

Description.—Fruit ; small, round, and a little flattened. Skin ; smooth and shining, pale yellowish green in the shade, but clear yellow, with sometimes a faint tinge of red or orange next the sun. Eye ; small and closed, surrounded with a few small plaits, and set in a very shallow basin. Stalk ; an inch long, slender, and inserted in a shallow cavity, which is lined with delicate russet. Flesh ; white, crisp, brisk, and juicy, with a vinous and slightly perfumed flavour, but becoming mealy and tasteless if kept only a few days after being gathered. It is in the greatest perfection when eaten from the tree.

The tree is hardy and healthy, but does not attain a large size. If worked upon the Paradise stock it may be grown in pots, when the fruit will not only be produced earlier, but in greater abundance than on the crab or free stock.

2. SUMMER GOLDEN PIPPIN.

[SYN.: *Summer Pippin; White Summer Pippin.*]

Description.—Fruit ; below medium size, two inches and a quarter broad at the base, and two inches and a quarter high ; ovate, flattened at the ends. Skin ; smooth and shining, pale yellow on the shaded side, but tinged with orange and brownish red on the side next the sun, and strewed over with minute russety dots. Eye ; open, set in a wide, shallow, and slightly plaited basin. Stalk; thick, a quarter of an inch long, completely imbedded in a moderately deep cavity, which is lined with russet. Flesh ; yellowish, firm, very juicy, with a rich vinous and sugary flavour.

This is one of the most delicious summer apples, and ought to form one of every collection, however small. It is ripe the end of August, and keeps about a fortnight.

The tree is small, about one-third of the ordinary size ; it succeeds well when grafted on the doucin or paradise stock ; and is an early and abundant bearer. The Summer Golden Pippin is very frequently grown in Herefordshire, and attains a larger size than it appears to do elsewhere. It is most prolific, and since the fruit so quickly loses its freshness, it is well to remember that when not sold, or given away, it will make a delicious apple jelly.

When grown on the " pomme paradis" of the French, it forms a beautiful little tree, which can be successfully cultivated in pots.

PLATE III.

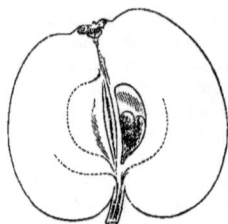

3. COURT OF WICK.

[SYN.: *Fry's Pippin ; Golden Drop ; Knightwick Pippin ; Phillips' Reinette ; Wood's Huntingdon ; Weeks' Pippin ; Yellow Pippin.*]

This variety is said to have originated at Court of Wick, in Somersetshire, and to have been raised from a pip of the Golden Pippin.

Description.—Fruit ; below medium size, roundish ovate, regular and handsome. Skin ; when fully ripe of a fine clear yellow, with bright orange, which sometimes breaks out in a faint red next the sun, and covered all over with russety freckles. Eye ; large and open, with long acuminate and reflexed segments, set in a wide, shallow and even basin. Stalk ; short and slender, inserted in a smooth and even cavity, which is lined with thin russet. Flesh ; yellow, tender, crisp, very juicy, rich and highly flavoured.

It is one of the best and most valuable dessert apples. The rich and delicious flavour of the fruit is not inferior to that of the Golden Pippin. In season from October to March.

The tree attains the middle size, is very healthy and hardy, and bears abundantly. It will succeed on almost every soil, and is not subject to attacks of blight and canker. In some places, as on the Hastings Sand, the colour of the fruit becomes a fine clear orange with a somewhat carmine cheek on the side next the sun ; the same rich tint is observed in some localities in Herefordshire. This richness of colour with its fine flavour, is doubtless the cause of this fruit being attacked with such avidity by birds and insects. On the sunny side of the tree it is often an exception to find a Court of Wick apple untouched by these marauders.

PLATE III.

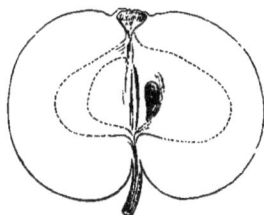

4. DEVONSHIRE QUARRENDEN.

[SYN.: *Quarrington; Red Quarrenden; Sack Apple.*]

The Devonshire Quarrenden is supposed to be a very old variety, but there is no record of it previous to 1693, when it is mentioned by Ray. It seems to have been unknown to Switzer, Langley, and Miller; and, except by Mortimer, it is not noticed by any subsequent writer till within a very recent period. The only early catalogue in which it is mentioned is that of Miller and Sweet, of Bristol, in 1790. It does not seem to have been grown in any of the London nurseries until the beginning of the present century.

Description.—Fruit ; rather below medium size, oblate and sometimes a little angular in its outline. Skin ; smooth and shining, entirely covered with deep purplish red, except where it is shaded by a leaf or twig, and then it is of a delicate pale green, presenting a clear and well-defined outline of the object which shades it. Eye ; quite closed with very long tomentose segments, and placed in an undulating and shallow basin, which is sometimes knobbed, and generally lined with thick tomentum. Stalk ; about three-quarters of an inch long, fleshy at the insertion, deeply set in a round funnel-shaped cavity. Flesh ; white tinged with green, crisp, brisk, and very juicy, with a rich vinous and refreshing flavour.

A very valuable and first-rate dessert apple. It ripens on the tree in the first week of August, and lasts till the end of September.

The tree attains a considerable size, is particularly hardy, and a most prolific bearer. It succeeds well in almost every soil and situation, and is admirably adapted for orchard planting. It grows in perfect health and luxuriance in almost every latitude of Great Britain, from Devonshire to the Moray Frith. Throughout Herefordshire the Devonshire Quarrenden is the pride of many a cottage garden, and the richly-coloured fruit which loads its boughs is a source of considerable profit to the good housewife. The fruit too is generally larger in size in Herefordshire than in its native county.

PLATE III.

5. BORSDÖRFFER.

[SYN.: *Borstörff Hâtive; Queen's Apple; Red Borsdörffer; Borsdörff; Postophe d'Hiver; Pomme de prochain; Reinette d'Allemayne; Blanche de Leipsic; Reinette de Misnie; Grand Bohemian Borsdörffer; Garret Pippin; King; King George; King George the Third.*]

This apple, above all others, is the most highly esteemed in Germany. Diel calls it the "Pride of the Germans." It is believed to have originated either at a village of Misnia, called Borsdörf, or at a place of the same name near Leipsic. According to Forsyth, it was such a favourite with Queen Charlotte, that she had a considerable quantity of them imported from Germany for her own private use.

It is one of the earliest recorded varieties of the Continental authors. It is mentioned by Cordus in 1561, as being cultivated in Misnia. He also informs us it is highly esteemed for its sweet and generous flavour, and for the pleasant perfume which it exhales. Wittichius, in his "*Methodus Simplicium,*" attributes to it the power of dispelling epidemic fevers and madness! It does not seem to have been known in this country before the close of the last century. It was first grown in the Brompton Park Nursery in 1785.

Description.—Fruit; below the medium size, roundish oblate, rather narrower at the apex than the base, handsomely and regularly formed, without ribs or other inequalities. Skin; shining, pale waxen yellow in the shade, and bright red next the sun; it is strewed with dots which are yellowish on the sunny side and brownish in the shade, and marked with veins and slight traces of delicate, yellowish gray russet. Eye; large and open, with long reflexed segments, placed in a rather deep, round, and pretty even basin. Stalk; short and slender, inserted in a narrow, even, and shallow cavity, which is lined with thin russet. Flesh; white with a yellowish tinge, crisp, and delicate, brisk, juicy, and sugary; and with a rich, vinous, and aromatic flavour.

The Borsdörffer is a dessert apple of the first quality; in season from November to January.

The tree is a free grower and very hardy, not subject to canker and attains the largest size. The bloom is also very hardy, and withstands the night frosts of spring better than most other varieties. It is very prolific. If grafted on the paradise stock it may be grown as an open dwarf, or an espalier. This variety in Herefordshire has not got beyond the garden of the amateur.

PLATE III.

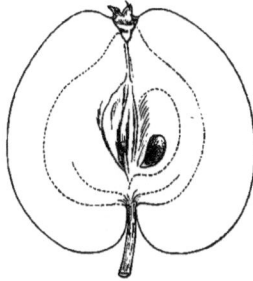

6. WORCESTER PEARMAIN.

This very handsome early apple was produced in the St. John's Nurseries, Worcester, in 1873, by Mr. Richard Smith ; and received a first-class certificate from the London Horticultural Society in 1874. It has been supposed to be a seedling from the Devonshire Quarrenden, but this is not known with certainty.

Description.—Fruit ; medium sized, two inches and three-quarters wide, and the same in height ; conical even, and very slightly angular towards the crown, where it is narrow. Skin ; very smooth and completely covered with a brilliant red, dotted with minute fawn-coloured dots ; here and there in some of the specimens, the yellow ground shows faintly through the red. Eye ; small, closed with long segments, forming a cone, set on the apex of the fruit, with a few plaits round it. Stalk ; three quarters of an inch long, deeply inserted in a russety cavity. Flesh ; very tender, crisp, juicy, sweet and sprightly, with a very pleasant flavour.

The great beauty of the fruit, and its usefulness, both for dessert and culinary purposes, cannot fail to render it a general favourite. It is ripe in August and September.

The tree is hardy, begins to bear at a very early age, and is very productive. When well trained, on the paradise stock, and laden with its bright red fruit, which has a peculiar rosy tint, it forms a very beautiful object.

PLATE III.

7. KERRY PIPPIN.

[SYN.: *Edmonton; Aromatic Pippin.*]

This is an Irish variety, as its name indicates, and nothing is definitely known of its origin. It was introduced through the instrumentality of Mr. Robertson, the nurseryman of Kilkenny.

Description.—Fruit; below the medium size, oval, sometimes roundish oval. Skin; smooth and shining, greenish yellow at first, but changing as it ripens to a fine clear pale yellow colour, tinged and streaked with red on the side next the sun, but sometimes when fully exposed, one half of the surface is covered with bright shining crimson, streaked with deeper crimson; it is marked on the shaded side with some traces of delicate russet. Eye; small and closed, with broad, erect, and acuminate segments, set in a shallow basin, which is generally surrounded with five prominent plaits. Stock; slender, three quarters of an inch long, obliquely inserted in a small cavity, by the side of a fleshy protuberance. Flesh; yellowish white, firm, crisp, and very juicy, with a rich, sugary, brisk and aromatic flavour.

The tree is a fine grower, hardy, and a good bearer, attaining about the middle size. It is well adapted for grafting on the paradise stock, and being grown either as a dwarf or espalier.

The Kerry Pippin is an early dessert apple of the highest excellence. It is in season during September and October.

PLATE III.

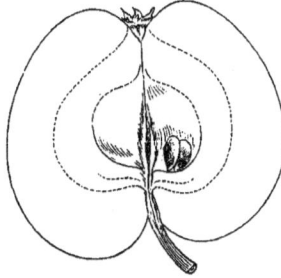

8. HEREFORDSHIRE SPICE APPLE.

There are several varieties of apples known by the name of Spice Apple, and this differs from all of that name which have been described in pomological works. We have therefore for the sake of distinction called it the Herefordshire Spice Apple. Forsyth describes one which in a MS. list in our possession he says he had from Mr. Jones of Fawley, Herefordshire; but he describes it as being of a yellow colour only. It is possible that it might have been a pale specimen for he says it is "a handsome middle-sized angular-shaped apple," ripe in January and keeps till March.

Description.—Fruit; below medium size, conical or ovate, uneven in its outline, being ribbed on the sides somewhat in the way of the Margil, and with ridges around the eye. Skin; smooth and shining, as if varnished; almost entirely covered with deep bright crimson, which is streaked and mottled with darker crimson, on the side next the sun; but where it is shaded, it is yellowish and only mottled with crimson. Eye; small and closed, with erect pointed segments set in a deep and plaited basin; the tube is funnel shaped, and the stamens are inserted in the middle of the tube; the styles form a concrete fleshy mass, which fills nearly the half of the tube. Stalk; very short, thick, and fleshy, set in a very shallow cavity. Flesh; tender, crisp, fine grained, sweet and with a pleasant sub-acid flavour. Cells of the core open and symmetrical.

An excellent apple, in use during October and November. It is grown abundantly in Herefordshire, and is fully appreciated.

PLATE IV.

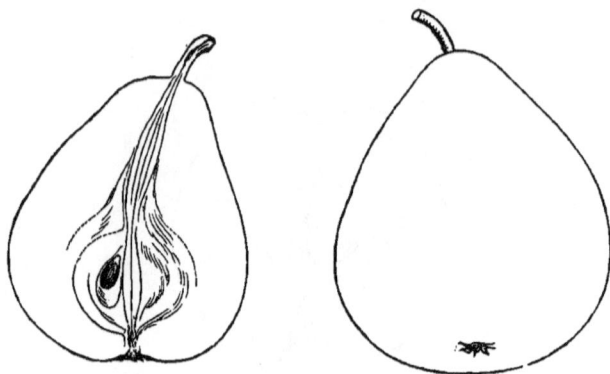

MONARCH. ┌ ┐

[SYN : *Knight's Monarch.*]

This pear was the favourite seedling of Mr. Thomas Andrew Knight. Of its parentage he was not quite sure, but he believed it to have originated from the seed of the Autumn Bergamot. "I named it the 'Monarch,' says Mr. Knight, " under conviction that for the climate of England, it stands without an equal ; and because it first appeared in the first year of the reign of our most excellent monarch " (William IV, 1830). *Trans. Hort. Soc., 2nd Series, Vol. I, p. 107.*

Description.—Fruit ; medium sized, roundish, two inches and three quarters long, by two and a quarter wide. Skin ; yellowish green, very much covered with brown russet, and strewed with grey russet specks. It has usually a tinge of brownish red next the sun. Eye ; small and open, set in a shallow undulating basin. Stalk ; three quarters of an inch long, inserted in a small cavity, frequently without depression. Flesh ; yellowish, buttery, melting, and very juicy ; with a rich, sprightly, sugary, and agreeably perfumed flavour. Season ; January and February.

The official report on the merits of the Monarch from fruit grown in the gardens of the Horticultural Society has already been given in full in the Introduction of this work (see page 43.) with Mr. Knight's comments upon it. Since this time its estimation has greatly increased and it is now universally admitted to be one of our most valuable pears.

The tree is very hardy, and forms a handsome pyramid, but its wood is slender and straggling. It is usually an excellent bearer, but in some situations it is capricious. Its greatest imperfection is the liability to lose its fruit from the wind before it is ripe. It requires therefore a

2 Althorp Crassane

PLATE IV.

sheltered situation ; but the fruit is so excellent that it deserves to have a proper place found for it in every garden.

The Monarch has indeed become so great a favourite that every one likes to think he has it in his orchard. Many of Mr. Knight's other seedling pears have thus been called the Monarch until in Herefordshire a complete confusion as to the real variety has taken place. For this reason, in addition to the coloured figure, an exact outline is also added of the coloured representation of the Monarch, as given in the " *Transactions of the Horticultural Society, 2nd Series, Vol. I, p. 103,* with the section at the head of the preceding page.

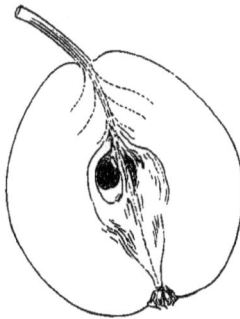

ALTHORP CRASANNE.

This well-known pear was produced by Mr. Thomas Andrew Knight, and the seedling first bore fruit in 1830.

Description.—Fruit ; rather above the medium size, two inches and three quarters wide, and two inches and a half high, roundish obovate, widest in the middle, and tapering gradually to the apex, which is somewhat flattened, but rounding towards the stalk. Skin ; pale green with a slight tinge of brown on the side exposed to the sun, and covered with minute russety dots. Eye ; rather large and open, placed in a shallow and slightly plaited basin. Stalk ; an inch and a half long, slender, curved, and not deeply inserted. Flesh ; white, buttery and juicy, with a rich and slightly perfumed flavour.

The Althorpe Crasanne is a dessert pear of the finest quality. " It possesses all the richness of the Crasanne with less grittiness, being perfectly melting." (See Introduction, p. 42.) Mr. Knight thought it the better of the two, " its rose-water flavour will please where musk offends." Season, October and November.

The tree is hardy, vigorous, and an excellent bearer. It succeeds best as a standard, and is found to produce fruit of a superior quality even in soils that are unfavourable to the growth of pears generally. Mr. Knight believed it to be as hardy as the Swan's Egg Pear.

PLATE V.

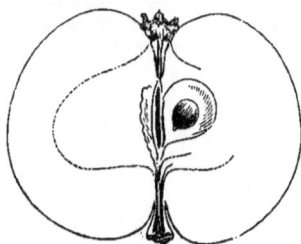

1. NEW NORTHERN GREENING.

A seedling from the Northern Greening, but superior to it in size, in flavour, and as a keeping apple.

Description.—Fruit ; rather above middle size, about two inches and a half broad, and two inches and one quarter high, roundish but more flattened than its parent. Skin ; smooth, of a dullish green and brownish red on the sunny side ; this colour frequently spreads over the greater part of the surface of the fruit, and often presents streaks of a darker colour. Eye ; small and neat, closed with long segments and set in a broad round and even basin. Stalk ; short and stout, inserted in a deep narrow cavity. Flesh ; greenish white, firm in texture, crisp, and very juicy. It is excellent in flavour, with a brisk vinous acidity, and is superior for tarts to all others of its season. The fruit is in best flavour in December, January, and February, and if allowed to remain late on the tree before it is gathered, it will be good in March.

The tree grows freely and bears well. It succeeds well as a bush or pyramid, and only requires for pruning, the summer pinching of the ends of the shoots.

Bew Northern Greening

2 Spring Grove Codlin

3 Stirling Castle

4 Wormsley Pippin

PLATE V.

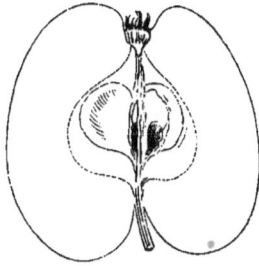

2. SPRING GROVE CODLIN.

This apple was a seedling of Mr. Andrew Knight, "produced by one of his judicious mixtures," said Sir Joseph Banks, after whose seat, Spring-grove, near Hounslow, Middlesex, it was named in 1810.

Description.—Fruit; above medium size, three inches wide at the base, and two inches and three quarters high; conical and slightly angular at the sides. Skin; pale greenish yellow, tinged with orange on the side exposed to the sun. Eye: closed with broad segments, and set in a narrow plaited basin. Stalk; short, inserted in rather a deep cavity. Flesh; greenish yellow, tender, juicy, sugary, brisk, and slightly perfumed.

A first rate culinary apple. It may be used for tarts as soon as the fruit is the size of a walnut, and continues in use to the beginning of October. (Figured and coloured in the *Trans. Hort. Soc.* Vol. I, p. 197.)

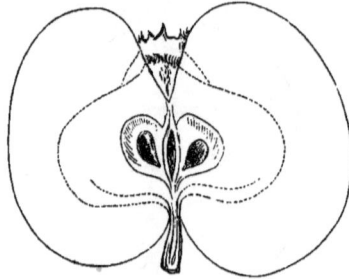

3. STIRLING CASTLE.

Messrs. Wm. Drummond & Sons, the nursery and seedsmen at Stirling, state that this apple was first brought to their notice about forty years ago by a Mr. Christie, an officer of Inland Revenue in Stirling, well known under the soubriquet of " Pookum Christie," who was a great amateur horticulturist. An " Auld Citizen" of Stirling, however, a man particularly well up in the archives of the " Sons of the Rock," informs us that the Messrs. Drummond are mistaken, and that the illustrious introducer of the Stirling Castle apple was a man known, some sixty years ago, as " Auld Johnnie Christie," a nursery gardener, in a very humble scale, out at Causewayhead, on the road to the Bridge of Allan. However this may be, the wide dispersion of the apple is probably due to the appreciation of its merits by Messrs. Drummond & Son.

Description.—Fruit ; rather large, sometimes very large, round and oblate, and when of moderate size even and regularly shaped. Skin ; clear pea-green, which becomes yellow when it ripens ; with a blush and broken stripes of pale crimson on the side next the sun, and several large dots sprinkled over the surface. Eye ; half closed, set in a pretty deep, wide, and saucer-like basin. Stalk ; an inch long, slender, inserted in a deep and wide cavity. Flesh ; white, very tender, juicy, and of the character of that of Hawthornden.

An excellent culinary apple ; "a gem of apples " a Herefordshire grower calls it, and says that "in addition to its other good qualities it is one of the best for making jelly." Rivers speaks of it as an improvement on Small's Admirable, and the improvement is certainly very great. It may be used in August and September but is best in flavour from October to December.

The tree is well adapted for bush or pyramid culture. It is an immense bearer and will certainly become a general favourite when it is better known.

PLATE V.

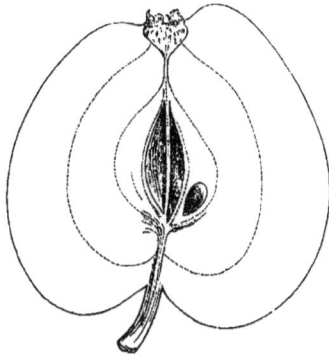

4. WORMSLEY PIPPIN.

[SYN.: *Knight's Codlin.*]

This apple was first brought into notice in 1811 by Mr. T. Andrew Knight. It was his favourite seedling apple, and the best he ever produced.

Description.—Fruit; large, three inches and a half broad in the middle, and narrowing both towards the base and the apex, with obtuse angles on the sides, which terminate at the crown in several prominent ridges. Skin; smooth, deep clear yellow, with a rich golden or orange tinge on the side next the sun, and covered with numerous small dark spots. Eye; large and open, with long acuminate segments, placed in a deep-furrowed and angular basin. Stalk; short, inserted in a deep and round cavity, which is thickly lined with russet. Flesh; yellow, tender, crisp, rich, sugary, brisk and aromatic.

A most valuable apple either for dessert or culinary purposes: it is in season during September and October.

As a culinary apple it is not to be surpassed, and even in the dessert when well ripened, Mr. Knight thought it resembled the New Town Pippin in flavour.

The tree is healthy and hardy; a free grower and a free and abundant bearer. It has been found to succeed in every latitude of Great Britain; the late Sir G. S. McKenzie found it succeed well as an espalier even in Rossshire. It ought to be cultivated in every garden however small.

The specimen of fruit coloured is rather above the average size.

Plate VI.

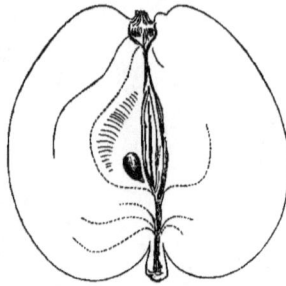

1. KESWICK CODLIN.

This excellent apple was first discovered growing amongst a quantity of rubbish behind a wall at Gleaston Castle, near Ulverstone, and was first brought into notice by one John Sander, a nurseryman at Keswick, who, having propagated it, sent it out under the name of Keswick Codlin.

Description.—Fruit ; above medium size, two inches and three quarters wide, and the same in height ; conical angular in its outline, the angles on its sides running to the crown, where they form rather acute ridges round the eye. Skin ; rather pale yellow on the shaded side, but deeper yellow with an orange or blush tinge on the side next the sun. Eye ; closed with long narrow segments, and set in a pretty deep and rather puckered basin. Stalk ; about a quarter of an inch long, downy, inserted in a deep cavity, which is marked with russet. Flesh ; pale yellowish white, very juicy, tender and soft, with a brisk and pleasant flavour, but becomes mealy after being kept for a month.

One of the most valuable of our early culinary apples. It may be used for tarts so early as the end of June, but it is in perfection during August and September.

In the Memoirs of the Caledonian Horticultural Society, Sir John Sinclair says : "The Keswick Codlin tree has never failed to bear a crop since it was planted in the episcopal garden at Rose Castle, Carlisle, twenty years ago (1813). It is an apple of fine tartness and flavour, and may be used early in Autumn. The tree is a very copious bearer, and the fruit is of good size, considerably larger than the Carlisle Codlin. It flourishes best in a strong soil."

1 Keswick Codlin.

2. Maux Codlin.

3 Lord Suffield

4 Hawthornden

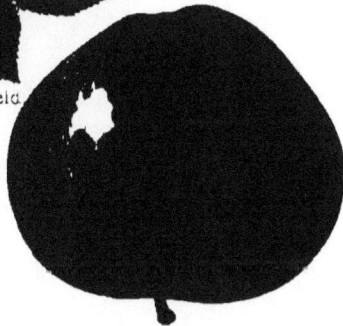

5 Tom Putt

PLATE VI.

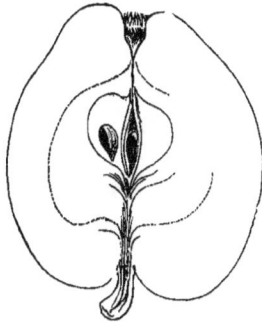

2. MANX CODLIN.

[SYN: *Irish Pitcher ; Irish Codlin ; Eve ; Frith Pippin.*]

Description.—Fruit ; medium sized, conical, slightly angular. Skin smooth, greenish yellow at first, but changing as it ripens to clear pale yellow, tinged with rich orange on the side next the sun, but sometimes when fully exposed, assuming a clear bright red cheek. Eye ; small and closed, set in a small, plaited, and pretty deep basin. Stalk ; three quarters of an inch long, more or less fleshy, sometimes straight, but generally obliquely inserted, and occasionally united to the fruit by a fleshy protuberance on one side of it. Flesh ; yellowish white, firm, brisk, juicy and slightly perfumed.

A very valuable early culinary apple of first-rate quality. It is ripe in the beginning of August and continues in use till November.

The tree is not large but very hardy and healthy, well adapted for growing as a bush on the paradise stock, or as an espalier. It is well suited for planting in exposed situations, and succeeds well in shallow soils. It is a very early and abundant bearer, often producing fruit when only two years old from the graft.

Plate VI.

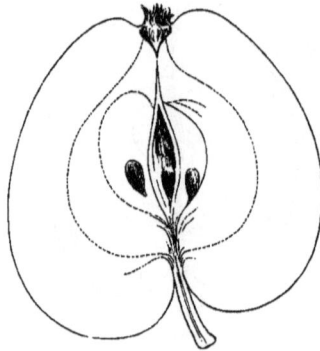

3. LORD SUFFIELD.

This apple was raised about forty-five years ago, by Thomas Thorpe, a hand-loom weaver, of Boardman Lane, Middleton, near Manchester, on the Middleton Hall Estate of the late Lord Suffield, and the apple was named from his Lordship, who was a very popular, benevolent man. In 1836, Thorpe sold the buds at threepence each, and trees thus obtained are now living.

Description.—Fruit ; large, ovate, even in its outline, with several obtuse angles on its sides. Skin ; thin, smooth, pale greenish yellow, with sometimes a tinge of red next the sun. Eye ; small, the segments being gathered together in a point, and placed in a plaited basin. Stalk ; slender, over half an inch long, inserted in a deep cavity. Flesh ; white, tender and firm, very juicy and briskly flavoured.

This apple has become the first favourite for early kitchen use, and in all modern gardens is rapidly displacing the early Codlins and the Hawthornden. Its fault is, that the skin is too fine and the flesh too tender to enable it to travel without being disfigured by bruises. It is in season in August and September.

The tree is hardy and a great bearer, but does not grow to a large size.

PLATE VI.

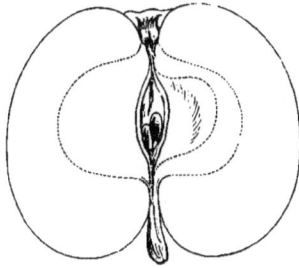

4. HAWTHORNDEN.

[SYN: *Hawthorndean ; White Hawthorndean ; Red Hawthorndean.*]

This variety was raised at Hawthornden, a romantic spot near Edinburgh, where Drummond the poet was born in 1585. The period when this apple was first produced is not known. It is first mentioned in the catalogue of Leslie and Anderson of Edinburgh ; but was not known about London until 1790, when it was introduced into the Brompton Park Nursery.

Description.—Fruit ; varying very much in size according to the soil, situation, or condition of the tree, generally above medium size, roundish and depressed, with occasionally a prominent rib on one side which produces an irregularity in its appearance. Skin ; smooth, covered with a delicate bloom, greenish yellow, with a blush of red on one side which varies in extent and colour according as it has been more or less exposed to the sun. Eye ; small and closed, with broad and flat segments, placed in a pretty deep and irregular basin. Stalk ; short, stout, and sometimes fleshy, inserted in a deep and irregular cavity. Flesh ; white, crisp and tender, very juicy, with an agreeable and pleasant flavour.

The Hawthornden has long been one of the most valuable and popular apples in cultivation. It is suitable only for kitchen use, and is in season from October to December.

The tree has always been considered as very healthy and vigorous, and unrivalled as an early and abundant bearer, but of late years in some situations it has lost its condition and only produced small and diseased fruit, as if it had exhausted the soils of its own particular requirements.

PLATE VI.

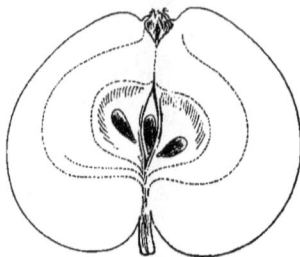

5. TOM PUTT.

[SYN : *Tom Potter.*]

The origin of this well-known apple, "Tom Putt" in Herefordshire and "Tom Potter" in Devonshire, has been lost. It is very generally cultivated in both counties.

Description.—Fruit ; about three inches wide and two and a quarter inches high ; roundish ovate, ribbed on the sides and terminated abruptly towards the eye in a narrow puckered apex. Skin ; smooth and shining, almost entirely covered with broken stripes and mottled blotches of deep bright crimson which becomes paler towards the shaded side, where the colour is lemon yellow. Eye ; set in a narrow puckered basin ; segments connivent ; tube funnel-shaped ; stamens median inclining to marginal. Stalk ; from a quarter to half an inch long, set in an uneven funnel-shaped cavity which is slightly russetty. Flesh ; white, very tender, sweet and with a pleasant acidity. Cells of the core all quite open.

An excellent culinary apple, juicy and high flavoured, when ripe it exudes a pleasant and powerful fragrance and has usually a beautiful colour. The tree is vigorous, and as " Tom Putt " in Herefordshire is very prolific and an annual bearer, but as " Tom Potter" in Devonshire, Ronalds speaks of it as "uncertain in bearing." In Herefordshire cottage gardens it is perhaps an equal favourite with the Blenheim Orange, and it certainly bears more regularly. It is in season from September till November.

Blenheim Orange

PLATE VII.

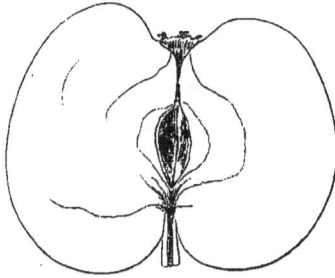

BLENHEIM ORANGE.

[SYN : *Blenheim Pippin ; Kempster's Pippin ; Woodstock Pippin ; Northwick Pippin.*]

This valuable apple was first discovered at Woodstock, in Oxfordshire, and received its name from Blenheim, the seat of the Duke of Marlborough, which is in the immediate neighbourhood. The exact date of its origin is not known. It is not noticed in any of the nursery catalogues of the last century, nor was it cultivated in the London nurseries until about the year 1818.

The following interesting account of this favourite variety appeared some years ago in the "*Gardener's Chronicle*" :—" In a somewhat dilapidated corner of the decaying borough of ancient Woodstock, within ten yards of the wall of Blenheim Park stands all that remains of the original stump of that beautiful and justly celebrated apple, the Blenheim Orange. It is now entirely dead and rapidly falling to decay, being a mere shell about ten feet high, loose in the ground, and having a large hole in the centre ; till within the last three years, it occasionally sent up long, thin, wiry twigs, but this last sign of vitality has ceased, and what remains will soon be the portion of the wood-louse and the worm. Old Grimmett, the basket maker, against the corner of whose garden wall the venerable relic is supported, has sat looking on it from his workshop window, and while he wove the pliant osier, has meditated for more than fifty successive summers on the mutability of all sublunary substances ; on juice, and core, and vegetable as well as animal, and flesh, and blood. He can remember the time when fifty years ago he was a boy, and the tree a fine full bearing stem, full of bud and blossom and fruit, and thousands thronged from all parts to gaze on its ruddy ripening orange burden : then gardeners came in the spring time to collect the much coveted scions, and to hear the tale of his horticultural child and sapling, from the lips of the son of the whitehaired Kempster. But nearly a century has elapsed since Kempster fell like a ripened fruit and was

PLATE VII.

gathered to his fathers. He lived in a narrow cottage garden in Old Woodstock, a plain, practical, labouring man; and in the midst of his bees and flowers around him, and in his "glorious pride" in the midst of his little garden, he realised Virgil's dream of the old Corycian: "*et regum equabat opes animis.*" The provincial name for the apple is still "*Kempster's Pippin,*" a lasting monumental tribute and inscription to him who first planted the Kernel from whence it sprung."

Description.—Fruit, large, being generally three inches wide, and two and a half high; globular and somewhat flattened, broader at the base than the apex, regularly and handsomely shaped. Skin, yellow, with a tinge of dull red next the sun, and streaked with deeper red. Eye, large and open, with short stunted segments, placed in a round, broad, and rather deep basin. Tube funnel shaped: stamens medium. Stalk, short and stout, rather deeply inserted, and scarcely extending beyond the base. Flesh, yellow, crisp, juicy, sweet and pleasantly acid. Cells of the core, open or closed: cell-walls roundish obovate.

A very valuable and highly esteemed apple, either for dessert or culinary purposes, but strictly speaking more suitable for the kitchen than the parlour, except for its very handsome appearance in size, and shape, and colour. It is in season from November to February.

The Blenheim Orange has a strong and vigorous habit of growth and forms a large and very beautiful standard. This is the best and most profitable form of its growth, and when it becomes fullgrown it usually bears regular and abundant crops; it is however apt even then, to bear well only on alternate years. As a dwarf or an espalier it does not bear so regularly, or so well.

The plate represents not only the fruit when fit to gather from the tree, with the leaves and blossom, but also shews the rich ripe tints it assumes when it takes the place of honour on the Christmas dinner table.

PLATE VIII.

MISCELLANEOUS CIDER FRUIT.

"THE AUTUMNAL SEASON PLESAUNCE BLITHE AFFORDS,
NOW THE SQUEEZ'D PRESS FOAMS WITH OUR APPLE HOARDS;
COME, LET US HIE AND QUAFF A CHEERY BOWL,
LET CIDER NEW, WASH SORROW FROM THE SOUL."

Gay.

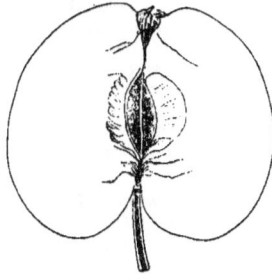

I. REJUVENATED FOXWHELP.

[SYN: *The Canon Apple; Crowe's Kernel.*]

This apple is one of peculiar interest in Herefordshire, and is therefore represented here. The epithet " new " will be used in treating of it but merely to prevent any confusion in discussing the question, as to whether it is " new," that is, a seedling; or whether it is the true " old " Foxwhelp restored to a flourishing rejuvenated form by a careful system of grafting and regrafting. This is a question that has been warmly discussed by the growers for some years past.

At first sight the distinction between them seems very marked; both the tree of the " new " Foxwhelp and its apple, are much more luxuriant than the " old " Foxwhelp. The apple of the " new " Foxwhelp is not only larger, but in its general character it is broad in shape—or in other words, its lateral is greater than its longitudinal diameter—whilst the apple of the " old " Foxwhelp is smaller and usually oblong in shape; but on a careful examination of the trees of either kind the apples are so similar in shape and appearance that it would be impossible to distinguish them if thrown together. The difference in size and shape is due simply to the improved vitality and luxuriance of growth of the tree. The points of similarity between them are very striking. There is the same brilliant colour; the same tough, leather-like skin; the same eye; the same long slender stalk set in its deep narrow channel; and to this it may be added they have the same period

PLATE VIII.

of arriving at maturity. Then again the chemical analysis shews no greater difference between them, than may be accounted for by the more watery juice of the fruit of the more free growing tree.

The history of the "new" Foxwhelp can be traced with some clearness. A farmer of the name of Yeomans living at Cowarne between 60 and 70 years ago, took an unusual interest in the "old" Foxwhelp, and both in that parish, and at Canon Pyon, to which he afterwards migrated, he grafted and regrafted it on healthy stocks, until he restored its luxuriance of growth. Another farmer, a Mr. Crowe, and Messrs. Skidmore, Miles, and Williams, wheelwrights of Canon Pyon, systematically but separately carried on the system of regrafting, beginning at Canon Pyon on seedlings of the "old" Foxwhelp. Their success had been well established by 1823 when Mr. Jay, of Lyde, got grafts, and afterwards Mr. Bosley, of Lyde, and Mr. Hill, of Eggleton, and thus from the centres of Cowarne and Canon Pyon, intelligent fruit growers got their supply of grafts, and we have the handsome, luxuriant, and useful fruit of this time. With this distinct history there can scarcely be a doubt but that the "new" Foxwhelp is simply the "old" historic variety rejuvenated by careful management, but the doubt in it has arisen from the absence in part or altogether of the true Foxwhelp flavour in the cider made from it, which is so remarkable and characteristic in the "old" Foxwhelp. As a matter of fact, its cider is more sweet and luscious than that made from the "old" Foxwhelp, and in flavour resembling far more the cider made from the Cowarne Red apple.

It must be remembered, however, that sometimes for years together, the cider from the "old" Foxwhelp itself gives but a faint suspicion of the true Foxwhelp flavour which is so highly esteemed, and moreover that it is only of late years, comparatively speaking,—that is, after the trees had become of considerable age—that the cider gained the pride of place it now so deservedly holds. In Evelyn's time, the "old" Foxwhelp was merely considered a first-class cider fruit. It must be left therefore for time to develope the true flavour of the Foxwhelp in its rejuvenated form.

The analysis of this apple by Mr. G. H. With, F.R.A.S., is as follows :—

Density of Fresh Juice	...	...	...	...	1·043
After 24 hours exposure	...	...	...	...	1·044

100 parts of the juice contains :—

Sugar	...	...	...	...	...	8·000
Tannin, Mucilage, Salts, &c.	...	...	...	...	4·301
Water	...	...	...	...	...	87·699

The rejuvenated Foxwhelp has intrinsic merits of its own, and for this cause alone it should be grown much more plentifully than it has been hitherto. Every orchard should possess it, and its owners may await with good faith the development of the true Foxwhelp flavour in its cider, as the trees grow older. Speaking of it as an apple, it may be said, that it is above the medium size, and its brilliant colour recommends it to every one. It sells well in September as a "pot fruit." It has a piquant acid rough flavour, which would not please all palates to eat raw, but as a cooking apple, it is excellent for pies and puddings; and "the apple of all others to make sauce for the Michaelmas goose, or for a roast leg of pork."

PLATE VIII.

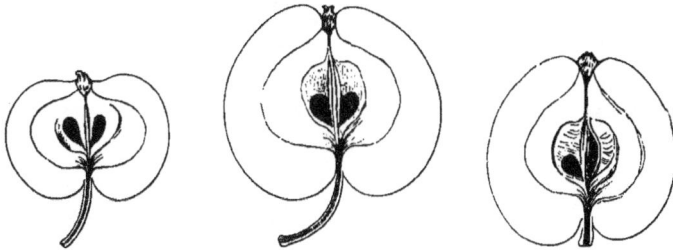

2. BASTARD FOXWHELP.   3. RED FOXWHELP.   4. BLACK FOXWHELP.

These several apples bear the Foxwhelp name. They have no special history, but the inference is that they are, what tradition supposes them to be, seedlings from the Foxwhelp.

---

2. BASTARD FOXWHELP.

There are two or three small apples called by this name, but that which is the most esteemed and grown is figured here.

Description.—Fruit, small and oblate, sometimes somewhat roundish, even and regularly formed. Skin, smooth and shining as if varnished, entirely covered with bright crimson, and striped with darker crimson on the side exposed to the sun; but on the shaded side it is greenish yellow striped with crimson; the stalk cavity only is lined with russet. Eye, very small and closed, with short connivent segments placed in a shallow saucer-like depression; tube, conical; stamens, marginal. Stalk, very long and slender at its insertion and throughout its length, but thicker at the end, inserted in a deep cavity. Flesh, yellowish stained with red, firm, unusually acid. Cells of the core, slightly open; cell-walls, orbicular.

The Bastard Foxwhelp bears well, and is much esteemed by some growers, who think they detect in the cider which it helps to make a slight Foxwhelp flavour.

---

3. RED FOXWHELP.

This apple is chiefly grown in the Bodenham and Marden districts. It is pretty, well-shaped, and very rich in colour. It is pleasant to eat, cooks well, and its growers value it as a cider apple.

Description.—Fruit small, roundish ovate, even and regular in its outline. Skin, uniformly

PLATE VIII.

very dark crimson, almost of a chestnut or mahogany colour over its whole surface, except a small portion on the shaded side, which is a little, but very little paler. Eye, small and slightly open, with short rather erect segments, and set in a shallow, plaited basin; tube, short conical; stamens, rather marginal. Flesh, yellow, deeply stained with crimson, both under the skin and at the core; very tender, pleasantly flavoured, and with a slight acidity. Cells of the core, open; cell-walls, ovate.

The want of size in the Red Foxwhelp, and its want of sufficient character too, will prevent its being generally grown. Its chemical analysis, however, shows it to be rich in sugar and mucilage.

---

4. BLACK FOXWHELP.

[SYN: *Monmouthshire Foxwhelp.*]

This apple is very widely grown through the County, and is to be found in the majority of "Apple heaps." Its definite ovate shape, smooth surface, and dull colour, make it quite unmistakeable. It bears very freely, and this perhaps is its best qualification, for the cider made from it is thin and poor.

Description.—Fruit, small, roundish ovate, inclining to short conical, even in its outline slightly angular towards the crown, where it is prominently plaited round the eye. Skin, smooth and rather shining, of a dark mahogany colour next the sun, but on the shaded side it is greenish yellow, covered with broad broken stripes of bright crimson. Eye, small and rather open, with rather connivent segments, and set nearly on a level with the surface, with only a very slight depression; tube, short conical; stamens, medium. Stalk, short, set in a shallow cavity. Flesh, yellowish sometimes, with a greenish tinge, briskly acid. Cells of the core, open; cell-walls, obovate.

The sooner the trees of the Black Foxwhelp are regrafted or cut down the better.

The report of the chemical analysis of these varieties by Mr. G. H. With, F.R.A.S., is as follows :—

	Bastard Foxwhelp.	Red Foxwhelp.	Black Foxwhelp.
Density of Fresh Juice	1·042	1·043 ...	1·038
Density after 24 hours exposure	1·042	1·500 ...	1·040
100 parts by weight yielded of :			
Sugar	7·780	10·010	6·400
Tannin, Mucilage, Salts, &c.	4·335	4·256	5·206
Water	87·885	85·734	88·394

PLATE VIII.

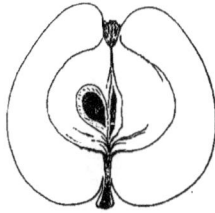

5. DYMOCK RED.

This apple takes its name from the village of Dymock, in Gloucestershire, on the borders of Herefordshire. It is an apple of considerable antiquity, and was probably produced towards the end of the seventeenth century. In Evelyn's time it bore a high reputation, and it well sustains its character in these days.

Description.—Fruit, roundish or oblate, even and regular in its outline; handsome. Skin, entirely covered with dark mahogany red, with streaks of bright pale crimson on the side next the sun, and somewhat paler, though of the same colour, on the shaded side; the whole surface is strewed with distinct russet dots, and mottled with patches and ramifications of cinnamon coloured russet. Eye, medium sized, with segments that are sometimes divergent and sometimes connivent; when the former, they are quite reflexed, and when the latter, they touch each other by their margins and close the eye, which is placed in a narrow, shallow, slightly plaited basin; tube, funnel-shaped; stamens, basal; stalk, very short, and often a mere knob, in a very narrow and shallow cavity. Flesh, yellowish, tender and soft, occasionally tinged with red, slightly sweet, and with a pleasant acidity. Cells of the core, closed; cell-walls, ovate.

Mr. With's analysis, season 1878, gives the following results:—

Density of Fresh Juice...	...	...	1·033
Ditto, after 24 hours' exposure	...	...	1·037
100 parts by weight of fresh juice gave of			
Sugar	...	... ...	12·100
Tannin, Mucilage, Salts, &c.		... ...	3·280
Water ... ... ·	...	... ...	84·620

The Cider made from this Apple, whether pure, or mixed with other fruit, is rich and excellent.

The Dymock Red Apple is grown chiefly in the neighbourhood of Ledbury, but from its high merits it deserves a far wider cultivation. The colour of the Apple on the plate should be of a much deeper and duller red, in sunny seasons it takes quite a mahogany tint.

PLATE VIII.

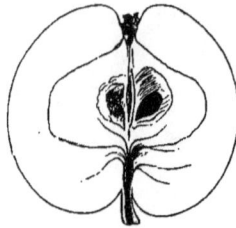

6. MUNN'S RED.

[SYN : *The Pretty Maid ; Greasy Apple.*]

This Apple derives its name from that of its producer, a householder at Canon Pyon. It is widely grown in Herefordshire, and attracts attention in most orchards by the remarkably bright and glossy colour of its fruit.

Description.—Fruit, round, sometimes slightly ovate, even and regular in its outline. Skin, bright red, approaching scarlet, mottled, and somewhat streaked with crimson over the whole surface. Eye, closed, with convergent segments, set in a pretty deep basin, which is sometimes even and saucer-like, and sometimes a little angular; tube, short, funnel-shaped ; stamens, median. Stalk, long, curved, and rather stout and woody, inserted in a very deep, round cavity. Flesh, yellowish, with a stain of red from the base of the eye round the carpels. Cells of the core, open ; cell-walls, elliptical.

The chemical analysis of this Apple, season 1878, by Mr. G. H. With, F.R.A.S., gives the following results :—

Density of Fresh Juice	...	...		1·0450
Ditto after 24 hours' exposure	...	...		1·0456
100 parts by weight yielded of				
Sugar	...	...	...	9·110
Tannin, Mucilage, Salts, &c.	...	...	...	4·718
Water	...	...	...	86·172

Notwithstanding this analysis, its cider is not deemed of first excellence.

PLATE VIII.

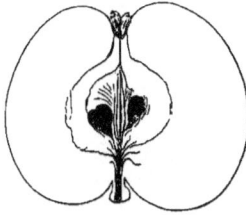

7. WHITE MUST.

[SYN : *White Musk.*]

This Apple is a very old variety. It is mentioned by Evelyn as "A great bearer, and its cider early ripe,"—and Philips says of it :—

> " But how with equal Numbers shall we match
> The *Musk's* surpassing Worth ! that earliest gives
> Sure hopes of racy Wine, and in its Youth,
> Its tender Nonage, loads the spreading Boughs
> With large and juicy Offspring, that defies
> The Vernal Nippings, and cold Syderal Blasts ! "

Description.—Fruit, roundish or oblate ; even and regular in its outline. Skin, smooth and shining, of an uniform pale straw colour, which is a little deeper where it is more exposed to the light. Eye, small and open, set in a narrow and rather deep basin, which is round and smooth ; segments, divergent ; tube, short conical ; stamens, basal. Stalk, short, and almost entirely within the cavity, and from which issues a ramifying patch of rough scaly brown russet, extending over the base. Flesh, yellowish, very tender, juicy, and pleasantly subacid. Cells of the core, closed ; cell-walls, obovate. This is a pretty apple, and, after being gathered, its skin becomes quite unctuous, and it gives off a powerful ethereal odour.

Mr. With's analysis, season 1878, gives these results :—

Density of fresh juice		1˙037
Ditto after 24 hours' exposure to air ...		1˙040
100 parts of weight of juice contained :		
Sugar		8˙030
Tannin, Mucilage, Salts, &c.		3˙580
Water		88˙390

The White Must apple still retains its useful qualities, and is largely grown in all the cider counties of England. It produces a deep-coloured, sweet, and pleasant cider ; but it has no great strength, and will not keep long.

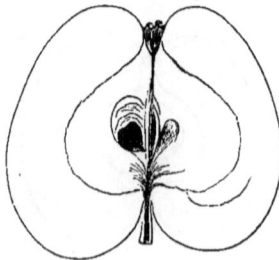

Plate VIII.

8. SAM'S CRAB.

[Syn: *Longville's Kernel.*]

This Apple, according to Mr. Lindley, "was originated in Herefordshire, where," he adds, curiously enough and very erroneously, "it is at present but little known." It is, on the contrary, well known in Herefordshire, widely distributed, and very highly esteemed as a very early dessert fruit. It is used also for cooking, and for cider.

Description.—Fruit, conical or roundish ovate, even and regular in its outline. Skin, beautifully streaked with crimson, and yellow on the side next the sun, and less so on the shaded side, where it is more yellow. Eye, closed, with connivent segments, set in a pretty deep, round, and somewhat plaited basin; tube, funnel-shaped; stamens, median. Stalk, about an inch long, slender, inserted in a deep cavity, which is tinged with green. Flesh, yellowish, tender, juicy, sweet, and of good flavour. It is tinged with red at the base of the eye, at the base of the stalk, and round the carpels. Cells of the core, open; cell-walls, ovate.

Mr. With's analysis, season 1878, gives these characters :—

Density of Fresh Juice...		1·037
Ditto after 24 hours' exposure to air		1·046
100 parts by weight of the juice contained :—		
Sugar		10·140
Tannin, Mucilage, Salts, &c.		4·370
Water		85·490

Sam's Crab is one of the most useful of all our early Apples. It requires a warm soil and sunny situation to bring its fruit to perfection In unfavourable situations it could hardly be recognised as the same Apple. When well ripened it has a rich aroma, and a juicy, sweet, and piquant flavour that is seldom equalled. It is a prime favourite with all Herefordshire school children (no mean judges of a good Apple), and it is equally attractive to birds and insects, who revel in its sweetness. There are undoubtedly two varieties of this Apple, or, as was quaintly expressed by a great admirer of the fruit, "There are two sorts of Sam's Crab: a basket full of one kind is eaten the same day, but the same basket full of the other kind lasts three or four days."

PLATE VIII.

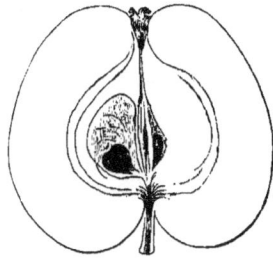

9. SACK APPLE.

[SYN : *Spice Apple : Fox's Kernel.*]

This Apple is one of our oldest historic varieties.

Description.—Fruit, below medium size ; conical, and uneven in its outline, being ribbed on the sides in the way of the Margil, and ridged round the eye. Skin, smooth and shining, as if varnished, almost entirely covered with deep bright crimson, which is streaked and mottled with darker crimson on the side next the sun ; but where shaded, it is yellowish and mottled with crimson. Eye, small and closed, with erect pointed segments, set in a deep and plaited basin ; tube, funnel-shaped : stamens, median ; the style very stout and thick at the base, filling nearly the half of the tube. Stalk, very short, thick, and fleshy, set in a very shallow cavity. Flesh, tender, crisp, fine-grained, sweet, and with a pleasant subacid flavour. Cells of the core, open ; cell-walls, ovate. In use during October and November.

The chemical analysis of this Apple, season 1878, by Mr. G. H. With, F.R.A.S, gives the following results :—

Density of fresh juice	1·036
Ditto after 24 hours' exposure to air ...	1.044
100 parts of juice by weight contained :	
Sugar	6.400
Tannin, Mucilage, Salts, &c.	5·220
Water	88·380

The Sack Apple is more used in the present day as a dessert or pot fruit, than for cider. It is an early apple, but keeps fairly well. It has a pleasant vinous aromatic flavour.

PLATE IX.

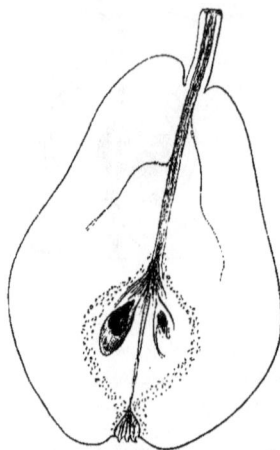

1. WILLIAMS' BON CHRÉTIEN.

[SYN: *Wheeler's; Bartlett; De Lavault; Williams'.*]

"In Amitermis vale, the Sabine boors
Added Bon-crêtiens to their former stores."
(*Rapin, "The Orchard," translated by Gardener.*)

This highly esteemed Pear was raised a short time previous to 1770, by a person of the name of Wheeler, a schoolmaster at Aldermaston, in Berkshire; from him it was obtained by Mr. Williams, the nurseryman at Turnham Green, Middlesex, and being by him first distributed, it received the name it now bears. In 1799 it was introduced to America by Mr. Enoch Bartlett, of Dorchester, near Boston, through whom it became generally distributed, and has ever since been known there by the name of the Bartlett Pear. It attains the highest perfection in America, and is esteemed as the finest and best keeping Pear of its season. It has even been brought back to England with its new name.

Description.—Fruit, large, obtuse-pyriform, irregular, and bossed in its outline. Skin, smooth, at first pale green, changing as it ripens to clear yellow, and tinged with streaks of red next the sun. Eye, open, set in a very shallow depression, but more generally even with the

PLATE IX.

surface. Stalk, an inch long, stout, and fleshy, and inserted in a shallow cavity. Flesh, white, fine-grained, tender, buttery, and melting, with a rich, sweet, and delicious flavour, and powerful musky aroma.

An excellent Autumn Dessert Pear ; in season in August and September. The tree is healthy and vigorous, but not a regular and abundant bearer. It succeeds best as a pyramid or standard on the pear, or quince stock, when the fruit is much better flavoured, though not so large as when grown on a wall. The fruit should be gathered before it is ripe, at intervals of a few days, that they may not all ripen together—for when ripe, it soon becomes mealy and decays. Its cultivation by the London market gardeners has become more limited than it was, in consequence of its fickleness in bearing.

An excellent coloured representation is given of this pear in the Transactions of the Horticultural Society (London,) Vol. II., p. 250, where, and in Vol. III., p. 357, its history and description are given.

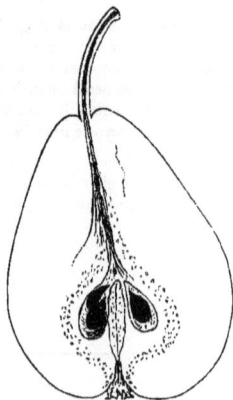

PLATE IX.

2. FORELLE.

[SYN: *Forellenbirne; Corail; Corille; Grain de Corail; Petit Corail; Truite; Trout Pear.*]

This is a very old variety, and a native of Germany, where it has been cultivated for some 200 years. Dr. Diel supposes it to have originated in northern Saxony.

Description.—Fruit, medium sized, elongated, obovate, but sometimes pyriform in shape. Skin, smooth, shining, of a fine lemon-yellow colour on the shaded side, and bright crimson on the side next the sun, covered with numerous crimson spots, which from their resemblance to the markings on a trout, have suggested the name it bears. Eye, small, set in a rather shallow basin. Stalk, an inch long, slender, inserted in a small shallow cavity. Flesh, white, delicate, buttery, and melting, with a rich sugary and vinous flavour.

This is a delicious flavoured pear when grown on a south wall, which it requires in Herefordshire. The tree grows freely, but is uncertain in bearing. The blossoms in Spring should be protected with sprays of fir, yew tree, or common larch. This pear ripens in November and its great beauty will always recommend it for the dessert table. A coloured drawing of it is given in the Transactions of the Horticultural Society (London,) Vol. V., p. 409.

PLATE IX.

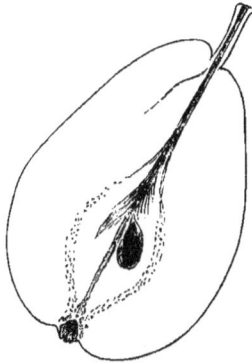

3. LOUISE BONNE OF JERSEY.

[SYN: *Beurré d'Avranches; Bonne d'Avranches; Louise d'Avranches; Bonne de Longueval; Bonne Louise d'Arandoré; William the Fourth.*]

This valuable pear was raised at Avranches about the year 1788 by M. Longueval, who at first named it simply "*Louise*" but subsequently added "*Bonne*," and it thenceforth became known as "*Bonne Louise d'Avranches.*" The original tree is still in existence in the garden where it was raised. It is cultivated with great success in the island of Jersey, and coming to England in this way, received the name it now bears.

Description.—Fruit, medium sized, pyriform, skin, smooth, yellow on the shaded side, but crimson next the sun, covered with crimson and russety dots. Eye, small and open, set in a rather deep basin. Stalk, three quarters of an inch long, obliquely inserted without depression. Flesh, white, buttery and melting, with a rich, sugary, and brisk vinous flavour.

A very delicious pear—ripe in October. The tree is hardy, and a very good bearer. Its best form of growth is as a pyramid on the quince stock and in this form it should take its place in every garden.

The colouring of this pear on the Plate was very true to the specimen, but it is only coloured so highly in very sunny seasons. An outline of it is given in the Transactions of the Horticultural Society (London,) Vol. II., p. 42.

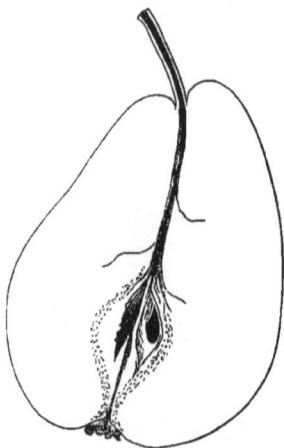

4. BEURRÉ D'AMANLIS.

[SYN: *D'Amanlis; Beurré d'Amalis; Delbart; Plombgastelle; Thiessoise; Kaissoise; Wilhelmine* of some, but not of Van Mons.]

This pear was introduced, says M. Prévost, into Normandy from Brittany so early as 1805. It is said to have been raised originally near Rennes.

Description.—Fruit, large, averaging three inches and a half long, by three quarters wide; obtuse pyriform, or obovate, uneven and undulating in its outline. Skin, at first of a bright green, tinged with brown next the sun, and marked with patches and dots of russet, but afterwards assuming a yellowish green tinge, and a reddish brown cheek as it ripens. Eye, open with stout segments, and set almost level with the surface. Stalk, long, slender, and woody, inserted in a small cavity. Flesh, greenish white, fine-grained, tender, juicy, melting, rich, sugary, and agreeably perfumed.

One of the best early pears; ripe the middle of September. The tree is hardy, a strong grower and an excellent bearer. It forms a handsome pyramid when properly attended to, and succeeds either on the pear or the quince stock. It is deservedly a general favourite.

PLATE IX.

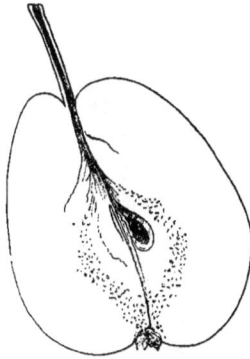

5. FLEMISH BEAUTY.

[SYN: *Belle de Flandres; Belle des Bois; Beurré des Bois; Beurré de Bourgoyne; Beurré Davy; Beurré Davis; Beurré d'Effingham; Beurré d'Elberg; Beurré Foidard; Beurré St. Amour; Beurré Spence; Bosch Peer; Mouille Bouche Nouvelle; Brilliante; Fondante des Bois; Gagnée à Henze; Impératrice des Bois.*]

This excellent pear was discovered by Van Mons, in the village of Deftinge, in Flanders, in 1810; and he brought it into notice by distributing grafts among his friends.

Description.—Fruit, large, and obovate. Skin, pale yellow, almost entirely covered with yellowish brown russet on the shaded side, and reddish brown on the side next the sun. Eye, open, set in a small shallow basin. Stalk, an inch long, inserted in a rather deep cavity. Flesh, yellowish white; buttery, melting, rich, and sugary.

A first-rate pear, ripe in September, but it requires to be gathered before it is ripe, or it will be inferior in quality. Best grown as a pyramid on the quince or pear stock.

PLATE X.

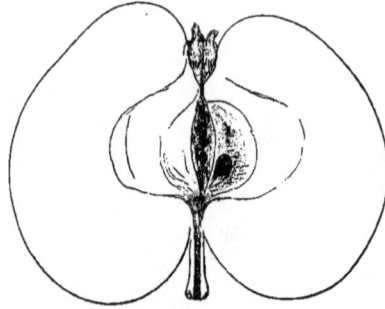

1. GLORIA MUNDI.

[SYN: *Baltimore; Belle Dubois; Mammoth; Monstrous Pippin; Ox Apple.*]

This Apple is believed to be of American origin, but some doubts exist as to where it was first raised, that honour being claimed by several different localities. The general opinion however is, that it originated in the garden of a Mr. Smith in the neighbourhood of Baltimore. It was introduced from America into France by Comte Lelieur, in 1804, and was brought over to this country by Captain George Hudson, of the ship Belvidere, of Baltimore, in 1817.

There is, however, some doubt as to whether it is a native of America. In the volume of the "*Allgemeines Teutsches Gärtenmagazin*" for 1805, it is said to have been raised by Herr Künstgärtner Maizman, of Hanover. If this account is correct, the apple must have been taken to America by some of the Hanoverian emigrants, which would account, moreover, for the claim of its origin by different localities there. Dittrich, vol. III, p. 41, has confounded the synonyms of the *Gloria Mundi* with *Golden Mundi*, which he has described under the head of *Moustow's Pepping*.

Description.—Fruit, immensely large, sometimes measuring four inches and a half in diameter; of a roundish shape, inclining to be angular on the sides, and flattened both at the base and apex. Skin, smooth, pale yellowish green, interspersed with white dots, and patches of thin delicate russet, and tinged sometimes with a faint blush of red next the sun. Eye, large, open, and deeply set in a wide and slightly furrowed basin. Stalk, short and stout, inserted in a deep and open cavity, which is lined with rough russet. Flesh, white, tender, and juicy, but not highly flavoured.

It is an excellent culinary apple, in season from October to December.

The tree grows freely, and bears abundantly. It was introduced into Herefordshire, some

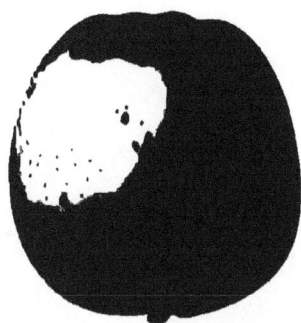

Potts' Seedling

3 Tower of Glams

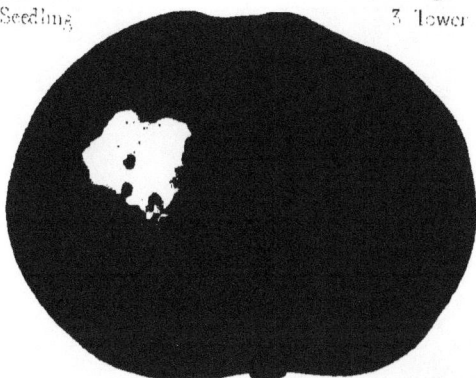

Gloria Mundi

4 Winter Hawthornden

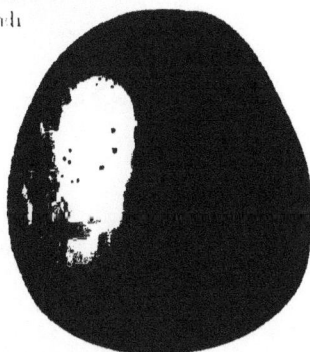

5 Nelsons Codlin

PLATE X.

years since, by a lady who met with the apple in London, got grafts through the Covent Garden salesmen, and brought them to Moccas. It grows best on a standard, but requires a sheltered situation, to prevent the heavy fruit from being blown off by the wind. The Woolhope Club, in one of the Autumn Forays, fell in with a full-grown tree at Byford, with all its boughs arched down with the weight of its fine fruit. The apples were all the more regular in shape, from not being so large in size. It was a sight not soon to be forgotten.

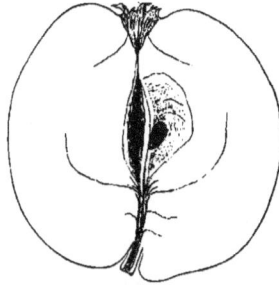

2. POTTS' SEEDLING.

This Apple was raised from the seeds of an American apple, at Ashton-under-Lyne, about the year 1849, by the late Mr. Samuel Potts, of Robinson Lane, Ashton.

Description.—Fruit, large, full, round, and upright, but not conical; sometimes irregular in shape, about three inches broad, and three and a half inches high. Skin, light green, not unlike *Lord Suffield*, becoming very yellow when ripe, with numerous small golden spots on the sunny side. Eye, shallow, with converging segments. Stalk, long, deeply inserted in a narrow cavity, and often connected with the apple by a fleshy protuberance on one side. Flesh, white, with a very agreeable acidity.

This is an excellent cooking apple. It is in its best flavour in October and November, but it will keep, with care, until January.

The tree is very hardy, robust in growth, with heavy roundish foliage, and bears abundantly. It forms a beautiful bush, or pyramid, and, from being so good a cropper, should become a very profitable market apple. The fruit should be gathered before it is too ripe, or, like the *Lord Suffield* apple, it will be bruised in carriage.

This apple is now described and represented for the first time.

PLATE X.

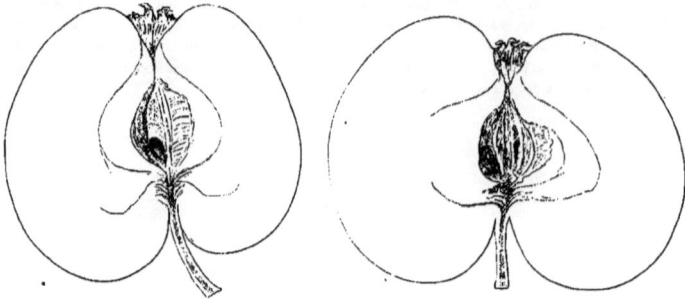

3. TOWER OF GLAMMIS.   4. WINTER HAWTHORNDEN.

3. TOWER OF GLAMMIS.

[SYN : *Glammis Castle ; Gowrie ; Carse of Gowrie.*]

The origin and history of this famous Scotch Apple seems to have been lost. It abounds in the orchards of Clydesdale, and the Carse of Gowrie, and this is all that could be learnt about it.

Description.—Fruit, large, conical, distinctly four sided, with four prominent angles extending from the base to the apex, where they terminate in four corresponding ridges. Skin, light green, becoming of deep sulphur yellow, tinged in some spots with green, and thinly strewed with brown russety dots. Eye closed, with broad ragged segments, set in a deep and angular basin. Stalk, an inch long, inserted in a deep funnel-shaped cavity, and only just protruding beyond the base. Flesh, greenish white, very juicy, crisp, brisk, and perfumed.

A first-rate culinary apple, in season from November to February. The tree grows well, and is an excellent bearer.

---

4. WINTER HAWTHORNDEN.

[SYN : *New Hawthornden.*]

Efforts have been made in vain to discover the history of this apple, although it must be a recent production.

Description.—Fruit, large, roundish ovate, and altogether not unlike the old Hawthornden in appearance. Skin, greenish, with a brownish red tinge on the side next the sun. Eye, closed. Stalk, long and slender. Flesh, white, very tender, juicy, and with a fine subacid flavour.

A first-rate culinary apple ; in season from December to January.

The tree forms an excellent pyramid, and is an abundant and early bearer.

PLATE X.

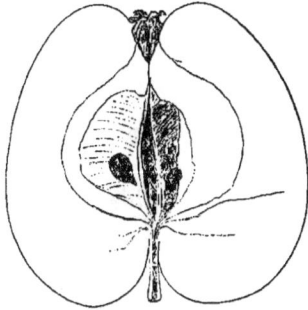

5. NELSON CODLIN.

[SYN : *Nelson; Nelson's Codlin; Backhouse's Nelson.*]

This much esteemed variety was first brought into notice by Mr. John Nelson, a noted Wesleyan preacher in the early days of Wesleyanism, who, while engaged in the work of evangelization in Yorkshire, used to distribute grafts among his friends ; from these circumstances it became known as the Nelson Apple. Mr. Hugh Ronalds, who received it from Mr. Backhouse, of York, published it in the *Pyrus Malus Brentfordiensis* as Backhouse's Lord Nelson, a name which the late Mr. James Backhouse disclaimed, and said he preferred that so excellent an apple should rather be the memorial of so excellent a man.

Description.—Fruit, large and handsome, conical, or oblong. Skin, greenish yellow, strewed with russety specks on the shaded side, but when exposed to the sun, of a fine deep yellow, covered with rather large dark spots, which are encircled with a dark crimson ring. Eye, open, with short segments, set in a deep, plaited, and irregular basin. Stalk, about a quarter of an inch long, inserted in a very deep and angular cavity. Flesh, yellowish white, delicate, tender, juicy, and sugary.

A very excellent apple ; of first-rate quality as a culinary fruit, and also valuable for the dessert. It is in season from September to January.

The tree is a strong, vigorous, and healthy grower, and bears abundantly.

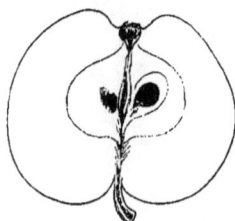

PLATE XI.

1. REDSTREAK.

[SYN: *Scudamore's Crab; Herefordshire Redstreak; Redstrake of King's Caple; Irchinfield Redstrake.*]

"Yours be the produce of the soil :
O ! may it still reward your toil !
But though the various harvest gild your plains,
    Does the mere landscape feast your eye ?
Or the warm hope of distant gains
    Far other cause of glee supply ?
Is not the Redstreak's future juice
    The source of your delight profound,
Where Ariconium pours her gems profuse,
    Purpling a whole horizon round."

Shenstone.

The Redstreak Apple has been the most fortunate of all cider apples for the renown it has obtained. It appeared at a time when great attention was paid to the Herefordshire orchards. It at once found a patron of remarkable energy and influence to propagate it: and its praises have been said and sung, in prose and verse, beyond all other kinds of fruit. It seems to have originated about the beginning of the 17th century, and was first brought into general notice by Lord Scudamore (see Introduction, *Life of Lord Scudamore*, p. 67). Evelyn is the first writer who notices it, as " The famous Red-strake of Herefordshire, a pure Wilding, and within the memory of

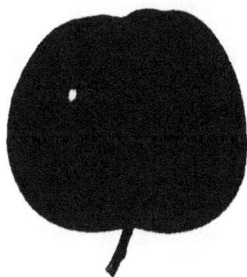

PLATE XI.

some now living, surnamed the *Scudamore's Crab*, and then not much known, save in the neighbourhood." Philips next took up its praise, and in his poem *Cyder* says :

> " Let every Tree in every Garden own
> The *Redstreak* as supream ; whose pulpous Fruit
> With Gold irradiate, and Vermillion shines.
> Tempting, not fatal as the Birth of that
> Primæval interdicted Plant, that won
> Fond *Eve* in hapless Hour to taste and die.
> This of more bounteous Influence inspires
> Poetic Raptures, and the lowly Muse
> Kindles to loftier strains ; even I perceive
> Her sacred Virtue. See ! the Numbers flow
> Easie whilst cheer'd with her Nectarious Juice,
> Her's, and my Country's Praises, I exalt.
> Hail Herefordian Plant ! that dost disdain
> All other Fields ! Heaven's sweetest Blessings, hail !
> Be thou the copious Matter of my Song,
> And thy choice Nectar ; on which always waits
> Laughter and Sport, and care beguiling Wit,
> And Friendship, chief Delight of Human Life,
> What should we wish for more."

The Redstreak Apple was thus brought into the highest popular favour, and its sweet and pleasant cider was deemed " a fitting present for princes." It completely supplanted the *Gennet Moyle*, which had previously been the favorite cider apple, and indeed, for the time being, all other Cider Apples were thrown into the shade. The extent to which its cultivation extended, is well shown by the following Extract from a *MS.* in the Bodleian Library, at Oxford, entitled *The History of Gloster*, or *The Antiquities, Memoirs, and Annals of ye Ancient City and Royal Dukedom of Gloster, from its Original to the present time*, by Abel Wantner, citizen of Gloster, 1714 :—

" Dimock and Kemply, before mention'd, are two of the most note'edst parishes in England for making the most and best rare *Vinum Dimocuum*, or that transcendant Liquor, called Redstrake Sider, not much inferior to the best French wines. And so plentifull that old Master Wyniat, of the Grainge, (a worthy gentleman, and a noble housekeeper,) hath caused but one Apple to be gathered from off each Apple Tree growing in his own Grounds, and with the Liquor thereof he hath made a whole hogshead of reare good Sider.—*Furley MSS., Vol. IV., fol. 196, p. 2.**

Its reputation however began to decline about the middle of last century. Its cider, though sweet and pleasant, had not much strength, and would not keep well. " Its liquor," Nourse describes as " of noble colour and smell, but withal very luscious and fulsome. They who drink it will find their stomachs pall'd sooner by it, than warmed and enliven'd." With the Herefordshire cider growers it must however be stated that its cider was thought from the very first to be inferior in strength and quality to that from many other kinds of fruit. " *Gennet Moyle* makes the best

---

*This Extract is copied into the Parish Register of Dymock, but the entry neither gives the date nor the authority for it, both of which have been obtained from the Bodleian Library, and the Extract itself verified.

PLATE XI.

Cyder in my Judgement, and such as I do prefer before the much commended Redstreak'd," says Dr. Beale (1656), and Evelyn, and the writers in the Appendix to the Pomona, say as much for several other apples, as the *Woodcock*, the *Hagloe Crab*, the *Underleaf*, the *Styre*, the *Must*, the *Bromsborrow Crab*, and some others. The soundness of this judgment was soon confirmed by experience, for by the end of the last century the *Redstreak* had quite lost all favour. Dunster, in his *Notes to Philips' Poems*, thought the true method of managing it was lost, for out of ten or twelve casks, seldom more than two or three proved good, and adds "it is now seldom made pure" (1791). And Marshall, in his *Herefordshire Orchard* (1796), says plainly, "The Redstreak Apple is given up." And Andrew Knight speaks of it (1811), as having survived its good qualities.

Description.—Fruit, medium sized, two inches and three quarters wide, and two inches and a quarter high ; roundish, narrowing towards the apex. Skin, deep clear yellow, streaked with red on the shaded side, but red, streaked with deeper red, on the side next the sun. Eye, small, with convergent segments, set in a rather deep basin. Stalk, short and slender. Flesh, yellow, firm, crisp, and rather dry. Specific gravity of the juice, 1079.

The tree seems naturally to have been very short-lived. It was low and shrubby, and rugged in growth. Evelyn says of it, "That as the best Vines of richest liquor and greatest burthen do not spend much in Wood and improfitable Branches, so nor does this tree."

The result of careful enquiries recently made for the true variety, was the discovery of one tree at King's Capel, which was however blown down in the Spring of 1878, and Mr. Reginald Wynniat, of the Grainge, has kindly ascertained for this paper, that there is still one tree remaining at Kemply, out of the many thousands growing at Dymock and Kemply in 1714. Redstreaks there are in abundance, in every parish, named from their colour, and this fact was also noticed by Evelyn. "The *Red-Strake* of King's Capel, and those parts is in great variety, some make Cider that is not of continuance, yet pleasant and good : others, that lasts long, inclining towards the *Bromsborrow Crab* rather than a *Red-Strake*."

As a distinct variety the *Redstreak* has now ceased to exist, and it may be added that the loss is not to be lamented.

PLATE XI.

THE NORMAN CIDER APPLES.

" Let foreign Apples in your Orchard live,
Hence will your fruit be always of the best,
And you with plenty of such kinds be blest."

Rapin, " Of Gardens."

" Fairest Apples Normandy adorn."

Rapin.

In the *Liber Landavensis*, the British Monks, St. Tèilo and St. Samson, are reported to have carried over a large quantity of apple trees from Monmouthshire to Armorica (Britanny), and to have planted there an orchard three miles in length (see Introduction *Early History of the Apple and Pear*, pp. 10-11.). In more modern times the advantage has been returned, and Norman Apples have been brought here. Lord Scudamore was the first to introduce them to Herefordshire; Mr. Foley and many others have imported them since his day. At the present time the Norman Cider Apples are in the highest esteem, and occupy a considerable share of the ground in all fresh plantations. They are so productive, as to cause it to be said, that since their introduction there has never been a complete failure in the apple crop. The saying is not strictly true, to wit the total failure in the years 1855 and 1858, but it affords, nevertheless, a great tribute to their hardihood and fertility. Another great proof of their extreme popularity consists in the fact that every unknown or nameless apple in the orchard, is at once called " Norman" : the orchards teem with them, and have done so now for many years. Marshall, in his *Herefordshire Orchards* (1796), first notices the tendency. He says, " At Ledbury I was shown a Normandy Apple, said to have been imported immediately from France. On seeing and tasting the fruit, I found it to be no other than this *Bitter Sweet*, which I have seen growing as a neglected Wilding in an English hedge" (p. 245). The habit has increased of late years to such an extent as to create the greatest confusion as to the right names, or want of names, of the many varieties of Cider Apples.

In the year 1862, an enquiry was commenced in France, by the Horticultural Society of the Seine-Inférieure into the best varieties of Apples and Pears adapted to the production of Cider and Perry. This Society was joined by others, and annual meetings were held for several years in succession, until in the year 1875 a very careful and elaborate *Report* was published at Rouen. It was edited by M.M. de Boutteville and Hauchecorne, and besides the full description and analysis of the different varieties of fruit, it gives very good coloured drawings of those most esteemed. This " Report," in addition to some little experience, enables the comparison to be made ; and it may safely be said that the so-called Norman Apples of Herefordshire, with very few exceptions are certainly not at this time the Apples of Normandy. The contrast is very marked and decided. It may be owing to the differences in soil, or climate, or situation, but it is there ; and the Pomona Committee of the Woolhope Club would be as completely at fault amongst the various apples that border the high roads of Normandy, as would M.M. de Boutteville and Hauchecorne in the more luxuriant orchards of Herefordshire.

PLATE XI.

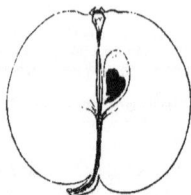

2. CHERRY NORMAN.

[SYN : *Hitterly.*]

This Apple corresponds with the Apple *Moulin-à-vent*, or *Douce-Morelle-Rouge*, of the Normandy orchards, in every particular, but in the season of its maturity. The Cherry Norman is an early apple here, September or October, but the *Moulin-à-vent* is not mature until December. The French "Report" speaks of it as a very old variety, cultivated chiefly in the Departments of the Orne and the Eure.

Description.—Fruit, round, and pretty regular in its outline, occasionally a little ribbed, and peculiarly rounded at the base, with a small and very narrow stalk cavity, on one side of which is a fleshy swelling uniting the stalk on one side to the fruit. Skin, clear straw yellow, with a large russet check on the side next the sun, and a few traces of russet extending to the shaded side; there is sometimes a crimson or reddish orange mixture among the russet of the sunward side. Eye, very small, with long convergent segments placed in a shallow depression, and set round with prominent plaits; tube, conical; stamens, marginal. Flesh, soft, spongy, bitterish, and sweet. Cells of the core, slightly open; cell-walls, roundish, obovate.

The Cherry Norman is much esteemed in our orchards. The tree grows well and freely, but it is apt to bear in abundance only every other year. It makes a cider of a deep colour, with a sweet, rich, and pleasant flavour. It is one of the best early fruits, and deserves a still more extended cultivation.

Chemical analysis by Mr. G. H. With, F.R.A.S., season 1878.

Density of Fresh juice ...	...	...	1·043
Density after 24 hours ...	...	...	1·046
In 100 parts by weight of juice :			
Sugar ...	...	...	...	12·830
Tannin, Mucilage, Salts, &c.	...	...	2·073
Water	...	...	...	85·097

PLATE XI.

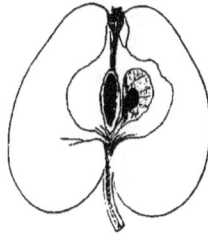

3. RED NORMAN.

This is an old variety in our orchards, which, although called Norman, does not correspond closely with any of the Norman apples of the French "*Report.*" It seems, however, to be most closely allied to that excellent variety *Martin-Fessard*, which is grown very much in the neighbourhood of Yvetot.

Description.—Fruit, conical, sometimes long conical, snouted, and puckered towards the apex. Skin, smooth, lemon yellow, with a faint blush of red on the side exposed to the sun; the surface sparingly strewed with minute russet points. Eye, very small, with convergent segments, set in a shallow, narrow, puckered basin: tube, very long and slender, funnel-shaped; stamens, marginal. Stalk, half an inch long, slender, and obliquely inserted, frequently with a swelling on one side at the base of the fruit. Flesh, greenish yellow, woolly, not very juicy, and sweet. Cells of the core, very large, and closed; cell-walls, ovate.

The *Red Norman* is much esteemed in the Herefordshire orchards, and very widely grown. The tree is vigorous and fertile, but of medium size. The apple juice is dark in colour, with a rich, sweet, and highly aromatic flavour.

Chemical analysis by Mr. G. H. With, F.R.A.S., season 1878.

Density of Fresh juice	1·044
Density after 24 hours	1·051
In 100 parts by weight of juice:	
Sugar	11·905
Tannin, Mucilage, Salts, &c.	3·942
Water	84·153

PLATE XI.

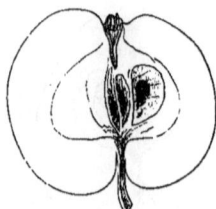

4. WHITE NORMAN.

The origin of this apple is unknown. It seems to correspond with the celebrated old Normandy variety, *Blanc-Mollet*, so much grown all through the orchards of the North-West of France. It is there also called *Pomme de Neige*, *La Blanche*, and various other local names in different districts. It has been described for nearly a century past (see French " *Report*," p. 147).

Description.—Fruit, small, roundish, and with obtuse angles on the sides, which are sometimes pretty prominent. Skin, perfectly white, or rather a very pale straw colour, clear and waxen-like, and with only a few large russet dots, distantly sprinkled over the surface ; the hollow of the stalk is lined with russet, which extends in ramifications a little way over the base. Eye, very small, with small convergent segments, set in a deep basin, which is plaited or slightly ribbed ; tube, deep, conical ; stamens, marginal. Stalk, long, and very slender, deeply inserted. Flesh, snow white, soft, and spongy, with a marked astringency or bitterness mixed with sweetness. Cells of the core, open, and very large for the size of the fruit ; cell-walls, elliptical.

The tree is of middle size, vigorous and fertile. It is very generally cultivated. The fruit is early, soft and juicy ; its juice being rather dark in colour, with a rich and rather bitter flavour.

Chemical analysis by Mr. G. H. With, F.R.A.S. :—

Density of fresh juice	...	...	...	1·040
Density after 24 hours	...	...	...	1·042
	In 100 parts by weight of juice :			
Sugar	...	...	...	10·770
Tannin, Mucilage, Salts, &c.	...	...		3·633
Water	...	...	...	...	85·597

PLATE XI.

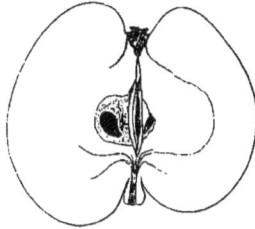

5. STRAWBERRY NORMAN.

The origin and history of this apple is not known; nor, though it bears a Norman name, does it seem to resemble any of the Norman apples given in the French "*Report*." It is probably a Herefordshire seedling.

Description.—Fruit, small, round, and flattened, uneven in its outline, being angular and considerably ribbed about the eye, which is deeply sunk. Skin, with a lemon yellow ground, covered with light crimson, which is thickly marked with broken streaks and mottles of bright and and darker crimson next the sun, and these extend for a considerable space to the shaded side of a paler tint; the base and stalk cavity are lined with cinnamon coloured russet. Eye, of medium size, with long, leafy, rather erect, and slightly divergent segments, set in a very deep and ribbed basin; tube, short, funnel-shaped; stamens, inclining to basal. Stalk, very short, quite embedded in the cavity, which is lined with russet, extending over the base. Flesh, yellowish, close, and spongy, with a sweet mawkish juice; it has a crimson stain at the base of the eye. Cells of the core, small, closed; cell-walls, obovate.

The tree grows freely, and bears well. The fruit is pleasant in taste, and its juice, when fresh, is sweet and rich, with something of the flavour of a ripe strawberry. It makes excellent cider, and is a variety that well deserves general cultivation.

Chemical analysis by Mr. G. H. With, F.R.A.S. :—

Density of fresh juice ...	...	...	1·043
Density after 24 hours ...	...	...	1·045
In 100 parts by weight of juice :			
Sugar	...	...	13·736
Tannin, Mucilage, Salts, &c.	...	...	1·071
Water	...	...	85·193

PLATE XI.

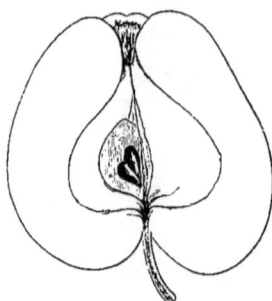

6. HANDSOME NORMAN.

[SYN: *Le Belle Normande; Bell Norman.*]

The common names given to this Apple support the traditions that it is really an introduced Norman Apple, and yet it is not to be found in the French "*Report.*" This is the more remarkable since its size, shape, and colour, make it very observable in the orchard, and a very distinct variety.

Description.—Fruit, conical, snouted towards the apex, very uneven and irregular in its outline, being angular, and having especially one very prominent rib, which makes the fruit one-sided; the base is rounded, and prominently swollen, so that the stalk is placed on an elevation of the surface. Skin, bright red on the side exposed to the sun, gradually fading towards the shaded side, where it is of a fine deep rich yellow; the whole surface is strewed with large russety specks, and the base surrounding the stalk has a patch of grey russet all over it. Eye, closed, with erect pointed segments, set in a deep, irregular, ribbed basin. Tube, long conical; stamens, marginal. Stalk, short, sometimes half an inch long, inserted in a small narrow cavity. Flesh, yellowish, spongy, and sweetish. Cells of the core, open; cell-walls, elliptical.

This tree grows freely, and carries an abundance of fine fruit. It is one of the most favoured varieties at the present time, and is very much planted through the county. The juice is of a rich red colour, with a fine bitter flavour, rich and full, like that of the *Cherry Norman.* It makes a rich and pleasant cider.

The chemical analysis of the juice given by Mr. G. H. With, F.R.A.S., gives the following results for the season of 1878 :—

Density of fresh juice	1·051
Density after 24 hours	1·052
In 100 parts by weight of juice :—	
Sugar	11·905
Tannin, Mucilage, Salts, &c, ...	4·038
Water	84·057

PLATE XI.

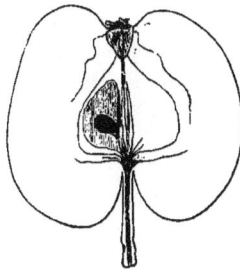

7. BLACK NORMAN.

This very distinct variety is again without any history, or any discoverable connection with the orchards of Normandy.

Description.—Fruit, roundish and flattened, obscurely ribbed, especially round the eye. Skin smooth and shining, unctuous to the touch after the fruit has been gathered; dull mahogany red on the side next the sun, and gradually becoming paler towards the shaded side, which is green and slightly mottled with red. Eye, closed, with long, leafy, convergent segments, set in a rather deep, irregular basin; tube, conical; stamens, median; stalk, long and slender, inserted in a deep, wide, funnel-shaped cavity, which is slightly russety; flesh, greenish, very tender, juicy, and brisk, with a faint sweetness; cells of the core, quite closed; cell walls, ovate.

The tree is hardy, and a good bearer. It is a favourite variety in the orchards. The juice, though without any especial flavour in itself, is rich and sweet, and is believed to give body to the cider made from it.

Chemical analysis by Mr. G. H. With, F.R.A.S :—

Density of fresh juice	...	...	...	1·036
Ditto after 24 hours	...	...	...	1·037
In 100 parts by weight of juice there is :				
Sugar	...	...	...	...	11·905
Tannin, Mucilage, Salts, &c.,		...	...	1·125
Water	...	...	...	...	86·970

PLATE XI.

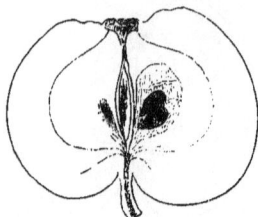

8. PYM SQUARE.

[SYN : *Izard's Kernel; Eggleton Red.*]

This apple originated at Eastnor Farm, near Eastnor Castle. Mr. Henry Izard, some forty years ago (c. 1839), when staying there as a boy, planted three pips of an apple he was eating, in a flower pot The seedlings were afterwards planted by Charles Bourne, the gardener, from Ledbury, in a waste corner of the garden. In due course they were grafted on young stocks. This plant grew very vigorously, and bore fruit the second year after grafting. The two others proved worthless. Bourne called it *Izard's Kernel,* but it afterwards got the name of *Pym Square,* under the mistaken idea that it was a Devonshire Apple introduced into Herefordshire. The origin of the name *Pym Square* is not known, but it is very peculiar, and since there does not seem to be any Devonshire apple of that name, it is retained here.

Description.—Fruit, of medium size, round, inclining to oblate, even and regular in its outline. Skin, smooth and shining, entirely covered with bright crimson, which is rather paler on the shaded side, and slightly mottled with yellow where the ground colour is visible. Eye, small and closed, with flat segments, and surrounded with small bosses or knobs on the margin of the depression ; tube, funnel-shaped ; stamens, marginal. Stalk, sometimes a mere knob, on the rounded base of the fruit ; at others, half an inch long, slender, and inserted in a deep narrow cavity. Flesh, yellowish, tinged with red under the skin, very tender and juicy, briskly and well flavoured. Cells of the core, open ; cell-walls, obovate.

The tree is hardy, grows strong in the wood ; and crops well. The apples are brilliant in colour, and good in flavour, so that in scarce seasons they will sell to advantage as pot fruit.

PLATE XI.

"For culinary purposes they are excellent" says Mr. Izard, "and as soon as the cook finds out their virtues, they are apt to prove bad keepers."

Analysis of *Pym Square* apple, by Mr. G. H. With, F.R.A.S., season 1878 :—Juice, plentiful and somewhat thin, of a very pale, brownish pink colour, with a delicate flavour, subacid and astringent.

Density of fresh juice ...	...	...		1·031
Density after 24 hours ...	...	...		1·035
100 parts by weight contain of				
Sugar	...	...	...	10·219
Tannin, Mucilage, Salts, &c.	...	...	...	2·499
Water	...	...	...	87·282

The *Pym Square* apple has spread from Eastnor into the neighbouring orchards of Herefordshire and Worcestershire, and is increasing still in favour. Its cider is excellent, and has made the voyage to India with great credit to itself.

This apple is described and figured here for the first time.

PLATE XII.

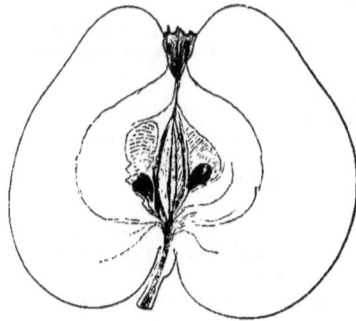

1. EMPEROR ALEXANDER.

[SYN: *Aporta; Alexander; Russian Emperor; Kief's Koy.*]

This beautiful apple is a native of the Southern Provinces of Russia, whence it was transported to Riga. It is called *Alexander*, in compliment to the Emperor Alexander the First. It was introduced into England by Mr. Lee, nurseryman, of Hammersmith, about the year 1805. On January 7, 1817, Mr. Lee exhibited some very fine specimens at the Horticultural Society, one of which measured five inches in diameter, four inches in depth, and sixteen inches in circumference. It weighed nineteen ounces. A plate of this apple is given in the Transactions (vol. II. p. 406), but it is not quite typical in shape, owing no doubt to its size. A better figure of it is given in Ronalds' *Pyrus Malus Brentfordiensis*, plate xxxv., figure 2.

Description.—Fruit, very large, heart-shaped. Skin, smooth, greenish yellow, with a few streaks of red on the shaded side, and orange, streaked with bright red next the sun; the whole strewed with numerous russety dots. Eye, large and half open, with broad, erect, and acuminate segments; set in a deep, even, and slightly ribbed basin. Stalk, an inch or more in length, inserted in a deep, round, and even cavity, which is lined with russet. Flesh, yellowish white, tender, crisp, juicy, and sugary, with a pleasant and slightly aromatic flavour.

This is one of the most beautiful and valuable apples, both as regards its size and quality. It is more adapted for culinary than dessert use, but when well coloured it presents a very noble appearance on the table. It is in season from September to December.

The tree grows well and vigorously. It is perfectly hardy, and a good bearer. The size of its fruit renders protection from the wind desirable. The leaves of the tree change in autumn to a fine red hue.

Alexander

Pom...n.

4. Beauty of Kent

PLATE XII.

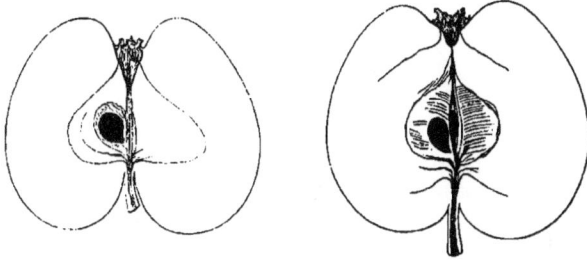

2. Cox's Pomona.   3. Cellini.

2. COX'S POMONA.

This beautiful apple was raised by Mr. Cox, of Colnbrook Lawn, about 1825.

Description.—Fruit, above medium size, sometimes large ; ovate and somewhat flattened and angular. Skin, yellow and very much streaked with bright crimson. Eye, slightly open, set in a deep and angular basin. Stalk, an inch long, and deeply inserted. Flesh, white, tender, delicate and pleasantly acid.

A first-rate and very handsome culinary apple. Season, October.

---

3. CELLINI.

This very useful apple originated with Mr. Leonard Phillips, of Vauxhall.

Description.—Fruit, rather above medium size ; roundish and flattened at both ends. Skin, rich deep yellow, with spots and patches of lively red on the shaded side, and bright red streaked and mottled with dark crimson next the sun, with here and there a tinge of yellow breaking through. Eye, large and open, with short, acute, and reflexed segments, and set in a shallow and slightly plaited basin. Stalk, very short, inserted in a funnel shaped cavity. Flesh, white, tender, very juicy, brisk and pleasantly flavoured, with a somewhat balsamic aroma.

This is one of the most useful and profitable apples. It eats well from the tree, but its chief use is as a culinary apple of the first quality. It is in season during October and November. The tree grows freely, comes very early into bearing, and is usually loaded with fruit. It is a fine showy and handsome apple, resembling the *Nonsuch*, from which it was probably raised.

PLATE XII.

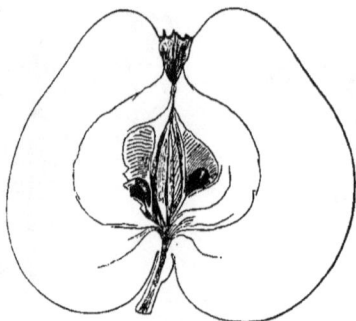

4. BEAUTY OF KENT.

The origin of this handsome apple is not known. It is first noticed by Forsyth in his *Treatise on Fruit Trees*, but it is not mentioned in any of the nurserymen's catalogues, either of the last or the early part of the present century. It was introduced to the Brompton Park Nursery about the year 1820, and is now very widely cultivated throughout the country. It is probably the *Rambour Franc* of the French pomologists. A very good representation of it is given in Ronalds' *Pyrus Malus Brentfordiensis*, plate xv, fig. 1.

Description.—Fruit, large, roundish ovate, broad and flattened at the base, and narrowing towards the apex, where it is terminated by several prominent angles. Skin, deep yellow, slightly tinged with green, and marked with faint patches of red on the shaded side, but entirely covered with deep red, except when there are a few patches of deep yellow on the side next the sun. Eye, small and closed, with short segments, and set in a narrow and angular basin. Stalk, short, inserted in a wide and deep cavity, which, with the base, is entirely covered with rough brown russet. Flesh, yellowish, tender, and juicy, with a pleasant subacid flavour.

A valuable and well-known culinary apple, in season from October to February. When well grown it is perhaps the most magnificent apple in cultivation. Downing says of it in America, "The fruit in this climate is one of the most magnificent of all apples, frequently measuring sixteen or eighteen inches in circumference."

The tree is a strong and vigorous grower. It is better adapted as a standard, than as an espalier or pyramid. It attains a large size, forms a very handsome tree, and bears freely. It is one of the most popular winter apples for culinary purposes, and one of the most desirable and useful either for a small garden or for more extended cultivation. It has a tendency to canker if not in congenial soil

Beurré Hardy

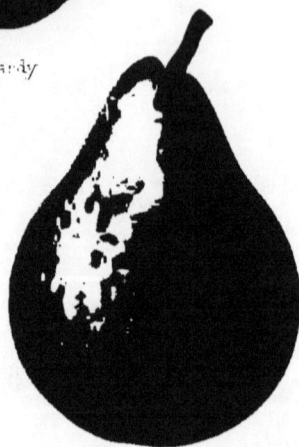

Fondante Duchess

Doyenné du Comice

PLATE XIII.

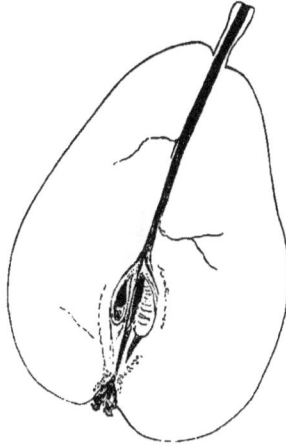

1. BEURRÉ HARDY.

[SYN : *Bonnet de Boulogne.*]

This fine pear was raised by M. Bonnet of Boulogne, the friend of Van Mons, about the year 1820. It was first distributed by M. Jamin, of Bourg-la-Reine, near Paris, who dedicated it to the late M. Hardy, director of the gardens of the Luxembourg.

Description.—Fruit, large, three inches wide and three inches and three-quarters long; oblong obovate or pyramidal, handsome and even in its outline. Skin, shining, yellowish green, thickly covered with large russet dots, and a coat of brown russet round the stalk and the eye. Eye, large and open, set in a shallow basin. Stalk, an inch long, stout and fleshy, with fleshy folds at the base, and inserted without depression. Flesh, white, melting, and very juicy, sweet, and perfumed with a rose water aroma.

A dessert pear of the first quality; in season in October and November. It forms a very handsome pyramid on the Pear, or on the quince stock. It is perfectly hardy, grows freely and bears well, and is worthy of general cultivation. The fruit should be gathered before it is ripe, at intervals, to prevent its getting mealy and to keep it fit for the table for a longer period.

The specimen represented on the coloured plate is perhaps rather above the medium size. It was grown on a Cordon tree at Holme Lacy.

PLATE XIII.

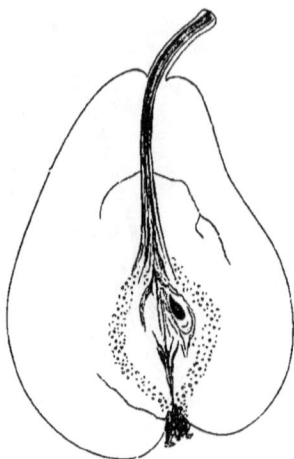

2. DOYENNÉ DU COMICE.

[SYN: *Comice.*]

This valuable pear was raised in the garden of the Comice Horticole, at Angers, and the original tree first bore fruit in 1849.

Description.—Fruit, large, three inches wide, and three inches and a half high; pyramidal or obovate, sometimes rather uneven in its outline. Skin, lemon yellow, with a greenish tinge, considerably covered with speckles and patches of pale brown russet, and particularly so round the eye and the stalk. Eye, small and open, with short pointed segments, set in a considerable depression. Stalk, half an inch to an inch long, fleshy at the base; sometimes curved, and inserted in a round narrow cavity; sometimes very short and stout, and obliquely inserted almost at right angles with the fruit. Flesh, yellowish white, very tender, buttery and melting, very juicy, rich, sweet, and delicately perfumed, with a sort of cinnamon flavour.

A most delicious pear, in season from the end of October to the middle of December. Its season is prolonged if it is gathered at intervals before it is ripe, and handled very gently.

The tree grows well, and bears pretty freely. It forms handsome pyramids on the quince. In this form it can be a little protected when in bloom, and its merits richly deserve it.

The specimen represented on the plate was grown on a cordon tree against the wall at Holme Lacy, and is larger than if it had been grown on a pyramid.

PLATE **XIII**.

3. PITMASTON DUCHESS.

[SYN : *Pitmaston Duchesse d'Angoulême ; William's Duchesse d'Angoulême.*]

This beautiful pear is of British origin. The seedling was raised by the late Mr. Williams, of Pitmaston, near Worcester, and obtained from crossing *Duchesse d'Angoulême* with *Glou Morceau*. It may be at this time some thirty or forty years old.

Description.—Fruit, large and handsome, four inches and a quarter long, and two inches and three quarters wide ; pyramidal, generally even or a little undulating in its outline, and sometimes rather prominently bossed. Skin, smooth and fine, of a pale lemon colour, thickly covered with patches of delicate cinnamon coloured russet, with a large patch round the stalk. Eye, large and open, set in a wide depression. Stalk, about an inch long, stout, and inserted either level with the surface, or in a small narrow cavity. Flesh, very tender and melting, very juicy, exceedingly rich, with a sprightly vinous flavour and delicate perfume.

A very handsome pear, of the finest quality ; in season from the end of October till the end of November. The tree is hardy, and grows freely. It is well adapted for pyramids, bushes, or espaliers. The fruit is too large to be grown as a standard.

PLATE XIV.

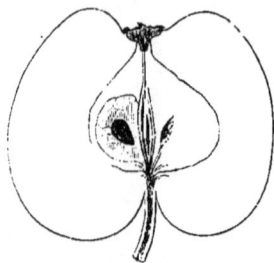

1. HEREFORDSHIRE PEARMAIN.

[SYN : *Royal Pearmain ; Pearmain Royal de longue durée.*]

This is a very old and well-known English apple. Rea is the first who notices it, under the name of *Royal Pearmain*, and he says, "it is a much bigger and better tasted apple than the common kind." In the Horticultural Society's catalogue this is called the *Old Pearmain*, and thus it is confused with the *Winter Pearmain*, as in some of the nursery gardens it is confused with the *Summer Pearmain*. It is figured in Ronalds' *Pyrus Malus Brentfordiensis*, plate xxii, fig. 4.

Description.—Fruit, large, three inches wide, and the same in height; Pearmain-shaped, and slightly angular, having generally a prominent rib on one side of it. Skin, smooth, dark, dull green at first on the shaded side, but changing during winter to clear greenish yellow, and marked with traces of russet; on the side next the sun it is covered with brownish red, and streaks of deeper red, all of which change during winter to clear crimson, strewed with many russety specks. Eye, small and open, with broad segments, which are reflexed at the tips, and set in a wide, pretty deep, and plaited basin. Stalk, from half an inch to three quarters long, inserted in a deep cavity, which is lined with russet. Flesh, yellowish, tinged with green, tender, crisp, juicy, sugary, and perfumed, with a brisk and pleasant flavour.

The *Herefordshire Pearmain* is a culinary fruit of high merit. It is in season from November to March, and will even keep to May.

The tree attains the middle size. It is a free and vigorous grower, very hardy, and bears well.

Wither Pearmain.

Adam's Pearmain.

Herefordshire Pearmain.

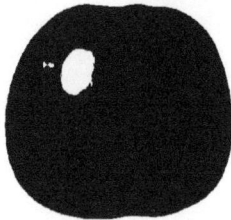

4. Loan Abbey Pearmain.

Winter Pearmain (Sussex)
or Duck's Bill.

5. Mannyt.

PLATE XIV.

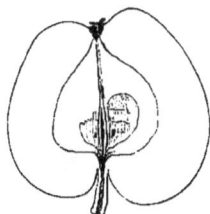

2. SCARLET PEARMAIN.

[SYN: *Bell's Scarlet Pearmain; Bell's Scarlet; Hood's Seedling; Oxford Peach Apple.*]

This apple is said to have been introduced by Mr. Bell, land steward to the Duke of Northumberland, at Sion, about the year 1800. It is figured in Ronalds' *Pyrus Malus Brentfordiensis*, plate viii, fig. 2.

Description.—Fruit, medium sized, two inches and a half wide, and two inches and a quarter high ; conical, regularly and handsomely shaped. Skin, smooth, tender and shining, of a rich deep bright crimson, with stripes of darker crimson on the side next the sun, and extending almost over the whole surface of the fruit, except where it is much shaded, and there it is yellow, washed, and striped with crimson, but of a paler colour, intermixed with a tinge of yellow, on the shaded side, and the whole surface sprinkled with numerous grey russety dots. Eye, open, with long reflexed segments, set in a round, even, and rather deep basin, which is marked with lines of russet. Stalk, from three quarters to an inch long, deeply inserted in a round, even, and funnel-shaped cavity, which is generally russety at the insertion of the stalk. Flesh, yellowish, with a tinge of red under the skin, tender, juicy, sugary, and vinous.

A beautiful well-shaped dessert apple, of first-rate quality ; in season from October to January. " Its good qualities are not exceeded by its great beauty " (*Transactions of London Horticultural Society*, vol. vii, p. 335). It has a delicious flavour when eaten ripe from the tree in September or October. The tree grows well but slenderly, and require a rich loamy soil, or it is apt to canker. It attains about a middle size, and bears well. It succeeds well on the paradise stock, on which it forms a good dwarf or espalier tree.

Plate XIV.

3. ADAMS' PEARMAIN.

[Syn : *Norfolk Pippin ; Hanging Pearmain ; Lady's Finger.*]

The history of this apple seems lost. The Mr. Adams, who may be presumed to have raised it, does not seem to have published any account of it.

Description.—Fruit, large, varying from two inches and a half to three inches high, and about the same in breadth at the widest part ; Pearmain-shaped, very even, and regularly formed. Skin, pale yellow, tinged with green, and covered with delicate russet on the shaded side ; but deep yellow, tinged with red, and delicately streaked with livelier red, on the side next the sun. Eye, small and open, with acute, erect segments, set in a narrow, round, and plaited basin. Stalk, varying from half an inch to an inch long, obliquely inserted in a shallow cavity, and generally with a fleshy protuberance on one side of it. Flesh, yellow, crisp, juicy, and sugary, with an agreeable and pleasantly perfumed flavour.

A dessert apple, of first-rate quality ; in season from December to February. It is a large and handsome variety, and worthy of great attention. The tree grows freely, producing long slender shoots, by which, and its cucullated ovate leaves, it is easily distinguished. It is an excellent bearer, even in a young state, particularly on the paradise or doucin stock ; and it succeeds well as an espalier.

PLATE XIV.

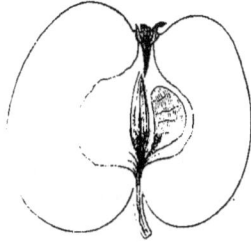

4. LAMB ABBEY PEARMAIN.

This variety is a seedling from the *Newtown Pippin*, the celebrated American apple. It was grown in 1804 by Mrs. Malcolm, the wife of Niel Malcolm, Esq., of Lamb Abbey, near Dartford, in Kent. It is figured in the *Transactions of the London Horticultural Society*, vol. viii, p. 267, and also in Ronalds' *Pyrus Malus Brentfordiensis*, plate xxi, fig. 2.

Description.—Fruit, small, roundish, or oblate oblong, regularly and handsomely shaped. Skin, smooth, greenish yellow on the shaded side, but becoming clear yellow when at maturity ; on the side next the sun it is dull orange, streaked and striped with red, which becomes more faint as it extends to the shaded side, and dotted all over with minute, punctured, russety dots. Eye, rather large and open, with long broad segments reflexed at the tips, and set in a wide, deep, and plaited basin. Stalk, from a quarter to half an inch long, slender, deeply inserted in a russety cavity. Flesh, yellowish white, firm, crisp, very juicy and sugary, with a brisk and rich vinous flavour.

A very valuable dessert apple, of first-rate quality, both as regards the richness of the flavour, and the long period during which it remains in perfection. It is in season from Christmas to April.

The tree is healthy, a free grower, and bears well.

PLATE XIV.

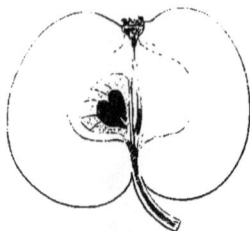

5. MANNINGTON'S PEARMAIN.

This excellent apple originated about the year 1770, in a garden at Uckfield, in Sussex, which belonged at that time to Mr. Turley, but which is now in the possession of his grandson, Mr. Mannington. The original tree grew up at the root of a hedge where the refuse from the cider press had been thrown. The original tree never attained any great size, and died about the year 1820. The great merit of its fruit however had been well recognized long before, and grafts had been freely distributed to many persons in the neighbourhood. It does not seem to have been much known beyond its own locality, until the Autumn of 1847, when Mr. Mannington sent specimens of the fruit to the London Horticultural Society, where it was pronounced to be a dessert fruit of the highest excellence. It thus received the name from Mr. Thompson of *Mannington's Pearmain*.

Description.—Fruit, medium sized, abrupt Pearmain-shaped. Skin, of a rich golden yellow colour, covered with thin brown russet on the shaded side, but covered with dull brownish red on the side next the sun. Eye, partially closed, with broad flat segments, set in a shallow and plaited basin. Stalk, three quarters of an inch long, obliquely inserted in a moderately deep cavity, with generally a fleshy protuberance on one side of it. Flesh, yellow, firm, crisp, juicy, and very sugary, with a brisk and particularly rich flavour.

This is one of the best and richest flavoured dessert apples, in season from November even till March. The fruit should be allowed to hang late on the tree to secure its richness of flavour, and its good keeping properties, for if gathered too soon it is apt to shrivel. The only objection to this is the difficulty of protecting the fruit from the birds.

The tree is very hardy, but does not attain a large size. It is an early and excellent bearer, even at two or three years from the graft.

PLATE XIV.

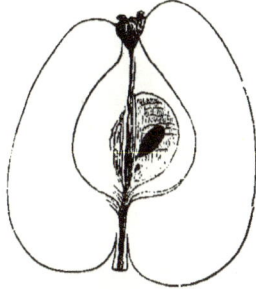

6. DUCK'S BILL.

[SYN : *Sussex Winter Pearmain; Sussex Scarlet Pearmain.*]

This apple is grown extensively in the County of Sussex where it is met with in almost every orchard. It is sometimes locally called *Winter Pearmain*, but is quite distinct from the *Old Winter Pearmain* of other parts of the country.

Description.—Fruit, large, three inches and a quarter wide, and about the same in height ; of a true *Pearmain* shape, somewhat five sided towards the crown. Skin, smooth and shining, at first of a greenish yellow, marked with faint streaks of dull red on the shaded side, and entirely covered with deep red on the side next the sun, and strewed all over with small russety dots. Eye, large and open, with short segments, set in a pretty deep and prominently plaited basin. Stalk, very short, not exceeding a quarter of an inch long, inserted in a deep funnel-shaped cavity, which is lined with russet. Flesh, yellowish, firm, crisp, juicy and sugary, with a brisk, piquant, and very pleasant flavour.

This highly esteemed apple is chiefly suited for culinary purposes, but can take its place on the dessert table with credit, when required to do so. It is a very handsome fruit and always fetches a high price in the markets. It is in season from December to April or May.

The tree grows well and freely. It is also very hardy, and a good bearer.

1 Cutillac.

3 Uvedale's St Germain

5 Bellissime d'Hiver.

PLATE XV.

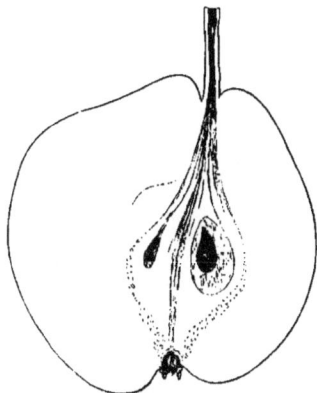

1. CATILLAC.

[Syn: *Cadillac; Quenillac; Chartreuse; Admirable de Chartreux; Bon Chrétien d'Amiens; Colillard; Monstreuse de Landes; Grand Monarque; Grand Mogol; Gratiole rond; Tête du chat; Gros Gilot; Besi de Marais; Bell Pear; Pound Pear.*]

The origin of this excellent Pear is lost. It is known to have been in cultivation for upwards of 200 years, and the number of its synonyms prove the esteem in which it has ever been held. It has been many times figured. Duhamel's "*Traité des Arbres Fruitiers (1768)* Pl. lviii., gives a good figure as does Brookshaw's "*Pomona Britannica* (1812,) Pl. lxxxvi.

Description.—Fruit : large, flatly turbinate. Skin : at first pale green, becoming after keeping a beautiful bright lemon yellow, with a tinge of brownish red next the sun, and covered with numerous large brown russety dots. Eye : open with short dry segments, set in a wide, even, and rather deep basin. Stalk : an inch and a half long, stout, curved and inserted in a small cavity. Flesh : white, crisp, gritty, with a harsh and somewhat musky flavour.

The tree is hardy, grows vigorously, and bears abundantly. It is best grown on a wall on the pear stock, when it is one of the most profitable pears that can be grown. A tree in full bearing is a small fortune to the owner, for it is one of the best culinary pears, and is so fine in size, as always to command the best price in the market. "It is considered by confectioners" says Brookshaw "as the best pear for stewing, preserving, and wet sweetmeats." It may be grown as an espalier in a very sheltered situation, but as a pyramid or an ordinary standard, it rarely succeeds, the size and weight of the fruit rendering the effect of wind most destructive to the crop. It is in season from December to April or even May.

PLATE XV.

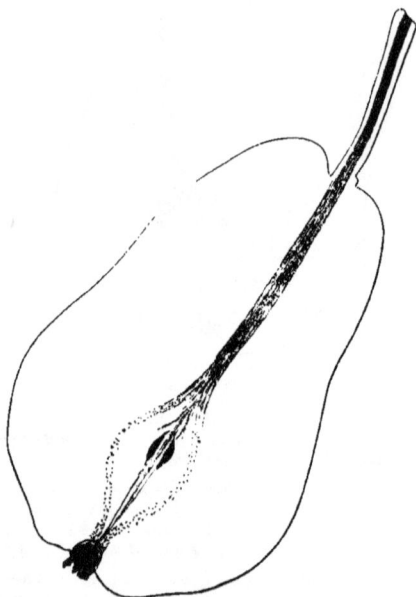

2. UVEDALE'S ST. GERMAIN.

[SYN : *Dr. Udale's Warden ; Union Pear ; Pickering Pear ; Pickering's Warden ; Piper ; Pound Pear ; Winter Bell Pear ; Chambers' Large ; Lent St. Germain ; German Baker ; Audusson ; Bolivar ; Royale d'Angleterre ; L'Inconnue à Compôte ; Belle Angevine ; Abbé Mongein ; Belle de Jersey ; Angora ; Beauté de Teruereu ; Comtesse de Teruereu ; Duchesse de Berri d'Hiver ; Gros Fin Or d'Hiver ; Berthebirne ; Grosse de Bruxelles.*]

This celebrated English Pear was raised by Dr. Uvedale, who was a schoolmaster, and lived at Eltham, in Kent, in 1690. Dr. Uvedale appears to have removed to Enfield where he continued his school. Miller, in the first edition of his Dictionary in 1724, speaks of him as " Dr. Udal of Enfield, a curious collector and introducer of many rare exoticks, plants, and flowers." Bradley in 1733 speaks of the pear as " Dr. Udale's great Pear, called by some the " *Union Pear*," whose fruit is about that length, one may allow eight inches." M. Leroy took much trouble to investigate

PLATE XV.

the history of this pear. He found that it had received the name of *Belle Angevine* from M. Audusson, a nurseryman at Angers. M. Audusson had obtained it from the Garden of the Luxembourg at Paris in 1821, under the name of *L'Inconnue à Compôte*. Beyond this, M. Leroy could not trace it. It had however been known in English gardens for upwards of a century before this time, and it is most probable therefore that it had been conveyd to Paris from England.

Description.—Fruit: very large, sometimes weighing upwards of three pounds; of a long pyriform, or pyramidal shape, tapering gradually towards the stalk and obtusely towards the eye, rather curved and more swollen on one side of the axis than the other. Skin: smooth, dark green, changing to yellowish green, and with dull brownish red on the exposed side, dotted all over with bright brown and a few tracings of russet. Eye: open, with erect rigid segments, set in a deep, narrow basin. Stalk: an inch to an inch and a half long, curved, inserted in a small close cavity. Flesh: white, crisp, juicy, and slightly gritty.

The tree grows vigorously and bears well if nourished well. For obvious reasons it should be grown only on a wall, and when grown, as it should be on the pear stock, will occupy a large space. It is an excellent culinary pear; it stews naturally of a rich red colour, and is very good when baked or preserved. It is in season from January to April, when it meets with a ready sale in the market at a high price.

PLATE XV.

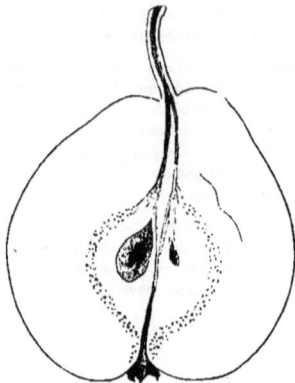

3. BELLISSIME d' HIVER.

[SYN: *Teton de Vénus ; Belle de Noisette ; Vermillion d'Espagne.*]

Nothing is to be found with reference to the origin of this fine pear, but it is a variety which has long been known and is well described by Duhamel (1758) and other old writers.

Description.—Fruit : very large, four inches wide and three inches and three quarters high ; roundish, turbinate. Skin : smooth and somewhat shining, of a fine deep green colour on the shaded side, and fine vermillion next the sun ; strewed all over with large brown russet dots. Eye : large and open, set in a rather deep basin. Stalk : an inch long stout and somewhat fleshy, particularly at the insertion, where it is placed in a rather deep cavity with a fleshy swelling on one side of it. Flesh : white, fine grained, crisp and tender, sweet, and with a musky flavour.

The tree is a free and vigorous grower, an excellent bearer, and succeeds well as a standard either on the pear, or the quince stock. It is one of the very best culinary pears and quite free from the disagreeable grittiness which is peculiar to baking pears generally. " Elle est beaucoup meilleure cuite sous la cloche que le Catillac. On peut même en faire d'assez bonnes compotes " says Duhamel, Vol. II., p. 235. In favourable situations, it is superior in size and in every other respect to the *Catillac.* It is in season from November to April.

1 Golden Harvey

2 Cox's Orange Pippin

PLATE XVI.

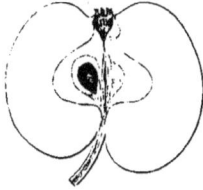

1. GOLDEN HARVEY.

[Syn: *The Harvey Apple ; Round Russet Harvey ; Brandy Apple.*]

The history of this apple is not known. It is doubtful if it existed before the 17th century. Evelyn mentions it, and says that "some persons prefer the cider from the *Hervey Apple* (being boiled) to all other ciders." The *Harvey Apple* and *Russet Hervey* are both mentioned by Worlidge, though he could scarcely have known its excellence for the dessert table, or he would have caused it to be cultivated in every part of England, and have made it to be "everywhere esteemed, as it is in Herefordshire, as the best fruit of its species." It is called "golden" from the bright yellow colour of its flesh.

Description.—Fruit: small, oblato-cylindrical, even and free from angles. Skin: entirely covered with rough, scaly russet, with sometimes a patch of the yellow ground colour exposed on the shady side, and covered with brownish red on the side next the sun. Eye: small and open, with very small, reflexed segments, set in a wide, shallow and slightly plaited basin. Stalk: half an inch long inserted in a shallow cavity. Flesh: yellow, firm, crisp, juicy, sugary, with an exceedingly rich and powerful aromatic flavour.

The *Golden Harvey*, or as it is more commonly called in Herefordshire, the *Brandy Apple*, in a warm and favourable season, is perhaps the most delicious of all dessert apples. It is fit for the table from December to May or June, but it requires however to be carefully kept, in jars, or in boxes in sand, for if exposed to light and air it will shrivel as do most of the late keeping russet apples.

The flesh of the *Golden Harvey* apple is firm and crisp. Mr. Edward Solly found that 1000 parts of the ripe fruit, consisted of organic matter 2140; inorganic matter 35; and water 7825; *Trans. of Horticultural Society,* London, 2nd Series, Vol. I., p. 62.) The juice is singularly rich

PLATE XVI.

and aromatic. Mr. Thomas Andrew Knight found its specific gravity to be 1085, exceeding that of any other apple he had examined. The cider made from it has so much strength, as to give the apple its favorite local name of *Brandy Apple*. It must be confessed however that its richness and flavour are most commonly lost in the manufacture. When however in a good season, the sweetness and aroma are happily retained, the cider is as rich as it is potent, and tradition tells that it has not unfrequently been exchanged, bottle for bottle, for the best old port wine.

The tree grows freely to a medium size with slender drooping branches, and is perfectly hardy; but it is rather fickle in bearing. On the paradise stock it is well adapted for dwarf training, forms a good espalier, and bears with more certainty. " No garden, however small, should be without it " says Lindley.

The *Golden Harvey* has been figured in Knight's " *Pomona Herefordiensis*" (1812,) Pl. xxii.; in Ronalds' " *Pyrus Malus Brentfordiensis* (1836,) Pl. xxiii.; and in Lindley, " *Pomological Magazine*," Pl. 39.

PLATE XVI.

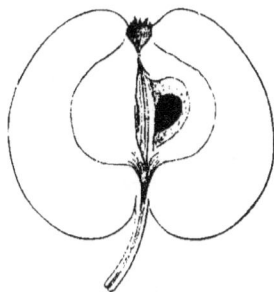

2. COX'S ORANGE PIPPIN.

This excellent apple is said to have been produced in 1830 from a pip of the Ribstone Pippin, by Mr. Cox, at Colnbrook Lawns, near Slough, Buckinghamshire.

Description.—Fruit : medium sized ; roundish-ovate, even and regular in outline. Skin : greenish yellow and streaked with red in the shade, but dark red where exposed to the sun, and this colour extends over three fourths of the whole surface. Over the coloured parts are patches and traces of ash-grey russet forming a smooth and firm crust. Eye : small and open, filled with stamens and with short erect segments, set in a somewhat shallow saucer-like basin which is entirely lined with russet. Stalk : half an inch long, somewhat fleshy, set in a moderately deep cavity, which has a slight swelling on one side, and is covered with russet extending over the base. Flesh : yellowish, very tender in the grain, crisp, juicy and sweet, with a fine perfume.

This beautiful apple is the best modern addition to the dessert table ; excellent in flavour, and very handsome. It is in season from October to February.

The tree is not vigourous in growth and for this reason is admirably adapted for dwarfs and pyramids. It succeeds best on the paradise stock when it bears freely. It should take its place in every garden, however small, it ranks in excellence with the *Ribstone Pippin, Golden Harvey,* and *Nonpareil,* and is more beautiful than any of them.

PLATE XVII.

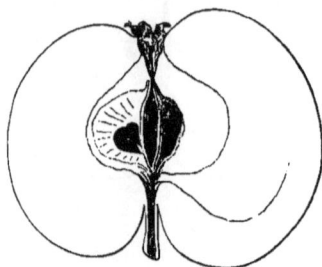

1. ECKLINVILLE SEEDLING.

This Irish apple was raised at Ecklinville four miles from Portaferry, and eighteen miles from Belfast, by a Scotch gardener named Logan, some 60 years since.

Description.—Fruit : large, roundish and flattened, even in its outline, but slightly angular round the eye. Skin : bright, rather deep lemon colour with a tinge of green, strewed but not thickly with large russet dots, and with a crimson blush on the side exposed to the sun. Eye : large with closed segments, deeply set in an angular basin. Stalk : half an inch long, slender, not protruding beyond the flat base of the fruit. Flesh : white, tender and full grained, with a brisk acidulous flavour.

This is an excellent and handsome culinary apple, in season from October to Christmas. The tree has a good habit of growth and is a great bearer. It is an excellent variety for the market, and is now extensively grown in the three kingdoms.

Waltham Abbey Seedling

5 Bedfordshire Foundling

Winkinsville Seedling

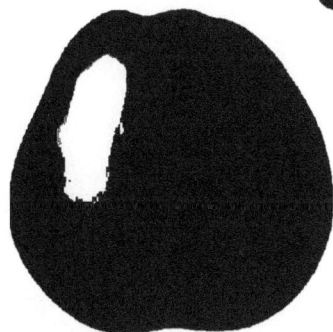

4 Hanwells Souring

5 White Spanish Reinette

PLATE XVII.

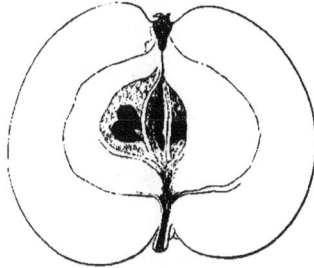

2. WALTHAM ABBEY SEEDLING.

This apple was raised about the year 1810 from the seed of the *Golden Noble*, by Mr. John Barnard, of Waltham Abbey in Essex; and was introduced by him at a meeting of the London Horticultural Society in 1821. It resembles its parent the *Golden Noble*, but is yet quite distinct from it.

Description.—Fruit: large roundish, inclining to ovate, in which respect it differs from the *Golden Noble*, which is quite round. Skin: pale yellow, assuming a deeper tinge as it attains maturity, with a faint blush of red where exposed to the sun, and strewed all over with minute, russety dots, and occasionally a few patches of thin russet. Eye : large and open, set in a shallow and even basin. Stalk: short, deeply inserted, and surrounded with rough russet. Flesh: yellowish, tender, juicy, sweet and pleasantly flavoured; when cooked it assumes a clear, pale amber colour.

The Waltham Abbey Seedling is a culinary apple of first rate quality ; it requires but little sugar when cooked, and has inherited from its parent, the *Golden Noble*, the great merit of becoming translucent. It is in season from September to Christmas. The tree is remarkable for its very small foliage, though the fruit is abundant and of a good useful size.

Plate XVII.

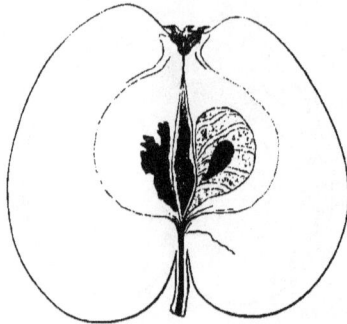

3. BEDFORDSHIRE FOUNDLING.

[Syn : *Cambridge Pippin.*]

The details of the origin of this apple are unknown. Its name gives the locality of its production; and the extent of its cultivation and age of many of the trees, lead to the inference that it is as old as the century. It is well figured by Ronalds, Pl. xxvii.

Description.—Fruit: large, three inches and a quarter wide, and three inches and a half high ; roundish ovate, inclining to oblong, with irregular and prominent angles on the sides, which extend to the apex and form ridges round the eye. Skin: dark green at first, changing as it attains maturity to a pale greenish yellow on the shaded side, but tinged with orange on the side next the sun, and strewed with a few fawn-coloured dots. Eye: open, set in a deep, narrow angular basin. Stalk : short, inserted in a deep cavity. Flesh : yellowish, tender, pleasantly, subacid, and with a somewhat sugary flavour.

An excellent apple of first rate quality ; in season from November to March. The tree is vigorous in growth and bears freely. It is a general favourite, and is grown extensively not only in the United Kingdom, but on the Continent, where it takes an honoured place at all Fruit Exhibitions.

PLATE XVII.

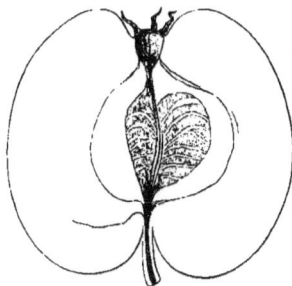

4. WHITE SPANISH REINETTE.

[Syn : *Camuesar ; Reinette blanche d'Espagne ; Reinette d'Espagne ; Reinette tendre ; Joséphine ; Belle Joséphine ; De Ratteau ; Concombre ancien.*]

The history of this valuable old variety is quite lost. It is said to be the national apple of Spain with the name "*Camuesar*," and has been known there from great antiquity. The date of its introduction to England is not known, but it was exhibited at the Horticultural Society by Mr. John Darby in 1829, from some very ancient trees then growing in Sussex. It is figured by Lindley, Pl. 110.

Description.—Fruit : very large, three inches and a half wide, and three inches and three quarters high ; oblato-oblong, angular on the sides and uneven at the crown, where it is nearly as broad as at the base. Skin: smooth and unctuous to the touch; yellowish green in the shade, but orange tinted with brownish red next the sun, and strewed with dark dots. Eye : large and open, set in a deep, angular and irregular basin. Stalk : half an inch long, inserted in a narrow and even cavity. Flesh : yellowish white, tender, juicy and sugary.

An apple of first rate quality suitable for dessert, but more especially for culinary purposes. It is in season from December to March or even April. The tree is healthy and vigorous in growth and an excellent bearer. It does best in a dry, warm and loamy soil, a characteristic which its name at once suggests. It is better adapted for dwarf, or espalier growth from the size and weight of the fruit. The general appreciation of this apple is shown by the wide extent of its growth and the great variety of its synonyms. " Its principal defect " says Dr. Lindley is its gigantic size.

PLATE XVII.

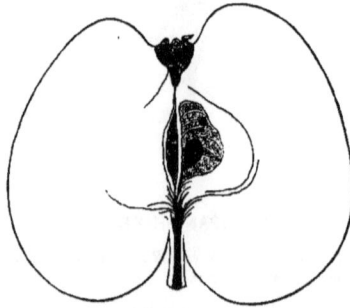

5. HANWELL SOURING.

This apple is said to have been raised at Hanwell, near Banbury, in Oxfordshire, and named from its strong acidity. It is figured in Ronalds' " *Pyrus Malus Brentfordiensis*," Pl. xxx.

Description.—Fruit : above medium size, three inches wide, and two inches and three quarters high ; roundish-ovate, angular or somewhat five sided, and narrow towards the eye. Skin : greenish yellow, and sprinkled with large russety dots, which are largest about the base, and with a faint blush of red next the sun. Eye : closed, set in a deep, narrow, and angular basin, which is lined with russet. Stalk : very short, inserted in an even funnel-shaped cavity, from which issue ramifications of russet. Flesh : white, firm, crisp, with a brisk and poignant acid flavour.

A culinary apple of first-rate quality, and very distinct character. It deserves general cultivation. It is in season from December to March, or April, and from the strong acidity it possesses it retains its flavour better than any other late keeping variety. The tree is vigorous in growth but a little uncertain in bearing. It makes an excellent pyramid tree.

Rutland Pear

Whit

3 black Huffcap   Yellow Huffcap

Longland   4 Red Pear

PLATE XVIII.

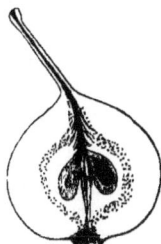

1. THE BARLAND PEAR.

"What tho' the pear tree rival not the Worth
Of *Ariconian* Products ? yet her freight
Is not contemn'd
. . . .
Chiefly the Bosbury whose large increase
Annual, in sumptuous Banquets claims applause."
(Philips' Cyder.)

[SYN : *Bosbury Pear ; Bareland, or Bearland Pear.*]

This pear from one of its common names may be supposed to have originated in the parish of Bosbury, near Ledbury, Herefordshire. The original tree is said to have grown in a field called *Bare Lands*, on an Estate called " Bosbury Farm," and to have been blown down about the end of last century. The variety was well established in the 17th century, and in great repute. Evelyn (1674) says of it "The *Horse Pear*, and the *Bear-land Pear* are reputed of the best, as bearing almost their weight of spriteful and vinous Liquor. They will grow in common Fields, gravelly and stony Ground to that largeness, as only one tree has been usually known to make three or four Hogs-heads." *(Evelyn's Pomona.)*

This fruit is well represented in Mr. Thomas Andrew Knight's *"Pomona Herefordiensis,"* Pl. xxvii.

Description.—Fruit : small, turbinate, pinched in near the stalk. Skin : bright green, very much covered with patches and large dots of thick, pale brown, or ash grey russet, but not so much so as entirely to obscure the green ground colour. Eye : large for the size of the fruit, open, with short erect segments; filled with the permanent stamens. Stalk : an inch long, slender, and inserted on the end of the fruit without any depression.

This variety has been much planted in Herefordshire and the adjoining counties. The trees have acquired an extraordinary size and height, and they are much distinguished by the beauty of

PLATE XVIII.

their form and foliage. The largest orchards of this variety are now to be found in the parishes of Dymock in Gloucestershire, and Newland in Worcestershire. Very few Farms on the Eastern side of Herefordshire are without Barland pear trees, shewing how extensively this favourite variety was at one time cultivated. Evelyn several times mentions the *Barland Pear*, "and as no trees of this variety" says Mr. Knight "are found in decay from age in favourable soils, it must be concluded that the identical trees which were growing when Evelyn wrote, still remain in health and vigour. The specific gravity of the juice is 1070."

The fruit Evelyn describes as "of such insufferable taste that hungry swine will not smell to it, or if hunger tempt them to taste, at first crash, they shake it out of their mouths": but of the Perry he speaks much more favourably. "There's a *Pear* in *Bosberry* and that neighbourhood, which yields the liquor richer the second year than the first, and so, by my experience very much amended the third year." Another writer says : "It hath many of the Masculine Qualities of Cyder. It is quick, strong, and heady, high coloured and retaineth a good vigour . . . . many years before it declineth . . . As it approacheth to the Apple Cyder in Colour, Strength and excellence in Durance, so the bloom cometh forth of a damask Rose Colour like Apples, not like other Pears." *Herefordshire Orchards* by J. Beale, 1730.

The juice is rich in colour and full in flavour, its chemical analysis by Mr. G. H. With, F.R.A.S., is as follows :—

Density of fresh juice ...	...	...	...	1·0421
Ditto after 24 hours exposed to air		...	...	1·0435
One hundred parts by weight contains :—				
Sugar ...	...	...	...	...	10·670
Tannin, Mucilage, &c....		...	...	...	2·763
Water ...	...	...	...	...	86·567
				100·000

Mr. Knight in his *Pomona* says "many thousand hogsheads of Perry are made from this fruit in a productive season ; but the Perry is not so much approved by the present as it was by the original planters. It however sells well, whilst new, to the merchants, who have, probably, some means of employing it with which the public are not acquainted ; for I have never met with it more than once within the last twenty years out of the district in which it is made, and many Herefordshire planters have applied to me in vain, for information respecting its disappearance. It may be mingled in considerable quantity with strong new port without its taste becoming perceptible ; and as it is comparatively cheap, it possibly sometimes contributes one of the numerous ingredients of that popular compound."

Barland Perry does not bottle well. It curdles in the bottles. It is usually drunk in Herefordshire as soon as made, where it is considered very wholesome and singularly beneficial in nephritic complaints.

PLATE XVIII.

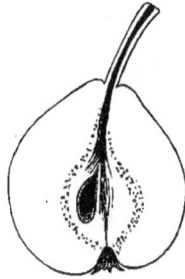

2. LONGLAND.

[SYN : *Longdon; Longland Pear.*]

The name of this pear, says Mr. Thomas Andrew Knight in the "*Pomona Herefordiensis*," was probably derived from the field in which the original tree grew—but nothing is really known as to the circumstances or date of its origin. It is certainly a very old variety. This Pear is well represented in the "*Pomona Herefordiensis*," Pl. xviii.

Description.—Fruit: roundish obovate, or doyenné shaped, even, regular, and rather handsome. Skin : very thickly covered with large russet freckles of a pale ashy colour ; the side next the sun has a bright pale red cheek, and on the shaded side it is a greenish yellow. Eye : large, open, and clove-like, set even with the surface, with a ring of permanent stamens round the mouth. Stalk : an inch long, straight and stout, perpendicular with the axis of the fruit, and very slightly depressed in a narrow cavity. Flesh : yellow, very astringent.

The specific gravity of the juice Mr. Knight found to be 1063.

The trees of this variety are very hardy and productive, since the blossoms are extremely patient of cold and unfavourable weather.—The Perry is very high coloured and without fine flavour. It is generally, however, free from sharp acidity, and more nearly resembles Cider than any other kind of Perry.—It does not answer for fining and bottling, but is excellent for ordinary use, either alone or mixed with apples, and its hardy, prolific character makes it a general favourite.

The chemical analysis of the fresh juice from this Pear (1879) made by Mr. G. H. With, F.R.A.S., is as follows :—

Density of the fresh juice	...	...	...	1·036
Ditto after 24 hours exposure to the air		...	...	1·041
One hundred parts of the juice by weight contained :—				
Sugar	...	...	...	...	...	8·400
Tannin, Mucilage, Salts, &c.,	....	...	. ...	4·187
Water	...	...	...	...	...	87·413

100

PLATE XVIII.

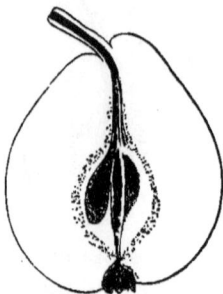

3. WHITE LONGLAND.

[SYN: *White Horse Pear.*]

The origin of this Pear seems to be unknown, but it is mentioned by Dr. Beale in his *"Herefordshire Orchards"* (1657).

Description.—Fruit : oblong, obovate, even and regular in its outline. Skin : very thickly sprinkled with large russet dots and tracings of russet, and with a solid patch surrounding the stalk ; on the exposed side it has a thin pale red cheek, and on the shaded side, it is yellowish green. Eye : open, with short, incurved segments, set in a shallow depression. Stalk : half an inch long, woody, inserted in a narrow and shallow cavity. Flesh : yellowish, firm, coarse-grained, briskly acid and sweet.

"*The White Horse Pear*," says Dr. Beale, "yields a juice somewhat near the quality of Cyder." It is a favourite Pear in Herefordshire, not so much for its Perry—indeed it is seldom or never used alone for this purpose—as for its cooking qualities. It is an excellent baking pear, somewhat coarse and rough in flavour, but with a natural deep rich red colour."

The chemical analysis of the fresh juice from this Pear (1879) made by Mr. G. H. With, F.R.A.S., is as follows :—

Density of the fresh juice	1·036
Ditto after 24 hours exposure to the air	1·039
One hundred parts of the juice by weight contained :—	
Sugar	8·580
Tannin, Mucilage, Salts, &c.	3·408
Water	88·012

100

PLATE XVIII.

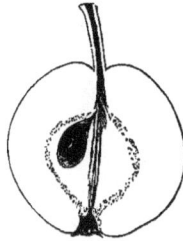

4. RED PEAR.

[Syn : *Red Horse Pear.*]

"The *Red Horse Pear* next the *Bosbury*." (Evelyn's *Pomona.*)

This Pear seems to have been well-known in the 17th century, but its origin is involved in obscurity.

Description.—Fruit : small and perfectly round, even and regular in its outline, sometimes inclining to turbinate. Skin : almost entirely covered with a rather bright red colour, except round the stalk, and where it has been shaded, and there it is yellow ; the whole surface is sprinkled with pale grey russet dots. Eye : open, having clove-like segments, and set level with the surface. Stalk : three quarters of an inch long, stout and straight with the axis of the fruit, set in a narrow shallow cavity. Flesh : pale yellow, firm, dry and gritty.

The tree is very hardy, early, and an excellent bearer.

The Perry is very good, and has a strong cider like character. It bottles well, and in good seasons makes an excellent saleable beverage.

Mr. With's analysis of the fresh juice (1879) is as follows :—

Density of fresh juice ...	...	...	...	1·0398
Ditto after 24 hours exposure to air...		...	...	1·0398
One hundred parts by weight contained :—				
Sugar	...	...		8·742
Tannin, Mucilage, Salts, &c.,	...	...		3·202
Water		...		88·056

100

PLATE XVIII.

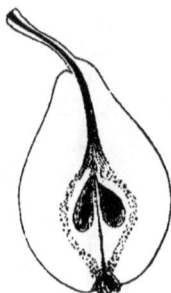

5. BLACK HUFFCAP.

[Syn: *Black Pear; Brown Huffcap.*]

The *Huffcap Pear* have been known from the 17th century, and were thought by Mr. Knight to have been included amongst the " Choke Pears," which abounded in Herefordshire at that time.

This Pear is represented in Mr. Knight's *Pomona Herefordiensis,* Pl. xxiv., under the name of *The Huffcap Pear,* as the best of all the varieties.

Description.—Fruit : oblong, obovate, sometimes elliptical, tapering gradually from the bulge both to the eye and stalk ; it is even and regular in its outline. Skin : olive green on the shaded side and entirely covered with dull rusty red on the side next the sun : the whole surface thickly sprinkled with large grey russet dots. Eye : prominently set; open, with erect segments. Stalk : three quarters of an inch long, woody, connected with the fruit by a thickened continuation of the flesh. Flesh : yellowish green, firm, and very gritty.

" The fruit is excessively harsh and austere," says Mr. Knight, " but it becomes very sweet during the process of grinding ; its Perry possesses much strength and richness, and has the credit of intoxicating more rapidly that made from any other Pear.

It maintains the same character at the present time and is therefore one of the most esteemed varieties.

The chemical analysis of the juice of this Pear (1879) made by Mr. G. H. With, F.R.A.S., is as follows :—

Density of fresh juice	1·048
Ditto after 24 hours exposure to air ...	1·051
One hundred parts by weight contained :—	
Sugar	11·225
Tannin, Mucilage, Salts, &c., ...	3·575
Water	85·200

PLATE XVIII.

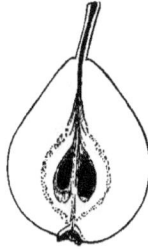

6. YELLOW HUFFCAP.

This Pear is figured in the *Pomona Herefordiensis*, Pl. xxiv.

Description.—Fruit : turbinate. Skin : entirely covered with rough, brown russet, but not so much so as to obscure the green ground colour, which is shown through the specks. Eye : open, small, with short, horny segments, set even with the surface. Stalk : three quarters of an inch long, inserted without depression. Flesh : yellowish with a greenish tinge.

The *Yellow Huffcap* is a very favourite pear in the neighbourhood of Ledbury. The tree is hardy ; it is earlier than the *Black Huffcap* ; and it bears freely, though usually in great abundance every second year. Its Perry is excellent. It is richer and has more body than the *Oldfield* Perry. " I always win the prize with the *Yellow Huffcap* Perry," says Mr. Hill, of Eggleton.

The chemical analysis of the juice of this Pear (1879) made by Mr. G. H. With, F.R.A.S., is as follows :—

Density of fresh juice	1·0462
Ditto after 24 hours exposure to air ...	1·0490
One hundred parts by weight contained :—	
Sugar	11·244
Tannin, Mucilage, Salts, &c.	2·290
Water	86·466
	100

PLATE XIX.

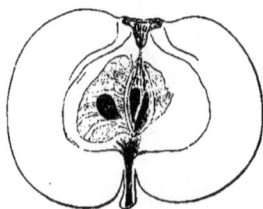

1. SUMMER STRAWBERRY.

The origin of this apple is not known. It is much cultivated in all the Lancashire and Northern Orchards of England, and is gradually extending into the special apple districts.

Description.—Fruit : rather below medium size, two and a half inches broad, and an inch and three quarters high ; oblate, even and regularly formed. Skin : smooth and shining, striped all over with yellow and blood-red stripes, except on any portion that is shaded, and then it is red. Eye : prominent, not at all depressed, closed with long flat segments, and surrounded with prominent plaits. Stalk : three quarters of an inch long, inserted in a round narrow cavity, which is lined with russet. Flesh : white, tinged with yellow, soft, tender, juicy, brisk and pleasantly flavoured.

A good dessert apple, ripe in September and October, but when long kept it becomes dry and mealy.

Margaret

ly Strawberry

PLATE XIX.

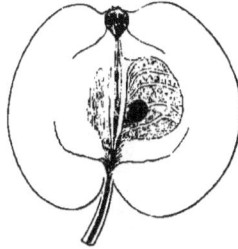

2. IRISH PEACH.

[SYN : *Early Crofton.*]

The origin of this apple is nowhere given. It was introduced into England by John Darby, Esq., of Addescombe, and Mr. Robertson, of Kilkenny. This fruit is represented by Lindley, Pl. 100, and by Ronalds, Pl. viii.

Description.—Fruit : middle sized, two and three quarters wide, by two inches and a quarter high ; roundish, somewhat flattened, and slightly angular. Skin : smooth, pale yellowish green, tinged with dull reddish brown, and thickly dotted with green dots on the shaded side, but fine lively red, mottled and speckled with yellow spots on the side exposed to the sun. Eye : small and closed, set in a rather deep and knobbed basin, which is lined with thick tomentum. Stalk : short, thick and fleshy, inserted in a pretty deep cavity. Flesh : greenish white, tender, and crisp, abounding in a rich, brisk, vinous and aromatic juice, which at this season is particularly refreshing.

An early dessert apple of the finest quality. It is ripe during the first week in August and lasts all through that month. It is a very beautiful apple and one of most excellent summer apples. It has an abundance of rich refreshing juice, and is so sweet, that it readily ferments on the ground, when it has fallen off ripe in hot weather, intoxicating the wasps and bees attracted to it by its aromatic scent.

The tree is hardy and vigorous. It grows freely and bears abundantly.

Plate XIX.

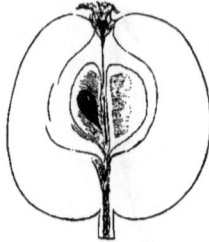

3. MARGARET.

[Syn : *Early Red Margaret ; Early Margaret ; Early Red Juneating ; Red Juneating ; Striped Juneating ; Early Striped Juneating ; Striped Quarrenden ; Summer Traveller ; Eve Apple* in Ireland ; *Marget Apple ; Maudlin ; Magdalene ; Marguerite ; Lammas.*]

This is a very old English apple. It is without doubt the *Margaret* of Rea, Worlidge, Ray, and all our early pomologists. It is figured by Lindley, Pl. 46.

Description.—Fruit : middle sized ; roundish ovate, and narrowing towards the eye, when it is angular. Skin : greenish yellow on the shaded side, but bright red next the sun, striped all over with dark red, and strewed with grey russetty dots. Eye : half open and prominent, with long, broad, erect segments, surrounded with a number of puckered knobs. Stalk : short and thick, about half an inch long, inserted in a small and shallow cavity. Flesh : greenish white, brisk, juicy, and vinous, with a pleasant and very refreshing flavour.

The *Margaret* is a first rate early dessert apple, with a fresh, juicy, vinous flavour. It is ripe in the beginning of August and should not be allowed to ripen on the tree or it will very quickly become mealy.

The tree is small in size, and well adapted for growing dwarf either for potting, or for training as a pyramid, or espalier, when grafted on the paradise or *pomme paradis* stock. It bears freely and is very hardy, though somewhat liable to canker in light soils.

PLATE XIX.

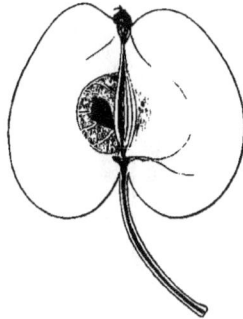

4. EARLY STRAWBERRY.

[SYN : *American Red Juneating.*]

This bright beautiful apple is said to have originated near New York ; and is supposed to be a seedling from the *Margaret* from which it is very distinct. The comparison is seen at once on the plate.

Description.—Fruit : middle sized, broad at the base and narrowing towards the eye, where it becomes angular, two inches and a quarter broad by two inches high. Skin : smooth, almost covered with a clear bright red colour on a yellow ground as seen on the shaded side, but finely striped with a deeper red on the sunny side. Eye : small, with converging segments, and set in a shallow basin. Stalk : thin, from one to two inches in length, and inserted in a deep narrow cavity. Flesh : white, or stained with red near the skin ; tender, with a pleasant sub-acid flavour and an agreeable aroma.

This apple is a very favourite early dessert apple in America, and has not very long been introduced into England. It is very attractive in appearance and excellent in flavour, but like all early apples does not keep long.

The tree is small, but very hardy and bears abundantly. Grafted on the paradise stock it makes a good pyramid, and when grown in an Orchard house forms a very attractive object.

PLATE XX.

1. YORKSHIRE BEAUTY.

There is no published account of the origin of this apple. Its name gives the only clue to the locality of its production ; but trees from 20 or 50 years old are scattered here and there through the orchards of Herefordshire and Worcestershire.

Description.—Fruit : large, three and a half inches wide, and three inches high ; roundish and flattened, with angles on its sides, which are often very marked. Skin : bright yellow, with a bright red blush on the side exposed to the sun, with occasional russet marks. Eye : open, with short erect segments, inserted in a contracted and angular basin. Stalk : very short, inserted in a russety cavity. Flesh : tender and juicy, with an agreeable acidity.

This is a first-rate culinary apple, and very handsome. It follows the early Codlins and Lord Suffield apple, and is in season from the end of August through September to the beginning of October. Its great size, its fine quality, and its beauty, make it the favourite fruit in the market for its short season. At the Apple Exhibition of the Woolhope Club, held October 10th, 1878, it was the most beautiful culinary variety exhibited.

Loddington

PLATE XX.

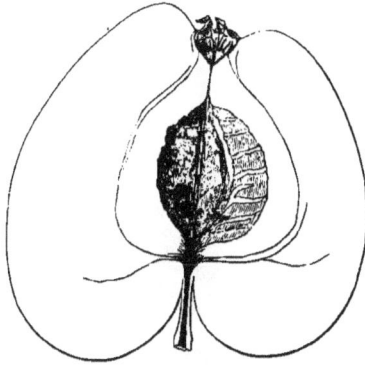

2. LODDINGTON.

[SYN : *Loddington Seedling ; Stone's Apple ; Killick's Apple.*]

This apple takes its name from the Village of Loddington, near Maidstone. The original tree grows on the farm of Mr. Stone, at Loddington. It is stated to have been brought from a nursery garden at Bath about 60 years since by a niece of Mr. Robert Stone, who had accompanied her uncle there. Mr. Stone soon saw its great merit, and gave grafts to his neighbours. It thus got the name of Stone's Apple, and spread through the orchards of the adjoining villages. In 1877 it was exhibited by Mr. Lewis A. Killick, of Langley, before the Fruit Committee of the Royal Horticultural Society, and was awarded a first-class certificate. Its cultivation is now becoming more and more extended through the country. It is figured in the " Florist and Pomologist," Plate 467, as *Stone's Apple.*

Description.—Fruit : large, varying from three to three and three-quarter inches in diameter, roundish, slightly flattened, and narrowing abruptly towards the eye ; it has obtuse ribs on the sides, which become more distinct towards the eye, where they form ridges round the crown. Skin : smooth and shining, grass-green at first, with a brownish red cheek ; but after being gathered it becomes a fine lemon yellow, with a pale crimson cheek, marked with broken streaks of dark crimson ; the surface is strewed more or less with minute russet spots. Eye : closed, with convergent leaf-like segments, set in a deep and prominently plaited or ribbed basin. Stalk : half an inch to three-quarters long, slender for the size of the fruit, and inserted in a deep, wide, funnel-shaped cavity, which is lined with pale, thin, ashy russet extending over the base of the fruit. Flesh : very tender, with a pleasant subacid flavour.

LODDINGTON is an early culinary apple, of great excellence, and a good market variety. It comes into use in September after the early Codlins, Lord Suffield, and Yorkshire Beauty, and will keep till Christmas. It is not so tender in the skin as Lord Suffield, and does not bruise so readily in travelling. The habit of the tree is compact and medium sized. It bears freely every year, and when in bearing, keeps on forming an abundance of fruit spurs without making much wood : for these reasons it answers well to graft on the branches of large trees of unproductive kinds.

PLATE XXI.

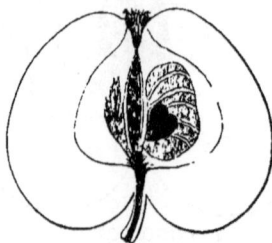

1. SYKE HOUSE RUSSET.

[SYN : *Englische Spitalsreinette ; Prager.*]

This apple originated at the village of Syke House, in Yorkshire, about the commencement of the present century. Diel translates this name in his nomenclature by a curious error of translation " Sik-House," and calls it " English Hospital Reinette," from the supposition that it had originated in the garden of an hospital, or that the flavour of its fruit was especially agreeable to invalids. It is figured by Lindley, Pl. 81, and by Ronalds, Pl. xxxviii.

Description.—Fruit : below medium size, two inches and a quarter broad by one inch and three-quarters high; roundish oblate. Skin : yellowish green, but entirely covered with brown russet, strewed with silvery scales ; sometimes it has a brownish tinge on the side which has been exposed to the sun. Eye : small and open, set in a shallow basin. Stalk : half an inch long, inserted in a shallow cavity. Flesh : yellowish, firm, crisp and juicy, with a rich, sugary, and very high flavour.

One of the most excellent dessert apples for winter use, and in season from the end of October to February.

The tree grows freely, is very hardy, and an excellent bearer. It attains a full middle size, and may be grown as a standard, or when grafted on the paradise stock as an espalier or a pyramid.

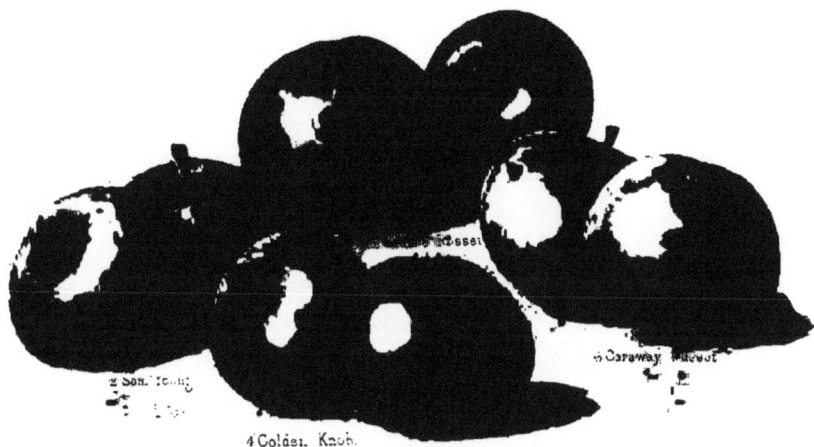

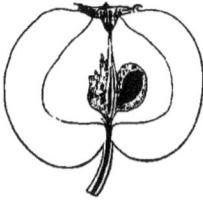

PLATE XXI.

2. SAM YOUNG.   3. CARAWAY RUSSET.

2. SAM YOUNG.

[SYN : *Irish Russet.*]

This apple is of Irish origin, and was first introduced to public notice by Mr. Robertson, the nurseryman of Kilkenny, about the beginning of the present century. It is figured by Lindley, Pl. 130.

Description.—Fruit : small, roundish oblate, two inches and a half wide, and an inch and three-quarters high. Skin : light greenish yellow, almost entirely covered with grey russet, and strewed with minute russety dots on the yellow part, but tinged with brownish red on the side next the sun. Eye : large and widely open, set in a wide, shallow and plaited basin. Stalk : short, not deeply inserted. Flesh : yellow, tinged with green, firm, crisp, tender, juicy, sugary, and highly flavoured.

A delicious little dessert apple of the first quality ; in season from November to February.

The tree is small, flat-headed in shape, and hardy ; it bears abundantly, and is well adapted for dwarf growth on the paradise stock. Dr. Lindley says of it, " It is one of the finest apples the English cultivate."

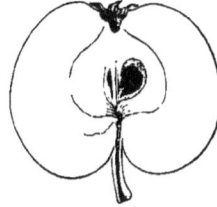

3. CARAWAY RUSSET.

The origin of this apple is not given by any of the leading Pomological Authorities.

Description.—Fruit : below middle size, two and a half inches wide, and two inches high ; oblate, even and regular in its outline. Skin : covered with a very thin coat of pale brown russet, which is dotted with darker russet, and on the same side the colour is inclining to orange. Eye : wide open, with broad reflexed segments, set in a pretty deep, wide, and saucer-like basin. Stalk : short and rather slender, inserted in a deep cavity. Flesh : yellowish, firm, crisp, rich, juicy and sweet, with a fine aroma.

A dessert apple of great excellence, and in season from November to February.

PLATE XXI.

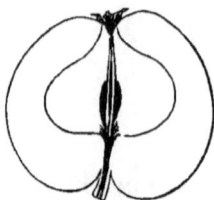

4. GOLDEN KNOB.

[SYN : *Kentish Golden Knob.*]

The precise history of this variety is not recorded. It has been extensively cultivated in the Kentish orchards, says Mr. Lewis A. Kellick, where there are many trees, some 100 years old. It is figured by Ronalds, Pl. xxxii.

Description.—Fruit : below middle-size, roundish. Skin : pale green, becoming yellowish green as it attains maturity, much covered with russet ; on the side next the sun it becomes yellow with an orange tinge. Eye : open generally, with long segments. Stalk : very short. Flesh : firm, with a brisk, juicy, sweet and pleasant flavour.

A good late keeping dessert apple, in season from December to March. It is one of the most remunerative sent to the London market, and at Christmas will bring as good a price as the *Ribston Pippin*, the *Golden Harvey*, and other choice sorts.

The tree is very hardy, grows to a very large size, with a drooping umbrageous character, and bears freely. It will succeed as a standard, but is well adapted on the paradise stock for dwarf growth as a pyramid or espalier.

PLATE XXI.

5. OLD NONPAREIL.

[SYN: *Nonpareil; English Nonpareil; Hunt's Nonpareil; Lovedon's Pippin; Reinette Nonpareil; Nonpareil d'Angleterre; Due d'Arsel; Grune Reinette.*]

It is generally believed that the *Nonpareil* came originally from France, though, as Switzer says, " It is no stranger in England . . . There are trees of them about the Ashtons in Oxfordshire of about a hundred years old, which (as they have it by tradition) were first brought out of France and planted by a Jesuit in Queen Mary or Queen Elizabeth's time."

It is strange that an apple of such excellence, and held in such estimation as the *Nonpareil* has always been, should have received so little notice from almost all the early continental Pomologists. It is not mentioned in the long list of the *Jardinier François* of 1653, nor even by De la Quintinye, or the *Jardinier Solitaire.* Schabol enumerates it, but it is not noticed by Bretonnerie. It is first described by Duhamel, and subsequently by Knoop. In the Chartreux Catalogue it is said "elle est forte estimée en Angleterre." Among the writers of our own country, Switzer is the first to notice it. It is not mentioned by Rea, Worlidge, or Ray, neither is it enumerated in the list of Leonard Meager. In America it is little esteemed. It has been cultivated in England for more than two hundred years. It is figured by Lindley, Pl. 86 ; and by Ronalds, Pl. xxxiv.

Description.—Fruit: medium sized, roundish, broad at the base and narrow towards the apex, apt to be more full on one side than the other. Skin: yellowish green, covered with large patches of thin grey russet, and dotted with small brown russetty dots, with occasionally a tinge of dull red on the side next the sun. Eye: rather prominent, very slightly if at all depressed, half open, with broad segments, which are reflexed at the tips. Stalk: an inch long, set in a round and pretty deep cavity, which is lined with russet. Flesh: greenish white, delicate, crisp, rich, and juicy, abounding in a particularly rich, vinous, and aromatic flavour.

One of the most highly esteemed and popular of all our dessert apples. It is in season from January to May.

The tree grows well and freely, but scarcely reaches the middle size. It is an excellent bearer, but rather a tender tree. It will succeed well as a standard, but on the paradise stock it is well adapted for dwarf growth in a pot ; as a pyramid ; or an espalier.

In the Northern counties and in Scotland it does not succeed as a standard, and when grown there under the protection of a wall the fruit is wanting in the fine delicate rich flavour of that grown for the South.

PLATE XXI.

6. WHITE NONPARIEL.   7. BRADDICK'S NONPARIEL.

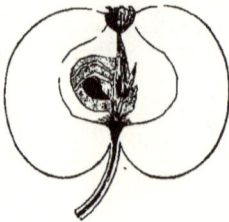

6. WHITE NONPAREIL.

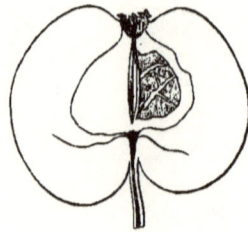

There is no notice of the origin of this variety.

Description.—Fruit: much resembling the *Old Nonpareil* in size and shape; oblate and obscurely ribbed, especially about the eye. Skin: greenish or yellowish green on the shaded side, and covered with a brownish red tinge on the side next the sun; the whole surface sprinkled with russet dots and a thin coat of grey russet, especially round the eye. Eye: closed, with broad, flat segments slightly recurved at the tips, set in a considerable basin, which is plaited and slightly angular. Stalk: half an inch to three-quarters long, slender, straight, and inserted in a deep, wide cavity. Flesh: greenish, tender, crisp, very juicy, sweet, and with a rich flavour, but not so rich as that of the *Old Nonpareil.*

A good dessert apple, in season from December to February. The tree has the ordinary growth and character of the Nonpareil race, and this variety is often greatly in favour with those who grow it.

7. BRADDICK'S NONPAREIL.

[SYN : *Ditton Nonpareil.*]

This variety was raised by John Braddick, Esq., of Thames Ditton. It is figured by Ronalds, Pl. xxxiv.

Description.—Fruit: medium sized; roundish and flattened, inclining to oblate. Skin: smooth, greenish yellow in the shade, and brownish red next the sun, russetty round the eye, and partially covered on the other portions of the surface with patches of brown russet. Eye: set in a deep, round and even basin. Stalk: half an inch long, inserted in a round and rather shallow cavity. Flesh: yellowish, rich, sugary, and aromatic.

One of the best winter dessert apples, and by many considered more sweet and tender than the *Old Nonpareil.* It is in season from November to April.

The tree is small and grows slenderly, but it is very hardy and an excellent bearer. It succeeds well on the paradise stock, and should be grown in every garden.

Plate XXI.

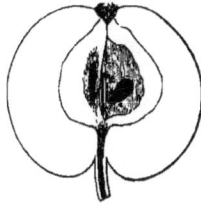

8. LODGEMORE NONPAREIL.

[Syn : *Lodgemore Seedling ; Clissold's Seedling.*]

This *Nonpareil* was raised by Mr. Cook, of Lodgemore, near Stroud, in Gloucestershire, about the year 1808. His garden was afterwards rented by Mr. Clissold, a nurseryman at Stroud, who propagated and sold it as a seedling of his own.

Description.—Fruit : about medium size, two and a half inches wide and nearly two inches high ; roundish ovate and regular in outline. Skin : rich golden yellow when fully ripe, dotted with minute grey dots, and with a blush of red on the side exposed to the sun. Eye : slightly closed, with broad, flat, leafy segments, and set in a narrow basin. Stalk : a quarter of an inch long, inserted in a narrow cavity. Flesh : yellowish, firm, crisp, juicy, sweet, and with a fine aroma.

A dessert apple of great excellence, and one of the very best late sorts. It is in season from February to June or even July.

The tree is hardy and bears freely.

Plate XXI.

9. ROSS NONPAREIL.

A variety of Irish origin, but without any published history. Mr. Roberston, of Kilkenny, has the credit of having first introduced it to notice. It is figured by Lindley, Pl. 90; and by Ronalds, Pl. xxxiv.

Description.—Fruit : middle sized, two and a half inches broad and two inches high ; roundish, even and regularly formed, narrowing a little towards the eye. Skin : entirely covered with thin russet, and faintly tinged with red on the side next the sun. Eye : small and open, set in a shallow, even basin. Stalk : an inch long, slender, inserted half its length in a round and even cavity. Flesh : greenish white, firm, crisp, brisk and sugary, charged with a rich and aromatic flavour.

This is one of the richest flavoured dessert apples.

The tree is very hardy and succeeds well upon most soils, and bears abundantly. Its great merit is that it is a free bearer on an open standard.

Beurre Butter's

2 Triomphe de Jodoigne

PLATE XXII.

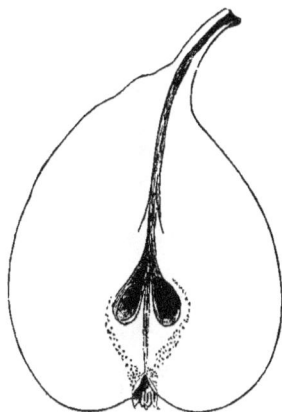

1. BEURRÉ SUPERFIN.

This excellent pear was raised at Angers, by M. Goubault, from seed sown about the year 1830, and it first bore fruit in 1844.

Description.—Fruit: above medium size, three inches wide, and a little more high ; obovate, or turbinate, somewhat uneven or bossed on its surface. Skin : thin, considerably covered with patches of cinnamon covered russet ; on the shaded side the ground colour is greenish yellow, which becomes more yellow at maturity, covered with small patches or veins of russet. Eye : very small and closed, with stiff, incurved, toothlike segments, and set in a deep, round and uneven basin. Stalk : over an inch long, fleshy at the base, and united to the fruit by fleshy folds. Flesh : yellowish white, fine grained, buttery and melting, very juicy, brisk and sweet, with a delicate perfume.

A very fine dessert pear, in season in the end of September and beginning of October.

The tree is a vigorous grower, hardy and prolific. It succeeds well as a standard on the pear stock, or as a pyramid on the Quince stock.

PLATE XXII.

2. TRIOMPHE DE JODOIGNE.

This fine pear was raised by M. Simon Bouvier, and fruited for the first time in 1843.

Description.—Fruit : large, obovate, regular and handsome. Skin : yellow, covered with numerous small russety dots and patches of thin brown russet. Eye : open, set in a slight depression. Stalk : an inch and a quarter long, curved and inserted without depression. Flesh : yellowish white, rather coarse, melting, juicy, sugary and brisk, with an agreeable musky perfume.

The *Triomphe de Jodoigne* is a very excellent pear when grown on soil that suits it, but on very heavy or indifferent soils it is apt to be coarse in flavour.

The tree is very hardy and very prolific. On the pear stock it grows as freely as the *Jargonelle*, and makes large extended espaliers. On the Quince stock it will grow well, but makes rather a straggling pyramid.

PLATE XXII.

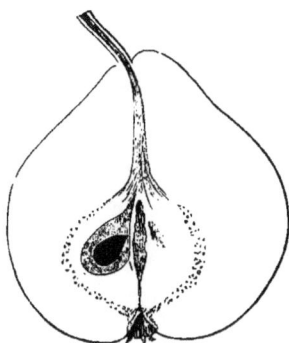

3. BEURRÉ DUHAUME.

There is no notice of the origin of this pear in any of the leading authorities.

Description.—Fruit : turbinate, evenly shaped. Skin : covered with brown russet, which only admits of a little of the yellow ground colour showing through on the side next the sun, where it has a red and orange cheek ; on the shaded side it is not so much covered with russet, and therefore shows more of the yellow ground colour through it. Eye : large, and quite open, set in a shallow basin, or almost level with the surface. Stalk : about half an inch long, very slender, and placed in a narrow round cavity. Flesh : firm, crisp and breaking, very juicy, sweet, rich and vinous, with a fine noyau flavour.

This is a pear of the first value, not unlike the *Passe Colmar* in the colour of the fruit and the texture of its flesh. It is, however, quite distinct from it.

The tree has a diffuse and bushy habit of growth. It succeeds well on the pear, or Quince stock, and bears freely on either when fully grown.

PLATE XXIII.

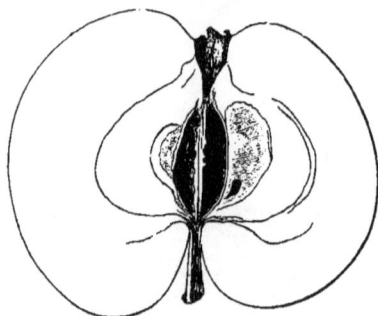

1. WARNER'S KING.

[SYN : *King Apple ; Salopian ; D. T. Fish ; Nelson's Glory ; Weavering ; Killick's Apple.*]

The origin of this valuable fruit is not given in any of the leading orchard authorities. " I have found this apple," says Mr. Lewis A. Killick in the *Journal of Horticulture*, "in almost every fruit growing district, and usually under different names. Although in Kent I cannot trace it under the name of *Warner's King* for many years, I am certain from my own examination it has been grown for nearly fifty years with a local name. In Middlesex I have seen it under the name of *Salopian.* I am growing it myself also as *Nelson's Glory* and *D. T. Fish.* One of our leading nurserymen imported a ' new' apple from Scotland, which acquired a great reputation three or four years since at the Crystal Palace on account of its size and appearance, but it proved to be *Warner's King.* Therefore I think we may fairly conclude that Mr. *Warner's King* was the *King* of many other people beside himself; but Mr. Warner recognising true loyalty linked his fortunes to those of the rising monarch." In a communication subsequently received direct from Mr. Killick, he says, "The apple I mentioned in *The Journal of Horticulture* as having been grown locally for 50 years, and that has turned out to be *Warner's King*, originated in an orchard in Weavering Street, near Maidstone, now in the possession of my brother, Mr. Austen Killick. The tree consists of about two thirds of Devonshire Quarrenden, and one third of ' Killick's Big Apple,' as it was called. I can trace it back nearly 40 years with certainty, and the tree was then

2 Golden Noble

3 D' Harvey

1 Warner's King

PLATE XXIII.

full grown, so that I have no doubt it must be over 50 years old, even if a graft had been introduced into the Old Quarrenden tree. We have grafted trees of it many years since, and it is now grown largely in the neighbourhood of Maidstone as *The Weavering*, or *Killick's Apple*. Most growers that now have planted *Waruer's King* are discovering their identity."

Description.—Fruit : very large, four inches wide, and three inches and a half high ; ovate. Skin : of an uniform clear deep yellow, strewed with russety dots and patches of pale brown russet. Eye : small and closed, with long accuminate segments, and set in a narrow, deep, and slightly angular basin. Stalk : half an inch long, deeply inserted in a round, funnel-shaped cavity, which is lined with thin yellowish brown russet. Flesh : white, tender, crisp, and juicy, with a fine brisk and subacid flavour.

A fine, handsome culinary apple, of first-rate quality. It is in season from November to March. This is now perhaps the most favourite culinary apple in cultivation, and usually takes the first place in its class at all fruit exhibitions.

The tree is very hardy, a good bearer, and not subject to disease. " It grows vigorously, and has a very upright mode of growth, and therefore in overcrowded orchards," says Mr. Killick, " instead of grubbing up the superfluous trees graft them with this variety, and in a few years the others would probably then receive the same fate." It succeeds admirably as an espalier, and should find a place in every garden of choice fruit.

PLATE XXIII.

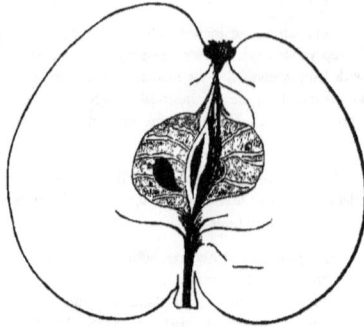

2. GOLDEN NOBLE.

[SYN : *Edelapfel gelber.*]

The origin of this apple is not known. It was first brought into notice by Sir Thomas Harr, of Stow Hall, Norfolk. His gardener procured it from a tree, supposed to be the original, in an old orchard at Downham, and communicated it to the Horticultural Society of London in 1820. It is probably therefore an old variety.

Description.—Fruit : handsome, large, round, and slightly narrowing towards the eye. Skin : smooth, clear bright yellow, without any blush of red, but a few small reddish spots, and here and there a small patch of thin russet. Eye : small, set in a round and deep basin, surrounded with plaits. Stalk : short, with a fleshy growth on one side of it, which connects it with the fruit. Flesh : yellow, tender, with a pleasant acid juice, perfectly melting, with a rich acidity, and baking of a clear amber colour.

A very attractive and valuable apple, in season from the end of September to the middle of December, and the best in its season.

The tree grows freely and bears abundantly. It should be grown much more generally than it is, as well for its own great merits as for its marketable qualities.

PLATE XXIII.

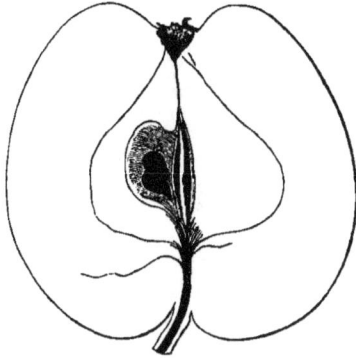

3. DR. HARVEY.

[Syn : *Harvey Apple.*]

This fine apple is one of the oldest English varieties. It is first mentioned by Parkinson as " a faire, greate, goodly apple ; and very well relished." Ralph Austin calls it "a very choice fruit, and the trees beare well." Indeed it is noticed by almost all the early authors. According to Ray, it is named in honour of Dr. Gabriel Harvey, of Cambridge : " *Pomum Harveianum ab inventore Gabriele Harveio Doctore nomen sortitum Cantabrigiæ suæ deliciæ.*"

Description.—Fruit : large, three inches wide and about the same high ; ovate inclining to be angular. Skin : greenish yellow, dotted with green and white specks, and marked with ramifications of russet about the apex. Eye : small, very slightly depressed, and surrounded with several prominent plaits. Stalk : short and slender, inserted in an uneven and deep cavity. Flesh : white, firm, crisp, juicy, pleasantly acid and perfumed.

A culinary apple of first-rate quality, well known and extensively cultivated in Norfolk and the north of England generally. It is in season from October to January.

The tree is large, hardy, and a great bearer. In the " *Guide to the Orchard*" it is said of this apple : " When baked in an oven, which is not too hot, these apples are most excellent ; they become sugary, and will keep a week or ten days, furnishing for the dessert a highly flavoured sweatmeat."

PLATE XXIV.

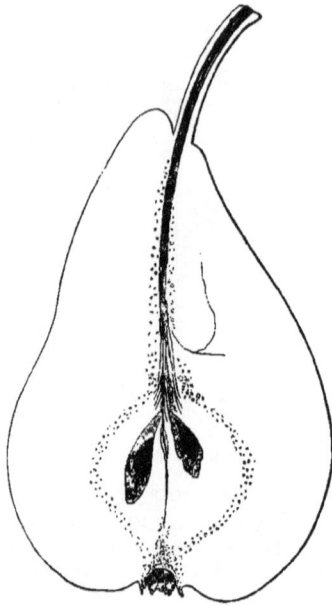

1. GÉNÉRAL TODTLEBEN.

This excellent pear was raised from seed sown at Gheling in Belgium in 1839, and the tree first produced fruit in 1855. M. Fontaine named it in honour of Général Todtleben, the gallant defender of Sebastopol.

Description.—Fruit : very large, four inches and a half long, and nearly four wide ; pyriform, ribbed round the apex, and often more swollen on one side than the other. Skin : yellow, covered with dots and patches of brown russet. . Eye : open, set in a wide furrowed basin. Stalk : an inch long, set in a small narrow cavity. Flesh : with a rosy tinge, very melting and juicy, slightly gritty, with a rich, sugary and perfumed juice.

A very delicious pear, in season from December to February. The tree is moderately vigorous, makes a good pyramid on the Quince, and bears abundantly.

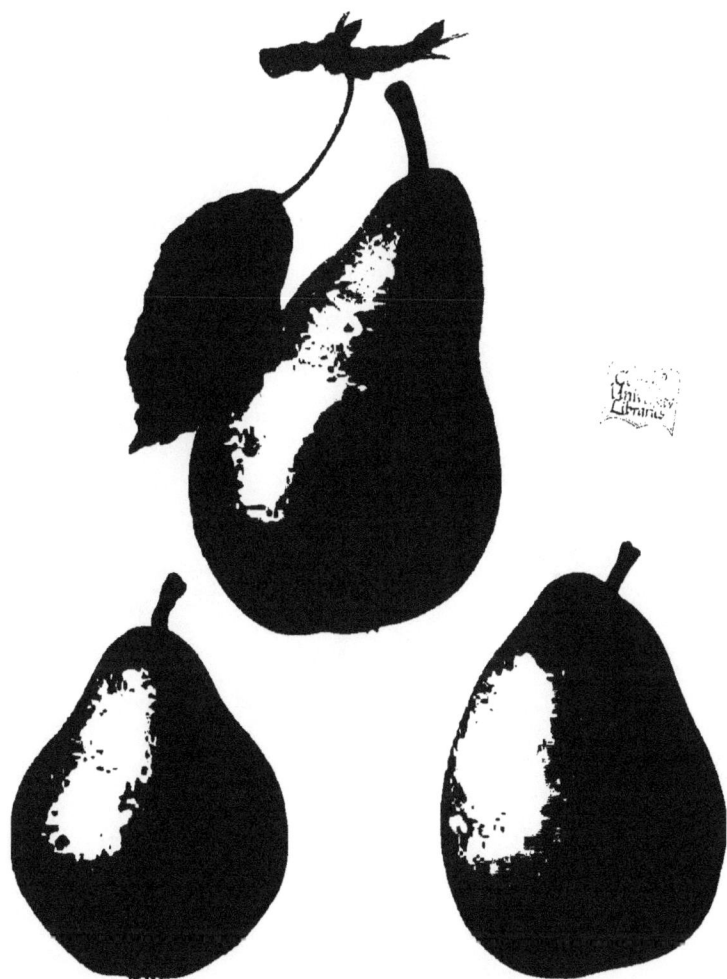

PLATE XXIV.

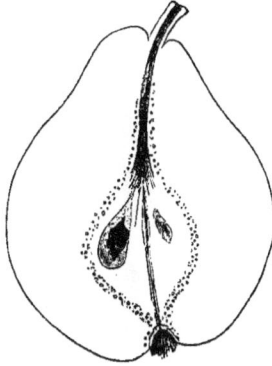

2. SOLDAT LABOURER.

[SYN : *Soldat Esperen ; Auguste Van Mons Soldat ; Beurré de Blumenbach.*]

A seedling grown by Major Esperen, and first fruited about the year 1820.

Description.—Fruit : large, three inches and three-quarters long and three inches wide ; oblong obovate, narrowing from the bilge both towards the eye and the stalk. Skin : pale lemon yellow, marked here and there with tracings of russet, and completely covered with minute russet dots. Eye : large, slightly closed, with long accuminate segments, and placed in a shallow depression. Stalk : an inch long, inserted by the side of a fleshy swelling in a narrow cavity. Flesh : yellowish white, buttery, melting and very juicy, rich and sugary.

This excellent pear has somewhat of the flavour of the *Autumn Bergamot.* It is in season through November.

On the Quince stock it forms a very good pyramid, and bears profusely.

PLATE XXIV.

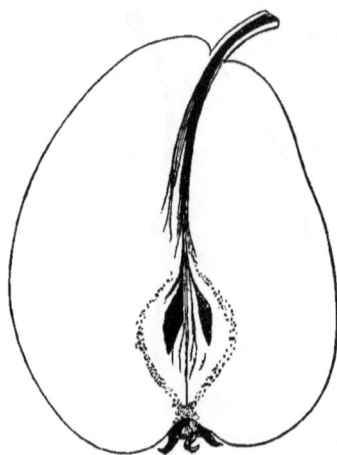

3. BEURRÉ BACHELIER.

[Syn : *Bachelier ; Chevalier.*]

This pear originated about the year 1845 with M. Bachelier of Cappelle-Brouck, in the department of Nord France.

Description.—Fruit : large and obovate, somewhat irregular in its outline.   Skin : greenish yellow, strewed with russety dots.   Eye : small and closed, set in a shallow basin.   Stalk : short.   Flesh : buttery and melting, rich, juicy, sugary, and aromatic.

A large, handsome, and very excellent pear.   In season through December.   When it ripens, this pear quickly becomes mealy.   It should therefore be gathered before it gets ripe, and its maturity looked for in time.

1 Newton Pippin

2 Arly Strad   3 Lunge oson

PLATE XXV.

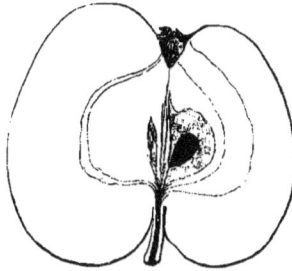

1. RIBSTON PIPPIN.

[Syn : *Glory of York ; Travers' Pippin.*]

The original tree of this celebrated Apple grows in the park at Ribston Hall, near Knaresborough, Yorkshire, on the banks of the river Nidd ; and if tradition is to be believed, the pip from which it grew came from Normandy. The following account of its origin is supposed to have been written by Sir John Goodricke, who died in 1789 :—" Three pippins (pips ?) were sent to Sir Henry Goodricke from Normandy, about the year 1709, only one of them succeeded, and from that tree all the *Ribston Pippins* have descended." Another account states that " Sir Henry Goodricke, father of the last Sir John Goodricke (owner of Ribston) being at Rouen in Normandy, preserved the pips of some fine-flavoured apples and sent them to Ribston. They were sown, and the young trees in due time planted in the Park (now George Garth). Out of seven trees planted, five proved decided " . . . (illegible—probably " failures and are all ") . . . " dead. The other two proved good apples. They are " . . . (illegible—probably " bearing ") . . . " yet. They never were grafted, and one of them is the celebrated original *Ribston Pippin* tree." And, lastly, in an undated letter from Miss Clough it is stated, " The *Ribston Pippin* came from Normandy about the beginning of the last century. My great grandfather, Sir Henry, had a friend abroad who sent him over three pippins (pips) in a letter, which being sown, two came to nothing. The present old tree at Ribston is the produce of the third, and has been distributed into all parts."

The original tree stood until the year 1810, when it was blown down by a violent gale of wind. It was afterwards supported by stakes in a horizontal position, as is shown by the woodcut vignette which has been kindly lent to the Club by the Editor of the *Gardener's Chronicle* to appear

PLATE XXV.

here. It continued to produce fruit until it lingered and died in the year 1840. On clearing away the old wood, a portion of which is still carefully preserved near the gardener's house, a healthy young sucker was found to be growing from the root about four inches below the surface of the ground. This young shoot has been well cared for but it does not grow well, and the tree is not so large as it was some years since. The branches died off very much after the frost of 1860. In the year 1875 the tree bore two pecks of very fine fruit. In 1876 and 1877 it did not bear so freely, and in the bad seasons of 1878—9 and 80 it did not ripen any fruit. The tree is still so far flourishing that it continues to bear fruit. It throws up suckers freely, so that when the present tree dies altogether, one of these with care would soon make a nice tree and still perpetuate the original.

The owner of Ribston Hall at this time is John Dent Dent, Esq., and much of the above information has been kindly supplied by Mrs. Dent. A tradition exists there that the gardener who raised the original tree, was a certain Robert Clemesha; but a second account states he was the father of Lowe, who, during the last century, was the fruit tree nurseryman at Hampton Wick.

Notwithstanding the great merits of the *Ribston Pippin* apple, now so universally admitted, it did not become generally known until the end of the last century; and it is not mentioned in any of the editions of Miller's Dictionary, or by any other author of that period; neither was it grown in the Brompton Park nursery grounds in 1770. In 1775 it appears in that collection, when it was grown to the extent of a quarter of a row, or about twenty-five plants; and as this supply seems to have sufficed for three years' demand, its merits must have been but little known. In 1788 its cultivation extended to one row, or about one hundred plants, and three years later to two rows. From 1791 it increased one row annually till 1794, when it reached five rows. From these facts we may pretty well learn the rise and progress of its popularity. It is now in the same nursery cultivated to the extent of about twenty-five rows, or 2,500 plants annually.

The extent to which the *Ribston Pippin* is now generally cultivated may be judged of from the fact, that in one single Nursery in the West of England, that of Messrs. Richard Smith and Co., at St. John's, Worcester, 3,000 plants are annually propagated by budding for trained plants, pyramids and standards.

Description.—Fruit: medium sized, roundish and irregular in its outline, caused by several obtuse and unequal angles on its sides. Skin: greenish yellow, changing as it ripens to dull yellow, and marked with broken streaks of pale red on the shaded side, but dull red changing to clear faint crimson, marked with streaks of deeper crimson, on the side next the sun, and generally russety over the base. Eye: small and closed, set in an irregular basin, which is generally netted with russet. Stalk: half an inch long, slender, and generally inserted its whole length in a round cavity, which is surrounded with russet. Flesh: yellow, firm, crisp, rich and sugary, and charged with a powerful aromatic flavour.

This delicious Apple needs no encomium. It is as highly appreciated as it is generally known. It is in greatest perfection during the months of November and December, but with good management it will keep until March.

The tree is hardy and vigorous in growth, and bears well when grown in a light, rich and dry soil. It is otherwise apt to canker. In all the Southern and middle counties of England it

PLATE XXV.

succeeds well as an open standard ; but in the North and in Scotland it requires the protection of a wall to bring it to perfection.   Nicol calls it "a universal apple for three kingdoms : it will thrive at John O'Groat's, while it deserves a place at Exeter or Cork."   It may be added, too, that it carries everywhere its delicious flavour.

" HONOUR'D AGE. "

The last remnant of the original *Ribston Pippin* Tree. Ribston Hall, Knaresborough.

PLATE XXV.

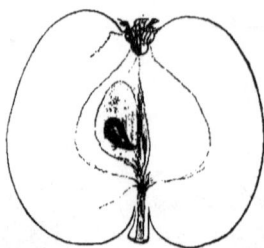

2. D'ARCY SPICE.

[Syn : *Baddow Pippin ; Spring Ribston.*]

The origin of this excellent Apple is involved in some obscurity. It was met with in the garden of the Hall, Toleshunt D'Arcy, in Essex, where a tree upwards of a century old is still standing. Some years ago Mr. John Harris, of Baddow, near Chelmsford, first introduced it to public notice under the name of *Baddow Pippin*, and Mr. Rivers afterwards renamed it *Spring Ribston*.

Description.—Fruit : medium sized, roundish or rather oblate, with prominent ribs on the sides, which terminate in four and sometimes five considerable ridges at the crown, very much in the character of the *London Pippin*. It is sometimes of an ovate shape, caused by the stalk being prominent instead of depressed, in which case the ribs on the sides and ridges round the eye are less apparent. Skin : deep lively green, changing as it ripens to yellowish green on the shaded side, but covered on the side next the sun with dull red which changes to orange where it blends with the yellow ground ; the whole surface is considerably marked with thin brown russet and russety dots. Eye : rather large and open, with short segments, and set in an angular basin. Stalk : very short, not more than a quarter of an inch long, and inserted in a shallow cavity. Flesh : greenish white, firm, crisp, juicy, sugary and with a particularly rich and vinous flavour, partaking somewhat of the *Ribston* and *Nonpareil*, but particularly of the former.

This is a dessert apple of first-rate character, in season from November, but possessing the desirable property of keeping well till April or May. It is not handsome in shape, nor attractive in appearance, but its excellent flavour will ensure its more extensive cultivation. The tree is hardy, grows freely, and bears well.

PLATE XXV.

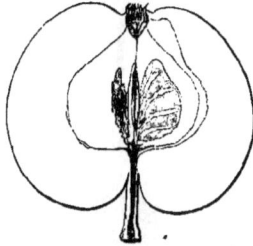

3. STURMER PIPPIN.

This Apple was raised by Mr. Dillistone, a nurseryman at Sturmer, near Haverhill, in Suffolk, and, it is said, was obtained by impregnating the blossom of the *Ribston Pippin*, with the pollen of the *Nonpareil*.

Description.—Fruit : below medium size, two inches and a quarter wide, by one inch and three quarters high ; roundish and somewhat flattened, and narrowing towards the apex, a good deal resembling the old *Nonpareil*. Skin : of a lively green colour, changing to yellowish green as it attains maturity, and almost entirely covered with brown russet ; it has a tinge of dull red on the side next the sun. Eye : small and closed, set in a shallow, irregular and angular basin. Stalk : three quarters of an inch long, straight, inserted in a round, even and russetty cavity. Flesh : yellow, firm, crisp, very juicy, with a brisk and rich sugary flavour.

This is one of the most valuable dessert Apples of its season : it is of first rate excellence, and highly desirable, as well on account of its excellent flavour, as from the fact of its arriving at perfection at a period when other favourite varieties are past. It is not fit for use till the *Ribston Pippin* is nearly gone, and continues long after the *Nonpareil*. The period of its perfection is from February to June.

The tree grows freely to a medium size. It is very hardy and an excellent bearer. It forms a good espalier and should have its place in every collection.

PLATE XXVI.

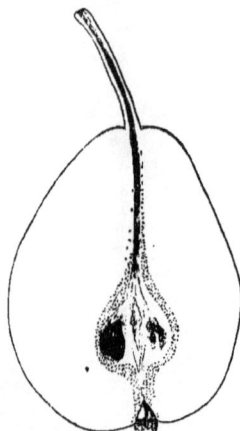

1. AUGUSTE JURIE.

This Pear originated from the seed of *Beurré Giffard*, sown on the 11th of August, 1851. It was grown at the École d'Horticulture at Écully, near Lyons, under the direction of the late M. Willermoz, and it was named in honour of M. Auguste Jurie, President of the Horticultural Society of the Rhone.

Description.—Fruit : above medium size, three and a half inches by two and a half inches wide, obtuse ovate. Skin : green, becoming yellowish green, as it ripens, with a thin speckled coat of russet on the side next the sun, and strewed all over with russet patches. Eye : closed, with tooth-like segments, and set even with the surface. Stalk : from an inch to an inch and a half long, inserted without depression. Flesh : crisp, rather granular, sweet, briskly flavoured, and with a fine melon flavour.

The fruit here represented and described was grown in the gardens of the Royal Horticultural Society at Chiswick. In France it grows more round in shape, with a shorter stalk, and, as described by the Abbé Dupuy, it has a much brighter colour. It is a valuable early pear, ripe in the middle of August, and well worthy of general cultivation. It received the certificate of the Royal Horticultural Society on August 24th, 1880.

The tree grows well and bears freely as a pyramid on the quince.

Auguste Jurie

Summer Beurré d'Aremberg

4 Jargonelle

6 Beurré Hard

PLATE XXVI.

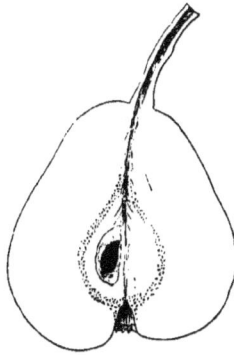

2. SUMMER BEURRÉ D'AREMBERG.

This excellent Pear was raised by Mr. Rivers, of Sawbridgeworth, and the tree bore fruit for the first time in 1863.

Description.—Fruit : small, two inches wide and the same high ; turbinate, even, and smooth in its outline.   Skin : entirely covered with a thin crust of cinnamon coloured russet.   Eye : wanting segments, very deeply set in a narrow hole.   Stalk : long, stout and fleshy, curved, and inserted without depression.   Flesh : yellowish or buttery, tender, melting, and very juicy, sweet, richly flavoured and with a musky aroma.

This is really a delicious Pear, one of the richest and best of its season.   It ripens about the middle of September, but should be gathered and put in the fruit room for a fortnight before being sent to table.

The tree grows freely, is very hardy and bears well.   It will succeed equally well as a wall tree, pyramid, or espalier.   If grown as a pyramid, or espalier, it should be double grafted on the pear stock.   "It is one of my favourite pears," says Mr. Ward, of Stoke Edith, "It deserves a wall with a south, or south-eastern aspect, and I give it this position.   It is but little known at present, but it will make its way, for it is really a sweetmeat when fully ripe.   It ought also to be successful as a standard on a warm soil and situation."

PLATE XXVI.

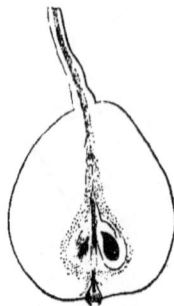

3. SUMMER DOYENNÉ.

[SYN : *Doyenné d'Été ; Doyenné de Juillet ; Jolimont ; Roi Jolimont ; Jolivet ; Duchesse de Berri d'Été ; St. Michel d'Été ; Brüsseler Sommer Dechantsbirne ; Julius Dechantsbirne.*]

The origin of this Pear has caused much discussion among pomologists, but there seems little doubt about it. In Van Mons' Catalogue at p. 28, we find "*Doyenné d'Été ; par nous.*" In the preface this expression "*par nous*" is stated to signify "*que ce fruit est un resultat de nos essais.*" This being the case, we cannot suppose that Van Mons would have claimed a fruit he did not raise. Diel acknowledges having received it from Van Mons in his "*Keruobstsorten,*" Vol. xix. ; and in his "*Systematisches Verzeichniss,*" 2 Fort. p. 90, he describes it under the name *Brüsseler Sommer dechantsbirne* with the synonyme "*Doyenné d'Été* V. M.*" This distinction of placing Van Mons' initials in conjunction with it, was, no doubt, to distinguish it from that other *Doyenné d'Été,* which he had described in the "*Kernobsorten,*" Vol. iii., p. 39, and which is a totally different fruit, of medium size, with no red on the sunny side, and which ripens in the end of August. This pear must have been raised by Van Mons at an early period, for Diel mentions it among his best pears in 1812.

An excellent coloured representation has been given of this pear in the "Florist and Pomologist," July, 1862, p. 104.

Description.—Fruit : small, two inches wide and one and three quarters high ; roundish, or roundish turbinate. Skin : smooth, and wherever shaded of a clear greenish yellow, changing as it ripens to a fine lemon yellow, and on the side next the sun covered with a red blush, and strewed with grey dots. Eye : small, half open, set in a shallow plaited basin. Stalk : three quarters of an inch long, not depressed. Flesh : yellowish white, half melting, and very juicy, sweet and pleasantly flavoured.

This is the earliest Pear known, and the best of the early Pears. It is ripe by the middle or end of July, but it should be gathered before it is ripe, or it will soon become mealy and quite insipid.

The tree is very hardy and bears abundantly. It succeeds well on the *Quince,* forms a handsome pyramid, and this is the best mode of growing it.

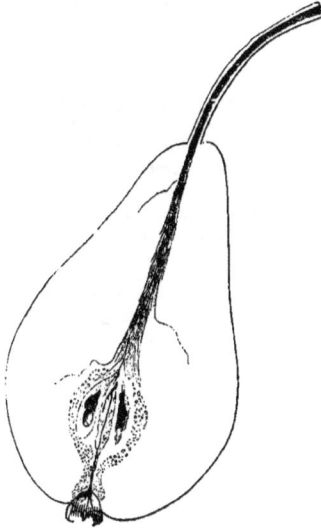

PLATE XXVI.

4. JARGONELLE.

[SYN: *Beau Présent*; *Belle Vièrge*; *Beurré de Paris*; *Chopine*; *Cueillette*; *Épargne*; *De Fosse*; *Grosse Cuisse Madame*; *Mouille Bouche d'Été*; *Sweet Summer*; *St. Lambert*; *St. Samson*; *De la Table des Princes.*]

The origin of this favourite old Pear is nowhere given, and it is not known at what period it was first introduced into this country. The first mention we have of it is by Switzer. It has been figured by Lindley in his *Pomologia Britannica*, Pl. 108, and in many other works.

Lindley states, that the name *Jargonelle* is derived, according to Ménage and Duchat from *Jargon*, anciently *Gergon*, in Italian *Gergo*, in Spanish *Gericonça*, all corruptions of *Græcum*; whence Merlet infers that the *Jargonelle* was the *Pyrum Tarentinum* of Cato and Columella; the *Numidianum Græcum* of Pliny, and the *Græcum* of Macrobius. If this conjecture be well founded, the kind to which the name belongs will be one of the most ancient in cultivation. It was certainly brought from France to England, though the French *Jargonelle* is not ours.

Description.—Fruit : large and pyriform. Skin : smooth greenish yellow, with a tinge of

PLATE XXVI.

dark brownish red next the sun. Eye: large and open, with short, stout, blunted segments, set in a shallow basin. Stalk: about two inches long, slender, and obliquely inserted without depression. Flesh: yellowish white, tender, melting, and very juicy, with a rich vinous flavour, and a slight musky aroma.

This well known Pear is ripe in August. It is one of the finest and best of our early Pears. The tree is strong, healthy, and vigorous, with long pendent shoots. It succeeds well as a standard, but in northern climates requires a wall. It grows too freely for a pyramid. Its best place is on a wall with plenty of space to extend itself, and there is no part of the kingdom where it will not grow in this way in the greatest perfection. As a standard its fruit is smaller, but it ripens it thoroughly even in the north if the situation is sheltered. In the city of Perth it may be seen wherever there is a space of ground sufficient to plant it. Never did Bourgeois of Rheims exhibit more partiality for his favorite *Rousselet*, than does the citizen of Perth for his adopted *Jargonelle*.

The wide extent on which it is cultivated is well shown by the fact that in the St. John's Nurseries, at Worcester, 3,000 plants are annually produced by budding; of which 1,500 are trained trees for walls or espaliers, and the remainder are equally divided between pyramids and standards.

PLATE XXVI.

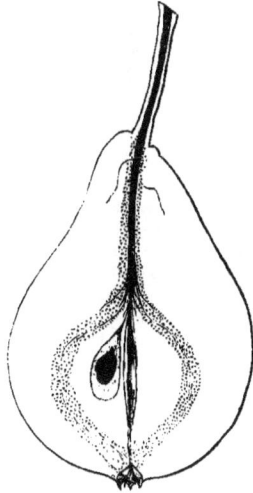

5. BEURRÉ GIFFARD.

[SYN : *Giffard.*]

This Pear was found as a wilding in 1825, by M. Nicolas Giffard. of Fonassières, near Angers, and it was first described by M. Millet in 1840.

Description.—Fruit : medium sized, pyriform, or turbinate. Skin : greenish yellow, mottled with red on the side next the sun. Eye : closed, set in a shallow basin. Stalk : an inch long, slender and obliquely inserted on the apex of the fruit. Flesh : white, melting and very juicy, with a vinous and highly aromatic flavour.

This is one of the very best early Pears, ripe the middle of August. The tree is hardy, grows freely, and bears well, but it must grow against a wall, or its fruit will be smaller than that which is represented on the plate, and which was grown on the Cordon Wall at Holme Lacy.

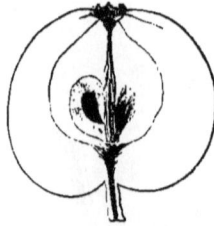

1. OSLIN.

[Syn : *Orglon; Orgeline; Arbroath Pippin; Original Pippin; Mother Apple; Golden Apple; Burrknot; Summer Oslin.*]

This is a very old Scotch Apple, supposed to have originated at Arbroath ; or to have been introduced from France by Monks of the Abbey, which formerly existed at that place. The latter opinion is in all probability, the correct one ; although this name, or any of the synonyms above given, are not now to be met with in any modern French lists. But in the "*Jardinier François*" which was published in 1651, an apple is mentioned under the name "*Orgeran*" which is so similar in pronunciation to "*Orgeline*," that it is not unlikely it may be the same name with a change of orthography, especially as our ancestors were not over particular in preserving unaltered the names of foreign introductions. This apple has been well figured in the *Pomological Magazine*, Pl. 5.

Description.—Fruit : medium sized, two inches and a half wide, and two inches high ; roundish oblate, evenly and regularly formed. Skin : thick and membranous, of a fine pale yellow colour, and thickly strewed with brown dots ; very frequently cracked, forming large and deep sinuosities on the fruit. Eye : scarcely at all depressed. Stalk : short and thick, inserted in a very shallow cavity. Flesh : yellowish, firm, crisp and juicy, rich and sugary, with a highly aromatic flavour, which is peculiar to this apple.

A dessert apple of the highest excellence ; ripe by the end of August, and continues through September, but does not last long. "This is an excellent apple," says Nichol, "as to flavour, it is out-done by none but the *Nonpareil;* over which it has this advantage, that it will ripen in a worse climate and a worse aspect." Lindley says of it "This delicious variety is the best, except the *Kerry Pippin,* of all the early Summer Apples."

The tree grows freely, but is subject to canker as it grows old ; it has an upright habit, and bears freely. The branches are generally covered with a number of knobs, or burrs, which when planted in the ground take root readily and produce a fresh plant.

PLATE XXVII.

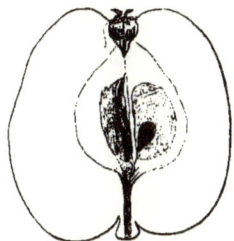

2. SACK AND SUGAR.

[Syn : Morris's *Sack and Sugar.*]

This Apple was raised the beginning of the present century, by Mr. Morris, a market gardener at Brentford, in Essex, and sometimes bears his name.

Description.—Fruit : below medium size, two inches and a quarter wide, and an inch and three quarters high ; roundish, inclining to oval, with prominent ridges round the eye. Skin : pale yellow, marked with a few broken stripes and streaks of bright crimson on the side next the sun. Eye : closed, with pointed segments overlapping each other, and set rather deeply in a round, angular and plaited basin. Flesh : white, very soft and tender, very juicy, sugary, and with a pleasant, brisk, balsamic flavour.

An excellent apple for dessert, or culinary use. It ripens in the end of July and beginning of August and continues during September.

The tree is hardy, a free and vigorous grower, and an immense bearer.

PLATE XXVII.

3. WHORLE PIPPIN.

[Syn: *Lady Derby; Thorle Pippin; Summer Thorle; Watson's New Nonsuch.*]

The origin of this Apple is not given in any of the leading works of authority. Its name has been supposed to be derived from its resemblance to the " Whorle," which was the propelling power, or rather impetus of the spindle, when the distaff and spindle were so much in use.

Description.—Fruit : below medium size, two inches and a quarter wide at the middle, and an inch and three quarters high ; oblate, handsome and regularly formed. Skin : smooth, shining and glossy, almost entirely covered with fine bright crimson, which is marked with broken streaks of darker crimson, but on any portion shaded, it is a fine clear yellow, a little streaked with pale crimson. Eye : scarcely at all depressed, large, half open, with broad and flat segments, which frequently appear as if rent from each other by an over swelling of the fruit ; set in a very shallow basin, which is often very russety, deeply and coarsely cracked. Stalk : a quarter of an inch long, inserted in a wide cavity. Flesh : yellowish white, firm, crisp, and very juicy ; with a brisk, refreshing and pleasant flavour.

A beautiful little summer dessert apple of excellent quality. It is ripe in August, and when the tree is loaded with fruit, as it usually is, it presents a very beautiful object.

The tree grows freely and makes an excellent pyramid. In Scotland and the North and East of England, it is to be met with in almost every garden and orchard, but in the South and West it is but little known.

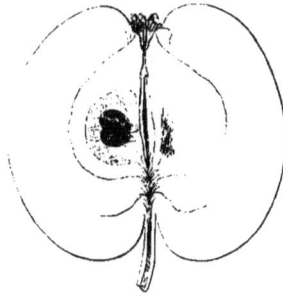

PLATE XXVII.

4. BOROVITSKY.

[SYN : *Duchess of Oldenburgh* ; Smith's *Beauty of Newark ; New Brunswick.*]

A Russian variety, whose origin is nowhere given. It was sent from the Taurida Gardens, near St. Petersburgh, to the London Horticultural Society, in 1824. It has been figured in the *Pomological Magazine*, Plate 10.

Description. Fruit : above medium size, sometimes larger, about three inches and a quarter wide, and two inches and a half high ; round, and sometimes prominently ribbed on the sides, and around the eye. Skin : smooth, greenish yellow on the shaded side, and streaked with broken patches of fine bright red on the side next the sun, sometimes assuming a beautiful dark crimson cheek ; it is covered all over with numerous russety dots, particularly round the eye, where they are large, dark and rough. Eye : large and closed, with long broad segments, placed in a deep and angular basin. Stalk : long and slender, deeply inserted in a narrow and angular cavity. Flesh : yellowish white, firm, crisp and very juicy, with a pleasant, brisk and refreshing flavour.

An excellent early dessert apple of the first quality, in season from the middle of August, to the end of September.

The tree is hardy, grows freely and bears well.

PLATE XXVIII.

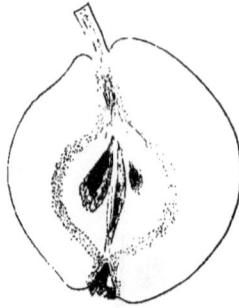

1. SECKLE.

[Syn : *Seckel; Sicker; New York Red Cheek; Shakespear; Lammas*, of the Americans.]

The *Seckle* Pear is an American production. Its precise origin is unknown. Downing, in his "*Fruits and Fruit Trees of America*," says, " Its affinity to the *Rousselet*, a well-known German Pear, leads to the supposition that the seeds of the latter Pear, having been brought by some of the Germans settling near Philadelphia, by chance produced this superior variety." It is first noticed by Coxe, an American pomologist, in his work " *View of the Cultivation of Fruit Trees.*" It was sent to the London Horticultural Society in 1819, by Dr. Hosack, of Philadelphia, with several other fruits.

The original tree is still in existence, and is over a hundred years old. It grows in a meadow in Passyunk township, about a quarter of a mile from the Delaware, opposite League Island, and about three miles and a half from Philadelphia. It is about thirty feet high ; the diameter of the trunk at a foot from the ground, is six feet ; and at five feet from the ground, it is four feet nine inches. The trunk is hollow and very much decayed ; the bark, half way round to the height of six feet, is entirely gone ; and the progress of decay has advanced so far, that it is feared in a few more years the tree will have ceased to exist. There are, however, young suckers growing from the root, by which the original stock will be preserved ; but it is to be regretted that means were not taken to preserve the original tree, by the very simple process of plaistering up the decayed portion. The property on which the tree stands belonged in 1817, according to Coxe, to Mr. Seckle (not Seckel), of Philadelphia, and hence the origin of the name.

" The following *morceau* of its history may be relied on," Downing adds, " it having been related by the late venerable Bishop White, whose tenacity of memory is well known. About 1765, when the Bishop was a lad, there was a well-known sportsman and cattle dealer in Philadelphia,

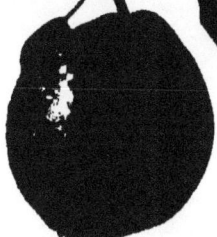

Stark

Hessle

Améris a

Bouillie de Mai

Souvenir du Congrès

Joie de s Capucins

PLATE XXVIII.

who was familiarly known as "Dutch Jacob." Every season, early in the autumn, on returning from his shooting excursion, Dutch Jacob regaled his neighbours with pears of an unusually delicious flavour ; the secret of whose place of growth, however, he would never satisfy their curiosity by divulging. At length the Holland Land Company, owning a considerable tract south of the city, disposed of it in parcels, and Dutch Jacob then secured the ground on which his favourite pear tree stood—a fine strip of land near the Delaware. Not long afterwards it became the farm of Mr. Seckel, who introduced this remarkable fruit to public notice, and it received his name." The *Seckle* Pear has been figured in the *Transactions of The London Horticultural Society*, Vol. III., Pl. ix., and also in the *Pomological Magazine*. Pl. 72.

Description.—Fruit : small ; obovate, regularly and handsomely shaped. Skin : at first dull brownish green, changing, as it ripens, to yellowish brown, with bright red on the side exposed to the sun. Eye : small and open, with very short segments, and not at all depressed. Stalk : half an inch long, inserted in a small, narrow depression. Flesh : buttery, melting and very juicy, with a rich and unusually powerful aromatic flavour.

One of the most valuable dessert Pears, surpassing in richness any other pear in cultivation. It is ripe in October, and from the beauty of its colour is as desirable as an ornament to the dessert table, as it is welcome there, for its delicious flavour.

The tree is very hardy, vigorous, and an abundant bearer. It succeeds well as a standard, and is admirably adapted for dwarf pyramids on the Pear stock, kept well root pruned. No garden, however small, should be without this excellent fruit.

PLATE XXVIII.

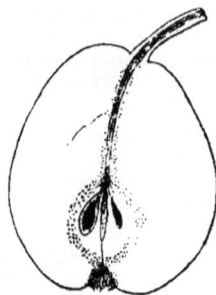

2. HESSLE.

[Syn : *Hessel; Hazel.*]

This Pear takes its name from the village of Hessle, near Kingston-upon-Hull, in Yorkshire, where it was first discovered, but its precise history is not known.

Description.—Fruit : below medium size ; turbinate.   Skin : greenish yellow, very much covered with large russety dots, which give it a freckled appearance.   Eye : small and open, slightly depressed.   Stalk : an inch long, obliquely inserted without depression.   Flesh : tender, very juicy, sweet, and with a high aroma.

An excellent Pear, in season in October.   There are two varieties of the *Hessle* Pear, or rather two *Hessle* Pears, known in Yorkshire.   The one figured here, the best of them, is of an olive colour thickly studded with large russety dots, but becomes a greenish yellow when ripe.   The other is an inferior Pear, larger, of a lighter colour, more yellow and with much fewer dots.   This Pear quickly decays in the centre.   It is juicy and pleasant if eaten at the right moment, but has no special merit.

This tree is very hardy, vigorous and a most abundant bearer.   It forms a fine standard succeeding in almost every situation, and particularly in the northern climates where the more tender varieties do not attain perfection.   In Yorkshire it is grown almost universally, and as a market pear it is one of the best and most remunerative to the grower.   The *Hessle* pear trees in Herefordshire were laden with fruit in the year 1880, when almost all other varieties failed.

PLATE XXVIII.

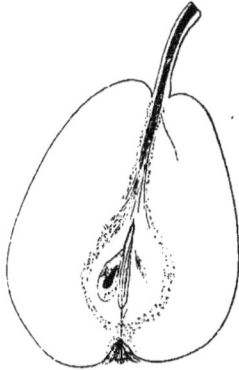

3. AMBROSIA.

[Syn : *Early Beurré; Summer Beurré.*]

It is related by Switzer that this variety was introduced from France "among that noble collection of fruit that was planted in the Royal Gardens in St. James's Park soon after the Restoration, but is now (1724) cut down." Nevertheless there is no record of it in any French author under this name. Jahn, in the "*Handbuch,*" considers it synonymous with Diel's *Braunrothe Pomeranzbirne*, which, Metzger says, is the same as *Orange Rouge* of the French authors, but it is certainly not the same as this last fruit.

Description.—Fruit : medium sized ; roundish obovate, and slightly flattened. Skin : smooth, greenish yellow, covered with small grey specks and slight marks of russet. Eye : closed, set in a considerable depression. Stalk : an inch and a half long, slender, inserted in an open cavity. Flesh : tender, buttery, melting, rich, sugary and perfumed.

A delicious summer dessert Pear of first-rate quality. It ripens in September, but keeps only a few days after being gathered.

The tree is hardy and vigorous in growth, and a good bearer. It succeeds well as a standard, either on the pear or quince stock.

PLATE XXVIII.

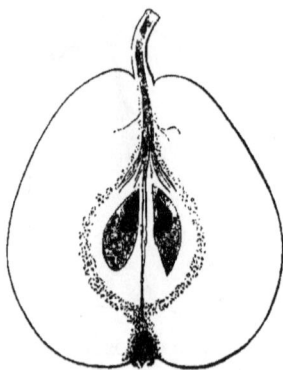

4. SUSETTE DE BAVAY.

This Pear was raised by Major Esperen, of Malines, in the early part of the present century (c. 1830-40). He named it in compliment to Madame de Bavay, the wife of a nurseryman at Vilvorde, near Brussels.

Description.—Fruit : medium sized, turbinate. Skin : yellow, covered with numerous large russet dots and traces of russet. Eye : open, placed in a shallow, undulating basin. Stalk : an inch long, inserted in a small cavity. Flesh : melting, juicy, sugary and vinous, with a pleasant perfume.

This Pear is in season in January and February, and valuable from its lateness.

The tree makes a good pyramid and bears freely.

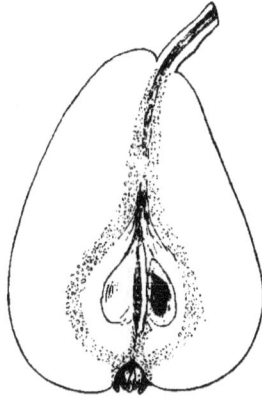

5. SOUVENIR DU CONGRÉS.

This Pear is said to have been produced by M. Morel, of Lyon-Vaise, in France, about the middle of the present century (1850).

Description.—Fruit : large, three inches and a half long, two inches and three quarters wide, and often much larger ; oblong, obovate, uneven and undulating in its outline. Skin : considerably covered with smooth cinnamon-coloured russet, with, here and there, patches of the yellow ground colour exposed ; on the side next the sun, there are streaks of bright crimson, and a warm glow of russet. Eye : large and open, deeply set. Stalk : an inch or more long, very stout, inserted either in a pretty deep cavity, or on the end of the fruit, in a slight cavity. Flesh : yellowish white, tender, very juicy and melting, with a rich, vinous flavour and a musky aroma.

A very handsome and excellent Pear ; in season from the end of August to the beginning of November. It has a great resemblance to *William's Bon Chrétien*, but is quite a distinct fruit.

The tree is hardy and bears well, making an excellent pyramid on the quince stock.

PLATE XXVIII.

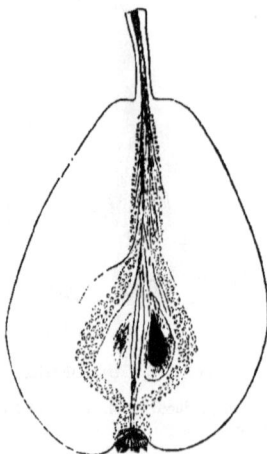

6. BEURRÉ BLANC DES CAPUCINES.

This French Pear has been confused by many French pomologists, with the *Amadotte*. M. Decaisne figures it as such in the "*Jardin Fruitier du Muséum*," but it is evidently not the *Amadotte* of Merlet, Miller, Forsyth, and Diel. This is a long pyramidal-shaped fruit, whilst the true variety is roundish and flattened. M. Tougard has an *Amadotte Blanc*, which he makes synonymous with *Beurré Blanc des Capucines*, and M. Decaisne made the mistake of adopting this as the variety described by Merlet. Jahn, following Decaisne, identifies *Beurré Blanc des Capucines* with the *Amadotte*, which he calls *Herbst Amadotte*; but Dr. Hogg has shown that there are, without doubt, two distinct varieties. Neither Tougard, Decaisne, nor Jahn, take notice of the remarkably high, musky flavour of the fruit; but on the contrary, the former says, "it is slightly acid and astringent ; and the latter, that "it has neither perfume nor flavour ;" characters which agree with *Beurré Blanc des Capucines*, but not with *Amadotte*. It is quite evident that Tougard, Prevost, Decaisne, and Jahn, have taken *Beurré Blanc des Capuchines* for the true *Amadotte ;* and the confusion still exists in the French orchards.

Description.—Fruit : large and handsome ; somewhat oval, even and regular in its outline. Skin : smooth, of a dull yellow colour, with a greenish tinge, strewed with flakes of russet, and with

PLATE XXVIII.

a russet patch round the stalk. Eye : small and half open. Stalk : upwards of an inch in length, woody, inserted in a small round cavity. Flesh : yellowish, coarse-grained and gritty, half melting or crisp, with a cold acidity.

This Pear is far inferior to the true *Amadotte*. It ripens in October, and must be eaten directly it is ripe, or it will rot at the core. It is not worth cultivating, and is inserted here to correct once again an error of Decaisne, which was first pointed out by Dr. Hogg—an error which still prevails in France—for it was sent as the true *Amadotte* by M. Benoit, of Hâvre, amidst a very fine collection of the best varieties of Pears, to the Apple and Pear Show of the Woolhope Club in October 1880.

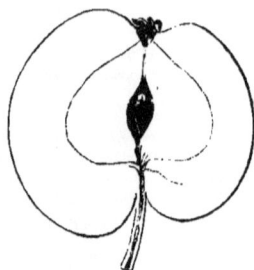

1. WHITE STYRE.

An old variety widely scattered through Herefordshire and Worcestershire, but it is without any known history.

Description.—Fruit : about medium size, round and obtusely ribbed. Skin : of a uniform lemon colour with patches and lines of russet over the surface, especially on the side next the sun, and in the stalk cavity ; the surface generally strewed with small russet dots. Eye : closed, with connivent segments, set in a pretty deep depression ; tube, short, funnel shaped ; stamens, median. Stalk : slender, half an inch long, set in a deep russety cavity. Flesh : yellowish, soft and tender ; the juice plentiful, moderately sweet. and with a delicate subacid flavour. Cells of the core, open.

The chemical analysis of the *White Styre* by Mr. G. H. With, F.C.S., F.R.A.S. (season 1880), is as follows :—

Density of fresh juice	1·033
Ditto after standing 24 hours	1·036

One hundred parts by weight contain of :—

Sugar	9·1
Tannin, Mucilage, Salts, &c.	3·5
Water	87·4

This Apple was formerly highly esteemed amongst the early cider fruits in Herefordshire, and is still valued in Worcestershire. It makes a light pleasant cider of a deep colour, with good keeping qualities, but it is without much flavour and with very little alcoholic strength. The fruit is therefore seldom used alone.

The tree is very hardy, bears abundantly, and seldom fails to bear. The sandy loams of Worcestershire with the blue clay (Lias) subsoil seems to suit it better than the clay loams of Herefordshire. The variety is old, and is not now propagated.

PLATE XXIX.

2. STYRE WILDING.

This fruit is without a history. It is widely grown, and many of the trees are more than a hundred years old.

Description.—Fruit : small, conical, bluntly angular and irregular in its outline. Skin : smooth and shining, greenish yellow on the shaded side and with a red cheek wherever exposed to the sun. Eye : closed, with connivent segments, set in a pretty deep, narrow and plaited basin ; tube, conical, sometimes inclining to funnel shape ; stamens, median. Stalk : very short, deeply imbedded in the cavity, which is russety, and generally with a fleshy swelling on one side of it. Flesh : soft and woolly, sweetish and scarcely acid. Cells of the core, open.

Mr. With's analysis of the *Styre Wilding* apple (season 1880), is as follows :—

Density of fresh juice	...	...	...	...	1·041
Ditto after 24 hours exposure		...	...	...	1·044

One hundred parts by weight of fresh juice yield :—

Sugar	...	...	...	...	14·121
Tannin, Mucilage, Salts, &c.		...	...	...	00·679
Water	...	...	...	...	85·200

This fruit ripens late. It is highly esteemed in some districts of the county, and is thought to give strength and flavour to the mixed fruit. With *Skyrme's Kernel* and the *Redstreak* it makes a very strong cider.

The tree is very hardy and bears profusely, so that the crop is usually very heavy though the fruit is so small. The apples often hang on the trees like ropes of onions. It is a sure bearer every other year, and the fruit keeps well.

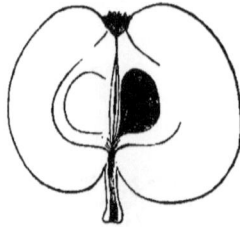

3. EGGLETON STYRE.

This Apple was raised from the kernel by the late Mr. William Hill, of Lower Eggleton, Ledbury, Herefordshire, in the nursery attached to the farm. The seedling first bore fruit about the year 1847, and the birds attacking the apples attracted attention to their sweet and rich flavour.

Description.—Fruit : medium sized, roundish with obscure ribs on the sides. Skin : rich yellow, orange next the sun, and covered with thin tracings and patches of russet. Eye : open with reflex segments like *Court of Wick*, set in an even basin. Tube : short, funnel shaped ; stamens, median. Stalk : slender, half an inch long, deeply inserted in a round cavity which is lined with russet, extending in branches over the base. Flesh : yellowish, tender, juicy, sweet and slightly acid. Cells of the core. open.

Mr. With's analysis of the *Eggleton Styre* juice (season 1880), is as follows :—

Density of fresh juice	1·049
Ditto after 24 hours exposure	1·050

One hundred parts by weight of juice yield :—

Sugar	10·591
Tannin, Mucilage, Salts, &c.	6·569
Water	82·840

The *Eggleton Styre* makes excellent cider alone, very sweet and rich, with a high red colour. It has been sold, fresh bottled, at 16/- a dozen. It fines better if mixed with *Redstreak, Cowarne Red, Pym Square, Cook's Kernel*, or *Strawberry Norman*.

This apple is second early, and is so sweet and aromatic, as to be very attractive to hares, rabbits, fowls, blackbirds, and fieldfares, who will choose it in preference to all others. The tree is hardy and bears freely. It is chiefly grown in the parish of Eggleton, and the neighbourhood, but it deserves a much wider growth.

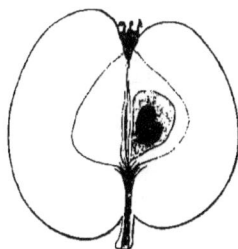

4. SKYRME'S KERNEL.

The Skyrmes are an old Herefordshire family, and a century or two since, one branch held an estate at Brockhampton, called the Upper House, for some generations. It passed to the Protheroes by marriage in 1788. Another branch of the Skyrmes lived at Dewsall, near Hereford. History is silent as to which of them grew the *Kernel* that bears the family name; but it may very probably have been reared at Brockhampton, for the trees there are some 100 or 150 years old; they are found in that district of the county, and may have spread from it. The apple is not mentioned by any of the old writers.

Description.—Fruit : small, about two inches wide and two inches high, ovate or slightly conical, even and regular in its outline, and somewhat snouted towards the apex. Skin : smooth and shining, almost entirely covered with broken streaks of brilliant crimson on a thin pale crimson ground on the side next the sun ; and lemon yellow, tinged with crimson, and marked with pale crimson stripes, on the shaded side ; the whole surface being strewed with distinct russet dots. Eye : small, set in a narrow, round and even basin ; segments, connivent ; tube, funnel shaped ; stamens, marginal. Stalk : short, or a fleshy knob, set in a deep wide cavity. Flesh : yellowish, firm, crisp but not very juicy, with an acid and rather harsh flavour. Cells of the core, closed.

Mr. With's analysis of *Skyrme's Kernel* (season 1880), is as follows :—

Density of fresh juice	1·034
Ditto after 24 hours exposure to air	1·037
One hundred parts by weight of juice yield :—	
Sugar	10·638
Tannin, Mucilage, Salts, &c. ...	3·662
Water	85·700

This Apple is very highly esteemed, and thought by many cider fruit growers to be second

PLATE XXIX.

only to the *Foxwhelp*, and to partake somewhat of its character. Its cider has a peculiar flavour of its own, which improves very much by keeping; but it is better mixed with other apples of its season, such as the *Styres, Styre Wilding, Strawberry Norman*, &c. If made by itself it is apt to turn dark coloured on exposure to air, even in the glass. *Skyrme's Kernel* is also used by its growers as a culinary fruit, and gives its special flavour to pies and puddings; but it does not come into the market in this character.

The tree is hardy, of large size, with wide-spreading growth.

PLATE XXIX.

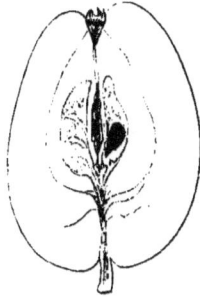

5. CIDER LADY'S FINGER.

The origin of this variety does not seem to be known, but from the age of the trees it was probably produced at the end of the last, or the beginning of the present century.

Description.—Fruit : medium sized, two and a half inches long by one inch and a half wide, oblong, even but not always regular in its outline, with a waist near the top. Skin : quite smooth, dull orange, or yellow on the shaded side, with a few broken stripes of red ; washed with thin red which is streaked with darker and brighter red, on the side next the sun ; the surface strewed with russet specks. Eye : small and prominently set ; open, with very short divergent segments, and surrounded with a few prominent plaits or little knobs ; tube, funnel shaped ; stamens, marginal. Stalk : very slender, short, inserted in a shallow cavity or merely in a slight depression, surrounded with russet. Flesh : yellowish, rather dry ; juice of a fine rich colour, with a sweet subacid and astringent flavour. Cells of the core, open.

Mr. With's analysis of the *Cider Lady's Finger* (season 1878), is as follows :—

Density of fresh juice	1·041
Ditto after 24 hours	1·045
One hundred parts by weight of the juice contain :—	
Sugar	13·242
Tannin, Mucilage, Salts, &c.	1·412
Water	85·346

This Apple is a valuable addition to the orchard. It ripens a little too early for the best class of cider ; but it is easy to manage and makes a very good cider. It is rich, strong, and brisk, often good enough to bottle—but too often it is apt to lose much of its richness in the hot weather of early autumn, from the difficulty of arresting the fermentation. This analysis fully bears out the esteem in which it is held.

The tree is hardy, grows well and bears profusely. It is rising in favour, and deservedly so. It is much grown in the orchards of the valley of the river Frome, and is becoming widely distributed through Herefordshire.

PLATE XXIX.

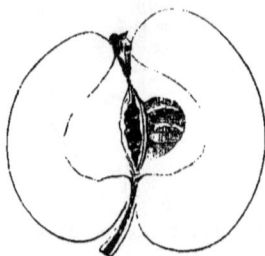

6. GENNET MOYLE.

"The Moile
Of sweetest honey'd Taste." (Philips' " *Cyder*.")

The *Gennet Moyle* was the favorite apple in the cider orchards of the fifteenth century, and continued to be so until Lord Scudamore's *Redstreak* supplanted it in popular esteem. Its history is lost; but its name signifies "a hybrid scion," from "gennet," a hybrid or mule; and "moyle" a scion or graft. It is still to be found in the old orchards of Herefordshire, but has now become scarce.

Description.—Fruit: round, somewhat prominently and obtusely ribbed on the side, and with ridges round the crown. Skin: of a clear lemon colour with a more or less russety cheek and with russet lines all over the side exposed to the sun. Eye: closed with convergent leafy segments, set in a puckered basin; tube, long, funnel shaped; stamens, marginal. Stalk: about half an inch long, inserted all its length in the cavity which is lined with russet. Flesh: with a yellowish tinge, tender, not very juicy, but rather dry, and with a very sweet, slightly acid flavour. Cells of the core, open.

Mr. With's analysis of the *Gennet Moyle* juice (season 1880), is as follows :—

Density of fresh juice	1.046
Ditto after 24 hours exposure to air...	1·053

One hundred parts by weight of juice afford of :—

Sugar	9'57
Tannin, Mucilage, Salts, &c.	5'43
Water	85'00

This sweet and fragrant apple, once so highly esteemed, is only to be found scattered here

PLATE XXIX.

and there in old established orchards. For its character we must therefore refer to the old writers. Dr. Beale says of it, " Our *Gennet Moyles* are commonly found in hedges, or in our worst soil, most commonly in *Irchenfield*, or towards Wales, where the land is somewhat dry and shallow. This Fruit is nice and apt to be discouraged by Blasts, and we do ordinarily expect a failing of them every other year. But this Fruit makes the best Cyder in my Judgment, and such as I do prefer before the much commended *Redstreak'd*. For this *Gennet Moyle* if it be suffered to ripen on the Tree, not to be mellow, but to be yellowish and fragrant, and then to be hoarded in Heaps under Trees a Fortnight or three Weeks before you grind them ; it is (at a distance) the most fragrant of all Cyder Fruit, and gives the Liquor a most delicate perfume. So for Tarts and Pyes it is much commended." " *Herefordshire Orchards* " (1730).

In Evelyn's *Pomona* the *Gennet Moyl* of one year is named first as a Summer Cyder, and of the fruit it is added, " The best Baking apple that grows ; and it keeps long baked ; but not so unbaked without growing mealy. It dries well in the Oven, and with little trouble."

PLATE XXIX.

7. BROMLEY.

[SYN : *Old Bromley.*]

A very old variety spread through the Orchards of Gloucestershire and Herefordshire, but not abundant in the latter county. It is without any known history.

Description.—Fruit: medium size, roundish and flattened, very uneven and angular on the sides ; and knobbed both at the crown and the base. Skin : bright yellow, much covered with firm broken streaks of crimson nearly over the whole surface, but especially where exposed to the sun ; russety all over the base, whence it extends in lines up the sides. Eye : closed, with broad, flat, convergent segments, set in a deep angular basin : tube, funnel shaped : stamens, basal. Stalk: straight and stout, from half to three quarters of an inch long, set in a deep cavity. Flesh: yellowish, firm and somewhat woolly in texture. Juice : pale, plentiful, fairly sweet, and with a brisk acidity. Cells of the core open.

Mr. With's analysis of the *Bromley* Apple (season 1880), is as follows :—

Density of fresh juice ...	...	...	...	1·033
Ditto after 24 hours exposure to air...		...	...	1·035

One hundred parts by weight of fresh juice yield :—

Sugar	...	...	...	...	...	12·10
Tannin. Mucilage. Salts, &c.		...	...	...	1·30
Water	...	...	...	...	...	86·6

The analysis of the juice of this fruit does not indicate any great merit as a cider apple, but it is held in high esteem in Gloucestershire where some think it makes a strong good cider next to *Skyrme's Kernel*, strong but not sweet. It cooks well; and as a sauce apple is unsurpassed. It is a late apple: keeps well; and will sell well in the market—all great merits, that no doubt enable it to maintain its place in the esteem of the growers.

The tree grows to a large size and spreads broadly, but it is considered rather shy in bearing, and for this reason has not been much propagated of late years.

PLATE XXIX.

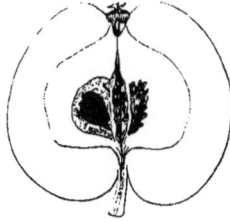

8. RED ROYAL.

This Apple is a favourite of some standing in the Gloucestershire Orchards. Its history has been lost.

Description.—Fruit : small, roundish, inclining to oblate and sometimes to ovate, bluntly angular on the sides. Skin : almost entirely covered with dark crimson, except on the shaded side where it is yellow ; the surface sprinkled with russety dots. Eye : quite closed with convergent segments ; tube, funnel shaped ; stamens, median. Stalk : short and slender, inserted in a rather deep cavity. Flesh : white and tender. Juice : plentiful, pale in colour, sweet but slightly bitter, and pleasantly subacid. Cells of the core, open.

Mr. With's analysis of the *Red Royal* (season 1880), is as follows :—

Density of fresh juice ...	...	...		1·035
Ditto after 24 hours exposure	...	...	...	1·037
One hundred parts by weight of fresh juice yield :—				
Sugar ...	...	...	...	...	13·70
Tannin, Mucilage, Salts, &c.	...	...	...	00·26
Water ...	...	...	...	...	86·04

This Apple is highly esteemed in East Gloucestershire, where it is thought to make cider of the first quality ;—of good colour and flavour, and very sweet and pleasant to drink. The analysis here given does not indicate so high a character, and it is very probable that this variety owes much of its favour to its brilliant colour, and the ready sale it meets with in the market as an edible, or culinary fruit.

The tree is hardy and bears well. It likes a high situation and a strong deep loam, well drained ; as indeed do most apples that make cider worth drinking.

PLATE XXX.

1. OLDFIELD.

This Pear is believed to have derived its name from an enclosure called the "Oldfield" near Ledbury, in Herefordshire. There is no notice of it in any early catalogues of fruits. Philips does not mention it, nor does it seem to have been known until the early part of the eighteenth century.

An excellent figure is given of it in the *Pomona Herefordiensis*, Pl. XI.

Description.—Fruit : small, round, even and regularly formed. Skin : of a uniform greenish yellow when ripe, covered with minute dots and a patch of russet round the stalk. Eye : open with incurved segments, set in a shallow depression, surrounded with plaits. Stalk : an inch long, slender, not depressed but tapering into the fruit at the base. Flesh : yellowish, firm, and crisp. Juice : pale, plentiful, sweet, and very astringent.

The chemical analysis of the juice of the *Oldfield Pear* (season 1880), by Mr. G. H. With, F.C.S., F.R.A.S., is as follows :—

Density of the fresh juice	...	...	...	1·057
Ditto after 24 hours ...	...	...	...	1·061
One hundred parts by weight of the juice contain :—				
Sugar ...	...	...	...	...	13·06
Tannin, Mucilage, Salts, &c.	...	...	...	3·71
Water ...	...	...	...	...	83·23

Mr. Thos. Andrew Knight gives the density 1·067 but states that it varies very much, like that of all other pears, according to the soil it grows on.

The Perry afforded by the *Oldfield* pear is rich and sweet with considerable strength, and ranks next to the *Taynton Squash* in general estimation. It fines readily in making ; keeps well ; and commands a high price in the market. It will keep and improve for 10 or 12 years in bottle.

The tree is very hardy and full blossoming. It is very late in season, and generally bears abundantly. The trees are large. The variety is very generally distributed through Herefordshire and is at this time in full luxuriance.

F. Bath.

Moorpark

Limston e

Hungar

Hy dan dywai

Charcley Green

White Bruash

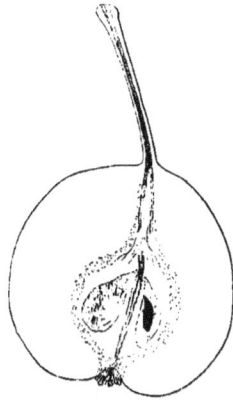

2. MOORCROFT.

[Syn : *Malvern Pear ; Malvern Hill Pear.*]

This Pear probably originated on the farm called "Moorcroft," in the parish of Colwall, near the western base of the Malvern Hills. There are many trees of considerable age there ; it is chiefly cultivated in that district, and has thus obtained the name of *Malvern Hill Pear.* Nothing, however, is positively known with regard to its origin.

Description.—Fruit : large for a Perry pear, pyriform, even and regular in its outline. Skin : greenish yellow on the shaded side, becoming quite yellow as it ripens, with a brownish tinge on the side next the sun ; the whole surface strewed with large ash grey freckles of russet. Eye : open, set in a saucer-like basin. Stalk : half to three-quarters of an inch long, rather stout inserted without depression. Flesh : crisp. Juice : abundant, pale, with a sweet *Jargonelle* flavour and some astringency.

Mr. With's analysis of the *Moorcroft*, (season 1880), is as follows :—

Density of the fresh juice	...	...	1·049
Ditto after exposure for 24 hours ...	...	...	1·050

One hundred parts by weight of the juice contain :—

Sugar	...	...	...	...	...	11·916
Tannin, Mucilage, Salts, &c.	...	...	...	2·384
Water	...	...	...	...	...	85·700

This analysis proves that the *Moorcroft Pear* possesses a very rich juice, capable of making Perry of considerable alcoholic strength. It ripens very early, about the same time as the *Barland,* following the *Taynton Squash,* and before the *Red Pear* and *Oldfield.* The pears are apt to decay soon, and care must be taken that they are used before this begins. It is usually mixed with other varieties to impart its excellent flavour and sweetness.

The tree takes a branching spreading form of growth, attains a large size, and is very hardy, but it can scarcely be called a free bearer.

PLATE XXX.

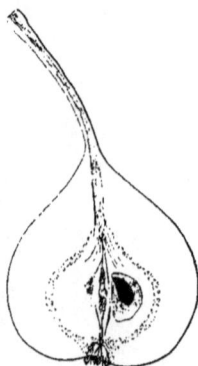

3. THURSTON'S RED.

[SYN : *Dymock Red.*],

The old family of the Thurstons held the estate of the Whitehouse, in the parish of Dymock, for several generations. Mr. Wm. Thurston now lives there, and has several fine trees of *Thurston Red* Pear. He was told by his father that Mr. John Hiatt, formerly of Merrables Farm, Dymock, a great fruit grower in his day, had grafted the young stocks there from the Whitehouse trees. This Whitehouse has the credit of being the birthplace of John Kyrle, the Man of Ross. It is believed also to be the place in which the Pear that bears the family name of *Thurston* originated. Some eight or nine trees have died there from old age ; so the variety is ancient though it has no history. It is now described and figured for the first time.

Description.—Fruit : small, turbinate, even in its outline, but often fuller on one side than the other. Skin : smooth, greenish yellow with a thin red cheek on the side next the sun ; has often a large patch of thin pale brown russet, especially round the eye, and a few spots here and there over the surface. Eye : small and open, set in a saucer-like basin. Stalk : slender, an inch and a quarter long, set on the point of the fruit without depression. Flesh : yellowish and firm. Juice : thin, deep straw colour, sweetish with an astringent aromatic flavour.

Mr. With's analysis of the *Thurston's Red* Pear (season 1880), is as follows :—

Density of fresh juice ...   .   1·035
Ditto after 24 hours exposure to air   ...   ...   1·036
One hundred parts by weight of the juice contains :—
Sugar   ...   .   9·20
Tannin, Mucilage, Salts, &c.   ...   .   2·84
Water   ...   .   87·96

This analysis is not favourable. It proves the juice to be thin and poor, and thus does not bear out the favourable character which many growers seem disposed to give it. The fruit clings

PLATE XXX.

to the tree, and keeps well, and hence is very useful for this quality. It is however a very local variety.

The tree is hardy with a nice upright growth and bears well. It is cultivated extensively at Pauntley in Gloucestershire, Newent, and the surrounding district. Pauntley Court was long in the possession of the Whittingtons, from whom came the celebrated Richard, thrice Lord Mayor of London—thus *Thurston's Red* Pear would seem to affect places of note.

PLATE XXX.

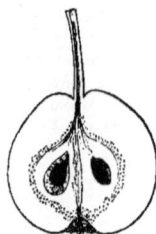

4. HOLMER.

[Syn : *Holmore Pear*—by printers error in the *Pomona Herefordiensis*.]

The original tree of this variety was found in a hedgerow on the estate of the late Charles Cooke, Esq., of the Moor, in the parish of Holmer, near Hereford. Mr. Thos. Andrew Knight judged it to be about eighty years old (c. 1730), and it is therefore now about a hundred and fifty years old.

A good figure is given of this Pear in the *Pomona Herefordiensis*, Pl. XX.

Description.—Fruit : small, roundish, turbinate, even and regular in its outline. Skin : pale green at first, but of a dull, greenish yellow when ripe ; thickly covered with russet dots, so as to form a kind of crust on the surface. Eye : open, full of stamens, having short divergent segments, and set in a very shallow, or in scarcely any depression. Stalk : from half to three quarters of an inch long, slender; inserted in a small hole with occasionally a slight swelling on one side. Flesh : yellowish, firm and crisp. Juice : plentiful, pale in colour, with a sweet subacid and very astringent flavour.

Mr. With's analysis of the *Holmer* Pear (season 1880), is as follows : —

Density of fresh juice	1.051
Ditto after 24 hours exposure to air	1·055
One hundred parts by weight of juice contains :—	
Sugar	11·9
Tannin, Mucilage, Salts, &c.	3·4
Water	84·7

Mr. Thomas Andrew Knight found the density of the juice of this Pear to be 1·066—so that the analysis is favourable for its spirit-producing power.

The Perry from this variety in a good season is of good flavour, very sweet and rich, and resembles that made from the *Red Pear*.

The tree is strong and vigorous, grows tall and bears abundantly. The fruit follows the *Moorcroft* and *Barland* in season. The pears ripen all together and perish very quickly, so that they must be sent at once to the mill. Notwithstanding these grave faults, Mr. Knight thought very highly of the *Holmer* Pear, and consequently it has been widely propagated.

PLATE XXX.

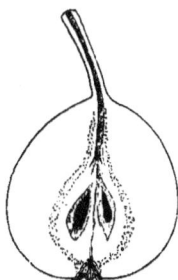

5. TAYNTON SQUASH.

" About Taynton (5 miles beyond Gloucester) Pears most abound, of which the best sort is that they name the *Squash Pear*, which makes the best Perry in those parts. These trees grow to be very large and exceedingly fruitful, bearing a fair round pear, red on one side and yellow on the other, when fully ripe, of a nature so harsh that Hogs will hardly eat them."—Evelyn's " *Pomona.*"

[SYN : *Teinton Squash; Red Squash.*]

The earliest mention of this Pear is by Evelyn, in the paragraph above given. There is no history of its origin, but its name and tradition lead to the belief that it was a native of the parish of Taynton. A *Red Squash* is mentioned by Worlidge, which may very possibly have been the same variety, since the great size and age of many of the trees sufficiently prove its antiquity.

A coloured representation of this Pear is given in the " *Pomona Herefordiensis,*' Pl. xiii.

Description,—Fruit : small, turbinate, even and regular in outline. Skin : dull, greenish yellow on the shaded side, and a clear red next the sun with a few interrupted streaks of deeper colour ; a thin light brown russet runs more or less over the fruit, often in thickly clustered dots, but not sufficiently deep to mar its bright colour. Eye : open, with stiff permanent recurved segments, giving it a star-like character, full of stamens set in a shallow depression, and surrounded with plaits. Stalk : slender, three quarters of an inch long, inserted without depression, with sometimes a fleshy lip on one side of it. Flesh : yellowish, abounding in juice of a rich sweet flavour, brisk, and very astringent, but sometimes very disagreeably harsh and rough.

Mr. With's analysis of the juice of the *Taynton Squash* Pear (season 1880), is as follows :—

Density of fresh juice ...	...	...	...	1·055
Ditto after 24 hours exposure to air		...	...	1·057
One hundred parts by weight of juice yield :—				
Sugar ...	...	...	...	...	13·471
Tannin, Mucilage, Salts, &c.	...	...		3·033
Water ...	...	...	...	...	83·496

The *Taynton Squash* is the earliest of all the Perry Pears and is ripe about the beginning or middle of September. It affords a Perry of the greatest excellence with a sweet rich distinctive

PLATE XXX.

flavour, peculiarly its own. The *Taynton Squash* is amongst Perry Pears what the *Foxwhelp* is amongst cider fruit, the first and the best. It is always sought after, and always commands a high price.

The trees are hardy and grow large and lofty with spreading branches. They bear freely. There is not a farm in Taynton parish without them, and they are scattered about widely, but nowhere in great abundance. There are eleven trees on Aylstone Hill, Hereford; ten trees at Eggleton; but usually they are less numerous; and indeed in most places the trees have died from age.

PLATE XXX.

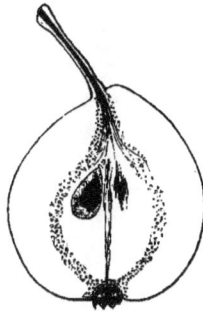

6. WHITE SQUASH.

[Syn : *Stanton or Staunton Squash ; Squirt Pear.*]

Nothing seems to be known of the origin of this Pear. Its synonym of *Stanton*, or *Staunton Squash*, may possibly indicate its origin to be a village of that name between Ledbury and Gloucester, but there are other villages called Staunton.

Description.—Fruit : medium sized, turbinate, even and regular in its outline. Skin : yellowish green when ripe, and strewn all over with small russety dots, with here and there a patch of russet, and always russety round the stalk and the eye. Eye : open, with short stunted segments set in a saucer-like basin. Stalk : an inch long, inserted without depression, and with a fleshy swelling on one side of it. Flesh : coarse and crisp. Juice : very abundant of a deep amber colour and harshly astringent.

Mr. With's analysis of the *White Squash* Pear (season 1880), is as follows :—

Density of fresh juice	1·046
Ditto after 24 hours exposure ... ... · ...	1·048

One hundred parts by weight of juice yield :—

Sugar	10·611
Tannin, Mucilage, Salts, &c.	2·259
Water	87·130

This Pear is rich and sweet, but it quickly decays, and becomes, with a fair outside, "rotten and squashy at the core." It makes a good family Perry if taken at the right moment, rich and sweet ; but it is "stubborn to fine," and its readiness to run into watery decay, makes its power of tub filling its chief merit.

The tree is of small size but a great cropper. It is "lucky for bearing," they say, and thus it maintains its place.

PLATE XXX.

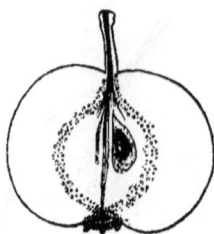

7. CHASELEY GREEN.

[Syn : *Hartpury Green.*]

This Pear is believed to have originated in the parish of Chaseley, a scattered village in the district formerly called Malvern Chase. It is also called *Hartpury Green* from the village of Hartpury in Gloucestershire, where it is much grown. It is without history.

Description.—Fruit : below medium size, two inches across and one inch and three-eighths high, round and flattened above and below. Skin : thick, of a fresh, pale green colour, becoming yellowish ; thickly studded with very distinct, thick, white, russet spots, like scales. Eye : very open and shallow with small upright segments ; set in a wide and shallow basin. Stalk : stout, from half to three quarters of an inch in length ; inserted without depression, but having often an irregular elevation of the fruit near it. Flesh : white, firm, more or less gritty. Juice : pale, mucilaginous, with a sweet, acid and astringent flavour.

Mr. With's analysis of the *Chaseley Green* Pear (season 1880), is as follows :—

Density of fresh juice ...	...	...	...	1·047
Ditto after 24 hours exposure	...	...	...	1·050

One hundred parts by weight of fresh juice yield :—

Sugar ...	...	...	...	...	8·4
Tannin, Mucilage, Salts, &c.	...	...	...	5·6
Water ...	...	...	...	...	86·0

The fruit of the *Chaseley Green* Pear, though capable of making a strong rough Perry, does not possess sufficient flavour to be used alone, except perhaps for home use. It resembles the *Holmer* pear very much in shape, appearance, and character, but is larger in size.

The tree grows well, and with upright growth, until its branches are bent down with the weight of fruit. It is a very prolific bearer. The variety is much grown in the lower valley of the Severn, both in Worcestershire and Gloucestershire ; but it is somewhat local, and has only as yet crept into Herefordshire in the neighbourhood of Ledbury, viz., at Eastnor, the Homend, and Eggleton, where there are many trees, and where it is called the *Hartpury Green* pear.

INDEX TO THE INTRODUCTORY PAPERS.

VOLUME I.